Drinking Behavior

Oral Stimulation, Reinforcement, and Preference

Drinking Behavior

Oral Stimulation, Reinforcement, and Preference

Edited by

Jan A. W. M. Weijnen
Tilburg University
Tilburg, The Netherlands

and

Joseph Mendelson
University of Kansas
Lawrence, Kansas

SPRINGER SCIENCE+BUSINESS MEDIA, LLC

Library of Congress Cataloging in Publication Data

Main entry under title:

Drinking behavior.

Includes bibliographical references and index.
1. Drinking (Physiology) 2. Drinking behavior. I. Weijnen. J. A. W. M. II. Mendelson, Joseph.
QP150.D74   599'.01'32   77-8299

ISBN 978-1-4684-2321-1   ISBN 978-1-4684-2319-8 (eBook)
DOI 10.1007/978-1-4684-2319-8

© 1977 Springer Science+Business Media New York
Originally published by Plenum Press, New York 1977
Softcover reprint of the hardcover 1st edition 1977

Contributors

Linda M. Bartoshuk, John B. Pierce Foundation Laboratory, New Haven, Connecticut

William J. Freed, The Laboratory of Clinical Psychopharmacology, National Institute of Mental Health, William A. White Building, St. Elizabeth's Hospital, Washington, D.C.

Richard M. Gold, Psychology Department, University of Massachusetts, Amherst, Massachusetts

Bruce P. Halpern, Department of Psychology and Section of Neurobiology and Behavior, Cornell University, Ithaca, New York

Don R. Justesen, Laboratories of Experimental Neuropsychology, U.S. Veterans Administration Hospital, and Department of Psychiatry, Kansas University Medical Center, Kansas City, Kansas

Nancy J. Kenney, Department of Psychology, University of Washington, Seattle, Washington

Robert G. Laforge, Psychology Department, University of Massachusetts, Amherst, Massachusetts

Joseph Mendelson, Department of Psychology, University of Kansas, Lawrence, Kansas

Douglas G. Mook, Department of Psychology, University of Virginia, Charlottesville, Virginia

Jan A. W. M. Weijnen, Department of Psychology, Physiological Psychology Section, Tilburg University, Tilburg, The Netherlands

Ronald F. Zec, Department of Psychology, University of Kansas, Lawrence, Kansas

Preface

This is the first book that is devoted entirely to a discussion of the effects on drinking behavior of sensory stimulation of the tongue and mouth. As Blass and Hall (1976) have recently pointed out, there has been an overrejection of the emphasis by Cannon (1932) and Hull (1943) on the peripheral origins (e.g., dry mouth, empty stomach) of the control of ingestive behavior. Thus most present-day investigators of drinking behavior have been concentrating on central mechanisms of control, to the neglect of the periphery. In this volume we have attempted to bring together much of the pertinent "peripheral" literature through originally written chapters that are concerned with the role of orosensory factors in the mediation of drinking and licking. Postingestive effects of fluids receive little attention. Indeed, two chapters deal with consummatory licking in the absence of intake of fluids. A good understanding of the consequences of orosensory stimulation on licking and drinking behavior requires insight into the functional anatomy of the tongue and mouth, the characteristics of the licking response and the problem of recording of licking behavior. Several chapters deal with these subjects.

It has not only been a pleasure but also a privilege to edit this volume. We have learned much from the expert treatment of the different aspects of licking and drinking behavior by the chapters' authors. We hope that further research will be stimulated—and will reveal the results of cross-fertilization among the several lines of approach to the study of licking and drinking reported herein; it is our wish that the volume will thereby become out-dated in the near future.

JAN WEIJNEN
JOSEPH MENDELSON

References

Blass, E. M., and Hall, W. G., 1976, Drinking termination: Interactions among hydrational, orogastric, and behavioral controls in rats, *Psychol. Rev.* **83:**356–374.
Cannon, W. B., 1932, "The Wisdom of the Body," Norton, New York.
Hull, C. L., 1943, "Principles of Behavior," A. C. Crofts, New York.

Contents

Chapter 2
The Recording of Licking Behavior
Jan A. W. M. Weijnen

Chapter 3
Classical and Instrumental Conditioning of Licking: A Review of Methodology and Data
Don R. Justesen

Chapter 4

Airlicking and Cold Licking in Rodents
Joseph Mendelson

Chapter 5

Current Licking: Lick-Contingent Electrical Stimulation of the Tongue

Jan A. W. M. Weijnen

Chapter 6

Temperature of Ingested Fluids: Preference and Satiation Effects (Pease Porridge Warm, Pease Porridge Cool)
Richard M. Gold and *Robert G. Laforge*

Chapter 7

Taste Modulation of Fluid Intake
Douglas G. Mook and *Nancy J. Kenney*

Chapter 8

Water Taste in Mammals

Linda M. Bartoshuk

Chapter 9

**Schedule-Induced Polydipsia: The Role of Orolingual Factors and a
New Hypothesis**

William J. Freed, Ronald F. Zec, and *Joseph Mendelson*

Introduction

Carl Pfaffmann

When the editors invited me to write an introduction to their book on *Drinking Behavior: Oral Stimulation, Reinforcement, and Preference,* I was delighted to accept for a number of reasons. First, of course, is my interest in any efforts that would clarify the sensory-oral determinants of ingestive behavior in relation to my own interpretation of the importance of such factors, particularly in the gustatory domain. Second, the task provided a good way to be brought up-to-date on recent developments. Third, the editors were so generous as to give me a free hand on length and substance of my comments. Who could resist? So in what follows I have freely chosen to discuss aspects pertinent to my interests, perhaps neglecting other, equally important aspects. I also felt free to include research from my own laboratory, especially of most recent vintage, on CNS mechanisms and their related behavioral functions. I have always felt that these aspects of behavioral acceptance and rejection provide the experimentalist with a microscale behavioral model for probing the nature of that most fundamental positive-negative polarity of all organismic reactions: approach and acceptance on the one hand, versus aversion and rejection on the other.

Investigators of food and fluid intake have often approached the topic from one of two points of view: (1) internal and regulatory, i.e., concern for the largely internal and visceral mechanisms that regulate and insure that electrolytes and metabolites in the internal milieu are properly maintained and replenished; or (2) external or stimulatory, i.e., concern for the processes by which the sensory and motor components of the oral and intake mechanisms at the entrance to the digestive tract both monitor and determine selection or rejection of foodstuffs according to their stimulatory properties. Such stimulation instigates closely-coordinated oral behaviors:

Carl Pfaffmann ● The Rockefeller University, New York, New York.

mastication or drinking, licking, lapping, and the ultimate consummatory act, swallowing. Once fluids or foods are in the digestive tract, digestion and related processes may have feedback relations with oral sensory-motor functions that affect sensory properties or reflexive strength. Whereas one might expect both sets of factors to be congruent, i.e., to lead to the same sets of consequences, there is no *a priori* reason that they be so. The distinction of homeostatic versus nonhomeostatic controls has often been evoked in attempts to account for the apparently maladaptive, even self-destructive tendency to eat or drink for the "pleasures of sensation" which is at odds with optimal physiologic benefit.

Drive theorists, at one time, predominantly emphasized the deprivational, bodily-need aspects of such motivated behavior, thereby putting the focus upon internal regulatory demands. Although this approach elucidated the motivation-by-deprivation control of ingestive behavior, the second class of oral-stimulatory and often positive-incentive features that control ingestion tended to be neglected until more recent decades.

The later emphasis on oral-centered features includes certain preparatory instrumental behaviors that often precede actual licking or ingestion such that one can speak of the appetitive or anticipatory and instrumental aspects of food and fluid selection, including preference behavior. In most mammals much instrumental behavior is subject to modification through experience and learning. In any case a very striking contribution to ingestive or drinking behavior is the oral stimulation that leads either to the sequence of acceptance and ingestion or of rejection (even regurgitation) of unwanted food or fluid.

Numerous investigators have remarked upon the general, often widespread occurrence of positive responses to certain classes of taste stimuli. Sugars or artificial sweeteners most commonly lead to acceptance and ingestion, whereas alkaloids or other bitter tasting substances are rejected. These responses are seen in newborn organisms prior to any feeding experience as consummatory mimetic responses. Such observations have led many to believe that certain gustatory receptor systems are biased (perhaps even "hardwired") toward specific reflexes or responses. Further, one is inclined to infer positive hedonic consequences; sugars are said to be palatable or pleasant, whereas quinine is said to be unpalatable. If instrumental responses are required in order to obtain the food object or stimulus, then animals will work for the palatable and will stop working for—or will work to prevent the presentation of—the unpalatable stimulus. By extension, one is led to encompass the concept of reinforcement in the repertory of oral behaviors. Thus, certain properties of oral stimulation, as stimulation *per se,* control and reinforce the acquisition of appropriate instrumental behavior. Certain aspects of oral behavior, such as licking, may display features of instrumental (operant) as well as consummatory responding. The distinction between instrumental and consummatory behavior becomes more subtle or even arbitrary as the analysis becomes more sophisticated, so that much effort has gone into clarifying just how reflexive and operant licking are related.

In my own earlier studies of preference and aversive behavior, with special emphasis on the gustatory control thereof, I reached the conclusion that the "Pleasures of Sensation" are intrinsic to certain peripheral sensory-neural mechanisms and their central connections (Pfaffmann, 1960). At that time, much of what was reviewed in the case of gustatory reinforcement could only be sketchingly documented. Even the categorization of taste as to quality was still unclear. Only recently have data been obtained on the gustatory afferents in our laboratory and those of others that re-affirm that taste receptors can be categorized into sweet, salty, bitter, and sour clusters of sensitivity with distinct afferent channels to the CNS by way of separate, labeled lines (Frank, 1974; Pfaffmann *et al.*, 1976; Sato *et al.*, 1975). Some of these sensory clusters are acceptable and lead to preference; others are unacceptable and lead to aversion.

The work of Norgren in our laboratory (Norgren, 1974; Norgren & Pfaffmann, 1975) has shown that there is a clear bifurcation of taste pathways in the central nervous system, which arises in the secondary gustatory nucleus of the pons (parabrachial area). One branch joins the classical thalamo-cortical circuits as elaborated by Benjamin and co-workers (Burton and Benjamin, 1971); the other, the newly discovered ventral pathway, passes ventrally and anteriorly, giving off fibers to the hypothalamus and, especially, to the amygdala and its central nucleus. The ventral pathway also includes a large visceral component by which afferent information presumably joins taste to its common limbic target. Other investigators have found that amygdalectomy or stimulation of the amygdala is implicated in the regulation of food intake and acceptance. In particular, Box and Mogenson (1975) have shown not only that bilateral electrolytic lesions of the central nucleus of the amygdala produce aphagia and adipsia, but that there is also a decreased reactivity to caloric or non-caloric sweet solutions. Much, however, remains to be elucidated about the behavioral functions of this system. Yet, it is now possible to be somewhat more specific about my statements of the CNS substrate, hypothesized in 1960, as follows:

> I would like to propose that sensory stimulation *per se,* together with its
> *ensuing central neural events* be considered as a prime determinant in
> the chain of events culminating in acceptance behavior, reinforcement,
> and hedonic effect. (Pfaffmann, 1960)

Throughout the literature on preference–aversion, oral stimulation, and drinking behavior, the existence and role of unlearned positive or negative values of gustatory or oral stimulation have been a recurring theme. The details of some of that literature and much else are covered in the chapters of this book.

Chapter 1 by Halpern is an encyclopedic review of the functional anatomy of the tongue and mouth of mammals including their sensory-motor endowments. He shows the precision with which the lick response can serve as an index of a behavioral decision, i.e., to lick or not to lick in a brief 110-msec interval. He further shows that taste buds are not purely lingual. Denervation of the lingual peripheral nerves does not cause complete gustatory desensitization. Miller's recent work (1976) extends our earlier finding on the persistence of gustatory preferences or aversions after combined *chorda tympani* and IXth nerve ablation. We noted then only the residual taste buds of the anterior hard palate, but not the rich palatal sensory apparatus, especially that of the *Geschmackstreife* (Pfaffmann, 1952). On the other hand, where electrical stimulation of the tongue is employed to reinforce operant licking, there is a clear effect of lingual denervation, suggesting that for this behavior lingual receptors are largely responsible (see Chapter 5).

The recording of licking behavior is not a simple matter when one examines methodology closely as is done in Chapters 2 and 3. Licking is more than a reflex and is subject to modification by operant contingencies. The distinction between *integrant* and *adjunct* licking is important for inter-preting many studies in which licking has been a behavioral measure. Integrant drinking ("wet drinking"), as when a thirsty rat drinks water from a pool or a laboratory rat drinks from a spout or cup, has properties that differ from the adjunct case, where licking is detected by one means and reinforced by another. Only in the adjunct ("dry licking") case is it possible to reinforce licks on an intermittent or delayed basis, or to program positive or negative reinforcement of the tongue lick without requiring concomitant ingestion of water. Integrant licking for fluids in conjunction with spatially remote, noncontingent, intermittent reinforcement by food pellets is the spe-cial case that is identified with schedule-induced polydipsia. This reviewer wonders whether polydipsia induced by schedules is not more likely to appear in spout licking than would be found in other, more "natural ways" of consuming fluids.

In Chapter 3 Justesen reviews methods and character of licking as a behavioral response. Rates of respondent (reflexive) licking by rats to differ-ing concentrations of a given volume of saccharin, sucrose, and saline are essentially invariant where a single reinforcement is used to elicit a burst of licking; however, when a rat engages in integrant licking, lick rate may be concentration dependent. The doctrine of invariance of lick rate within a burst first promulgated by Stellar and Hill is reviewed and rejected. Elicited and emitted licking have different properties. The former is faster, more stereotyped, and the emitted response is generally slower, less time locked, and more amenable to modification by training. The possibility that elect-rophysiological concomitants in the hypoglossal nucleus may only accompany elicited- or only operant-licking is entertained by Justesen but seems most unlikely to this reviewer. Regardless of whether tongue movement is voluntary or reflexive, one would expect to find activity in the nucleus of the final common path in all cases. Any useful difference would probably be in the temporal pattern of electrical discharge.

Chapter 4 on airlicking and cold-licking in rodents illustrates reinforcement by thermal stimulation. This striking phenomenon resists deafferentation of the anterior tongue. The crucial factor for maintaining airlicking behavior appears to be cooling of the whole oral cavity or roof of the mouth. Airlicking controls not only consummatory behavior but can serve as a reinforcer for acquisition of new instrumental responses. This stimulation appears to be a positive oral stimulus. Rats will airlick and hamsters will cold-lick even if they have never drunk water. Thus, both airlicking and cold-licking are primary reinforcers for water-deprived rodents, even for animals raised without the opportunity to drink water. Mendelson places these findings in a broad biological point of view when he states:

> Our results suggest that the cold receptors in rodents' tongues and/or mouths might help to promote the rodents' survival. Feedback from these receptors to the central nervous system is rewarding to animals whose bodies have been depleted of water. Thus, the orolingual cold receptors help thirsty rodents to detect water and increase the probability that they will drink it.

However, regardless of rewards that might occur in these experiments, results show that orolingual cooling constitutes a sufficient reward for maintaining licking behavior. For the water-deprived rodent, both water intake without the act of drinking and orolingual cooling without water intake are reinforcing events.

Chapter 5 by Weijnen deals with licking that is operantly reinforced by electrical stimulation of the tongue. Electrical stimulation may be more readily related to gustatory afferents than to hypothermic stimulation. Indeed, the gustatory receptors in electrophysiological studies are found to be extraordinarily sensitive to small anodal currents. Adachi (1969) reported that salt fibers of rats were highly sensitive to anodal currents but that sugar fibers were not so stimulated. But further study is needed to clarify just how strong and pervasive this differential gustatory effect may be. Anodal electrical polarization in man produces a distinctly sour or metallic taste; breaking the anodal current produces no new taste (Bujas, 1971). The cathodal current elicits a more complex taste, alkaline and bitter, or bittersweet. Breaking the cathodal current produces a brief yet clear sour taste often coupled with a sweet taste. Where there is an intermittency that could change as the animals lick, the parameters of electrical stimulation would change and conceivably affect the relative stimulating effect for the different taste qualities. The threshold for the reinforcing effect of electrical stimulation of the tongue is extremely low, on the order of 0.5 to 1 μA. Intensities of 50 to 100 μA are optimal values for current licking. These are well within the ranges of activation of the taste fibers in the *chorda tympani*, and it is unlikely that the trigeminal components are activated by currents at this level. Since bilateral section of the *chorda tympani* is reported to abolish the reinforcing effects of such stimulation but transection of the trigeminal portion of the lingual nerve has no effect, it seems that the gustatory afferents are indeed involved. But since current licking can be demonstrated consistently only in animals that have had experience with electrical stimulation of the tongue concomitant with water intake, the sensory result of electrical stimulation is probably novel if not unnatural. Since salt receptors predominate in the *chorda tympani* of the rat, this would probably be the approximate major taste quality for the animal with its intrinsic tendency to reinforce operant licking. This line of reasoning would fit in with Weijnen's view that "in rats under dipsogenic condition, the pairing of electrical tongue stimulation with water intake strongly elevates intrinsically reinforcing effects of lick-contingent tongue stimulation." Chapters 4 and 5 are consonant with this reviewer's conclusions on the intrinsic reinforcing effectiveness of appropriate gustatory and oral-sensory stimulation.

Chapter 6 on temperature of ingested fluids by Gold and Laforge follows naturally in sequence after airlicking and cold-licking by thirsty rodents. The next step in the chain is, of course, ingestion. Here a number of studies show that cold water is more satiating in the sense that thirsty animals drink less cold water than warm. Water intake increases directly with temperature up to 37°C. Although coldness of the source is a stimulus for licking, it appears to have a stronger satiating effect judged by its lesser intake. Subdiaphragmatic vagotomy, when bilateral, abolished the cold-water suppression of drinking.

That thirsty animals are more stimulated to lick by cold, and to bar press at a higher rate for cold water agrees with the demonstrated initial preference for cold water. But that cool water is at the same time more satiating seems paradoxical. It is suggested that within seconds after the onset of cool-water drinking, the cooling of certain parts of the body produces satiation of hypovolemic thirst; thus the cooler the water, the less rapidly it would have to be ingested in order to produce such satiation. After a drinking bout, the animal shifts to a preference for warm water and ingests more of it. Finally, as temperature rises much above 37°C, there is a falloff in water consumption. If coolness serves as a cue that identifies drinking water, as Mendelson suggests, it cannot be the sole cue sustaining drinking since warm water is also ingested by thirsty animals. That no prior experience is required to establish a cold-licking response indicates that coolness is a primary initiating unlearned cue perhaps like the taste of sugar or certain other taste qualities. Indeed, it might seem that the change in stimulus and effective value of coldness reflect a shift from oral control to other more viscerally located factors for which warm temperatures of ingested fluid ought to be less disturbing to the homeostatic machinery.

Mook and Keeney begin Chapter 7 on taste modulation of fluid and the control of intake by taste by examples (a) that seem opposed to homeostatic needs, or (b) that are irrelevant to need, and (c) where no need at all has been imposed. They eschew playing off "regulatory" versus "nonregulatory" factors in fluid intake. Rather they espouse an interactionist view, as do a number of other contributors to this volume, i.e., that control of ingestion requires a system that takes into account both the internal state of the organism and the commodities available in the environment so that demands of the one are met by appropriate selections in the other. That a hungry but nonthirsty rat will refuse to drink water but will accept a saccharin solution avidly—and the hungrier it is the more it will drink—exemplifies how the sweet taste can switch the rat's lapping from the control system for the maintenance of water balance to that for energy balance. The authors then review many other well-known cases. The mechanisms underlying appetitite for salt in salt-deprived rats appear to be quite different from those that generate salt preference by nondeficient rats. Salt appetite in the deprived animal is more focused and purposeful. Animals needing salt will press a lever, tolerate quinine adulteration, or overcome a learned taste aversion. The nondeprived animal will not bar press for delivery of NaCl (Lewis, 1960). However, when operant licking is reinforced by NaCl solution, the nondeprived animal will do so for salt solutions. Is the operant lick a more sensitive index of low-level sensory reinforcement than bar pressing (Pfaffmann *et al.*, 1967)?

The question of how sodium deficiency leads to the unusually vigorous response to the salt taste is still an open one. Changed sensitivity of the peripheral receptors, which earlier work (Pfaffmann and Bare, 1950; Nachman and Pfaffmann, 1963) ruled out, is now open to reconsideration. That sensitivity is not enhanced is still tenable. Indeed, recent work (Contreras, 1977) that is based on a more powerful technique for recording from single afferent taste-units shows that the taste receptors most reactive to NaCl, i.e., NaCl-best fibers, in fact become *less sensitive*. Other units, sugar-best or quinine-best, were unchanged after dietary salt deprivation. Although these sensory changes might cause a greater intake of salt solution if the animal strives to achieve the same sensory input as the more sensitive normal animal, there remain questions of the differing reinforcing effects of salt in a bar-press task and the guidance of behavior by a memory "search image" of where salt had been previously available in a state of no need (*cf.* Uexkull, as cited in Ch. VII, Krieckhaus, 1970).

In their analysis of the response to sweet stimuli, the authors note that although the intake pattern of the response to glucose appears identical to that for saline, the controlling mechanisms are not. In particular, they note that the descending limb of the bar-pressing response-function for sugar could be attenuated by periodic as opposed to continuous reinforcement, as first shown by Guttman. Such a change could be interpretable as reducing post-ingestive factors. Stomach preloads of glucose are known to inhibit subsequent glucose intake. Indeed Mook's own classic experiments with the esophagastomized animal, where intake by mouth bypasses the stomach and rules out post-ingestive factors, showed that intake continued to increase as concentration rose well into the higher ranges that would induce a decline in normal drinking. The form of the intake curves parallels the electrophysiological measures of taste afferent discharge to the same sugars (Hagstrom and Pfaffmann, 1959). This is probably one of the clearest cases arguing for a stimulatory role of gustation in intake.

Mook and Kenney have a predilection against purported hedonic properties of gustatory stimuli, preferring, for example, to invoke an identification role or releasing function for the sweet taste. Sweetness is said to be a cue for food but when it is redundant (i.e., when other cues identify the sweetened commodity as food) in their experiments, it has no effect on the choice an animal makes. Others, however (e.g., Strouthes, 1977), have contrary evidence. Ernits and Corbit's data, where neither water nor food deprivation is used, show enhanced drinking of sugar and salt and are accepted as strong evidence for a concept of palatability. But, they ask, does taste as a dipsogenic stimulus really operate in isolation? Even here, if the animal is obese, the Ernits and Corbit effect is not seen.

From a strictly methodological point of view, there can be no gainsaying Mook and Kenney's concern about concepts of hedonism or palatability in animal preparations. Too often this concept is a choice of last resort when all else fails. The basic circularity is ever present, for the criterion of hedonic effect is what the animal ingests or works for. What it does is just what one is attempting to explain.

In man, an independent rating of hedonic value can be given and is widely used in basic as well as in applied research on flavor. But for nonverbalizing organisms, can an independent measure be obtained, such as a grimace or the other mimetic responses to taste stimuli by human neonates that are described by Steiner? Steiner's studies (1973) of anencephalic neonates suggest that the basic reflexes of oral acceptance of sugar versus rejection of quinine include as well mimetic responses of pleasure or displeasure that are controlled by the brainstem. Harvey Grill in my laboratory (Grill and Norgren, 1977) has approached this problem in an animal model—a chronic decerebrate rat preparation—that is reasonably healthy and long lived. Aphagic and adipsic, the animal must be tube fed and carefully nursed, but it retains good postural control, grooms itself, and locomotes effectively. Its taste reactivity has been assessed by injecting solutions into the oral cavity via an indwelling catheter. Two major responses, both oral and mimetic, have been observed. Sugar, saline, and, surprisingly, acid solutions elicited similar labial and tongue movements and swallowing. Quinine elicited a clearly different mouth gape and fluid rejection at a threshold of $3 \times 10^{-5} M$ quinine. At $3 \times 10^{-4} M$ a second response appeared, rapid paw wiping of the snout and mouth followed by head shaking ("wet dog shake") and chin rubbing that is followed by fore and aft forepaw shuffling. All or part of these response patterns to bitter have been seen and described by others (e.g., Teitelbaum and Epstein, 1962). These same responses occurred in the chronic decerebrate animal at the same threshold values as in normal animals, showing the locus of neural control to be in the lower brainstem and medulla. Given these reflexes and fixed-action patterns, one might well conclude that the brain stem is the substrate of preferences and aversions.

Grill applied the taste–reactivity tests to normal rats after taste–aversion conditioning by associating toxicosis with sugar. Following conditioning, sugar elicited the full-blown aversive response previously seen to quinine, yet no aversive taste stimulus had been associated with the taste of sugar, only the visceral distress. After such conditioning, animals normally reject sugar in a preference situation, but when sugar is introduced into the oral cavity in an inescapable manner, the aversive response appears. Aversion appears to be a state of the organism that can properly be called negatively hedonic. Garcia and Hankins (1975) also report expressions of disgust in a variety of species examined after taste–aversion conditioning.

Such oral and mimetic responses (OMR) may be behavioral components of the total set of autonomic and endocrine response patterns alluded to by Powley (1977) as the cephalic-phase response of metabolism. The conditionability of OMRs has been demonstrated experimentally in at least one case by Grill (1975) and, from general observation, is seen in humans in a socially expressive context. Interestingly enough, gastrointestinal physiologists have not been hesitant to note the correlation in animals and man between magnitudes of such autonomic cephalic responses as salivary and gastric secretions, and ·palatability of food stimuli. There is thus hope indeed that some such response indicators will rescue the concept of palatability and hedonic properties from neglect or even exclusion on methodological grounds.

Although triggered by sensory aspects of food or oral stimulation, cephalic-phase responses are sensitive to the nutritional state of the organism. Nicolaides (1969) showed that the hypoglycemic response to oral stimulation with sucrose or saccharin varies significantly with the level of deprivation or repletion. The close relation of visceral and gustatory afferents in the nucleus of the solitary tract, in the dorsal pons and in the ventral pathway to the amygdala, especially the central nucleus (Norgren, Grill and Pfaffmann, 1977) provides the neuroanatomical substrate for such interactions.

Thus the task is not so much to isolate oral sensory factors from post-ingestional events or *vice versa* as primary determinants of ingestive behavior. Rather, the task is more to determine just what the interactions between the two are with the proviso that much more detailed knowledge is required of the visceral afferent system than is now available both as a sensory mechanism and in relation to the newly discovered CNS pathways. Especially intriguing is the anatomical finding that the ultimate destination of the ventral taste (and visceral) pathway is the central nucleus of the amygdala. Box and Mogenson (1975) report that lesions here produce transient aphagia and adipsia similar to but less severe than that seen after lateral hypothalamic lesions. Their preparations showed a *much reduced preference* for either caloric or noncaloric sweet solutions. Massive amygdalectomy had been previously reported to attenuate responses to both sweet and bitter solutions. Other lesions restricted to a more baso-lateral area are said to enhance sugar intake of strong solutions. Interference with the major limbic input via the central nucleus might be expected to cause a reduced response to a gustatory signaled solution. Inactivation of other amygdaloid and, indeed, limbic structures might cause enhancement if those structures modulate more complex interaction of oral-cephalic and postingestive factors. Further neuroanatomical study of the various components of the limbic system in combination with appropriate behavioral tests will be needed before a rationale for these and other com-plex interactions with brain lesions, which are reviewed by Mook and Kenney, can be developed.

Bartoshuk's interesting analysis of water taste shows the error of an ultrasimplistic model of sensory systems, such as gustation as mediated by rigid detectors in which the solvent is merely a vehicle for stimulating molecules. Adaptation level strikingly affects the receptors' properties and hence their discharge so that water as a stimulus may elicit sweet, sour, bitter or saltiness, depending on the substance that preceded it. A most important potential source of adaptation is saliva. Changes in salivary com-position in various states of deprivation and their effects on taste would bear further investigation.

In any case, adaptation level can affect the quality aroused by any particular taste stimulus, and systematic examination of these effects in electrophysiological studies bears promise of clarifying how the best stimulus profile of individual gustatory units may be modified. Similar psychophysical data from humans have strengthened classical, labeled-line views of taste coding. The electrophysiological data, combined with animal psychophysical tests of such adaptation, would contribute to the question of taste–quality coding. That water taste does control behavioral responses to fluids is amply demonstrated in Bartoshuk's study of the cat in which a massive water discharge presumably interferes with discrimination and preference for sweet which, however, can be revealed when the masking water taste is inhibited. Her work is clear warning not to consider water to be a neutral solvent in examining the gustatory determinants of licking and drinking. No sensory or perceptual experiment on vision would be carried out without great care in preadaptation to illuminance level and wavelength composition before beginning visual psychophysical studies. The same considerations apply to the chemical senses.

The final chapter by Freed, Zec, and Mendelson on schedule-induced polydipsia (SIP) assesses the role of many of the orolingual factors discussed in the preceding chapters in relation to this rather complex and puzzling phenomenon. When rate of water ingestion was controlled by decreasing the diameter of drinking-tube orifices, animals regulated volume of intake fairly precisely, whereas time of drinking to achieve the desired volume varied quite widely. The authors conclude that SIP is regulated by a volume-correlated feedback from water ingestion. They opt for an orolingual or hypothalamic origin of the feedback because of the rapidity of the effect. Consonant with this is the observation that schedule-induced airlicking does occur with no fluid ingestion, but then licking takes place for a much longer duration. Lingual denervation, which has a major effect on electric-current licking, does not seriously interfere with SIP. Presumably, partial desensitization of the tongue does not remove the remaining receptors of the extralingual oral tissue, which are sufficient to maintain the behavior.

That levels of SIP are inversely related to sugar content of dry, food-pellet reinforcers and that sweetened nonnutritive pellets reduce water intake in SIP appear to implicate oral- or cephalic-phase responses that could be triggered by nutritive or nonnutritive sweeteners. Such gustatory stimuli are known to change insulin secretion with a consequent change in blood-sugar levels. Freed, Zec, and Mendelson present an hypothesis that implicates insulin secretion as produced by the combined effects of intermittent priming by food and by stimuli associated (conditioned) with the testing situation such that insulin secretion might be excessive in relation to the amount of food actually ingested. The sweet pellet effect, i.e., inhibition of SIP, might be an important lead to follow. The series of experiments they suggest should clarify how nutritive and nonnutritive pellets that are sweetened with saccharin or with slowly metabolized sugars will affect blood glucose levels—and should shed light on this robust and well documented effect. In 1971 Falk wrote: "a decade of research has yielded no traditional physiological or behavioral explanation for schedule-induced polydipsia"; SIP in 1977 is still a challenge!

But there are other challenges! Most of the detailed analytical analyses of the role of sensory-oral factors in this book are based upon animal studies. Yet, we presume that much the same rationale applies to humans. Besides developmental studies in children (e.g., Nowlis and Kessen, 1976; Steiner, 1973), or psychophysical and hedonic scaling studies in normal adults, there are special populations such as the obese that merit further investigation (e.g., Cabanac, 1971; Grinker and Hirsch, 1972).

Another source of additional facts and findings might be those individuals with certain neurological syndromes. One such, who suffered a massive cerebral aneurism in late adolescence, showed remarkable recovery with no verbal or language deficiency except for a stabilized hemiplegia. She has an active professional career and social life, and participates in many recreational, even physical, activities. She reports that bitterness is a remarkably repulsive experience, that the thought of bitter tonic water gives her "the shudders". Hot coffee is painful to drink so that this beverage must be cooled to room temperature and also sweetened with five to six heaping teaspoons of sugar to counteract the bitterness before it becomes palatable. She also has a strong "sweet tooth" enjoying sweet soft drinks and fruit juices, candy, even sweet pickles, but not sour dills, or any other strong sour condiments, such as mustard, horseradish, etc. One may ask if this ensuing sensory-neural function reflects a state of "finickiness" or at least some part thereof.

Obese patients in Grinker and Hirsch's work show a clearly different set of hedonic judgments to sugar and saccharin than do most subjects of normal weight. A characteristic bitonic curve usually expresses hedonic value as a function of sugar concentration. For the normal in weight, pleasantness increases with concentration to a peak value of 0.3 M sucrose, then turns down, becoming less pleasant at higher concentrations. For the obese, pleasantness is a decreasing function throughout the entire range of concentrations. When obese subjects are fasted down to the so-called norm, the hedonic function for sweeteners shows the usual inverted U-shape.

This volume could stimulate and set the stage for more intensive human studies along these lines as well as for other animal studies, not merely to give relevance, but with the hope of uncovering new facts and elucidating species differences, and in particular in giving new insights and leads into those features of oral stimulation, reinforcement, and preference that the human being is uniquely able to experience and to report.

References

Adachi, A., 1969, Responses of the chorda tympani to electrotonic, electrolytic, and nonelectrolytic stimulations on the tongue, in "Olfaction and Taste, III" (C. Pfaffmann, ed.), The Rockefeller University Press, New York, p. 611 (Abstract).

Box, B. M., and Mogenson, G. J., 1975, Alterations in ingestive behaviors after bilateral lesions of the amygdala in the rat. *Physiol. Behav.* **15**:679–688.

Bujas, Z., 1971, Electrical taste, in "Handbook of Sensory Physiology, Vol. IV, Chemical Senses 2 Taste" (L. M. Beidler, ed.), Springer-Verlag, Berlin, Heidelberg, New York, pp. 180–199.

Burton, H. and Benjamin, R. M., 1971, Central projections of the gustatory system, in "Handbook of Sensory Physiology, Vol. IV, Chemical Senses 2 Taste" (L. M. Beidler ed.), Springer-Verlag, Berlin, Heidelberg, New York, pp. 148–164.

Cabanac, M., 1971, Physiological role of pleasure, *Science,* **173**:1103–1107.

Contreras, R. J., 1977, Changes in gustatory nerve discharges with sodium deficiency: A single unit analysis, *Brain Res.* **121**:373–378.

Falk, J. L., 1971, The nature and determinants of adjunctive behavior, *Physiol. Behav.* **6**:577–588.

Frank, M., 1974, The classification of mammalian afferent taste nerve fibers, *Chem. Senses and Flavor,* **1**:53–60.

Garcia, J., and Hankins, W. G., 1975, The evolution of bitter and the acquisition of toxiphobia, in "Olfaction and Taste V" (Denton, D. A. and Coghlan, J. P., eds.), Academic Press, New York, pp. 39–45.

Grill, H. J., 1975, Sucrose as an aversive stimulus, *Neuroscience Abstracts,* **1**:525.

Grill, H. J., and Norgren, R., 1977, The taste reactivity test. II. Mimetic responses to gustatory stimuli in chronic decerebrate and chronic thalamic rats. Submitted for publication.

Grinker, J., and Hirsch, J., 1972, Metabolic and behavioral correlates of obesity, in "Physiology, Emotion and Psychosomatic Illness (Knight, J., ed.), CIBA Fd. Symp.

Hagstrom, E. C., and Pfaffmann, C., 1959, The relative taste effectiveness of different sugars for the rat, *J. Comp. Physiol. Psychol.* **52**:259–262.

Krieckhaus, E. E., 1970, "Innate recognition" aids rats in sodium regulation, *J. Comp. Physiol. Psychol.* **73**:117–122.

Lewis, M., 1960, Behavior resulting from sodium chloride deprivation in adrenalectomized rats, *J. Comp. Physiol. Psychol.* **53**:464–467.

Miller, I. J., 1977, Gustatory receptors of the palate, *International Symposium on Food Intake and Chemical Senses,* Fukuoka, Japan, 1976, University of Tokyo Press, in press.

Nachman, M., and Pfaffmann, C., 1963, Gustatory nerve discharge in normal and sodium-deficient rats, *J. Comp. Physiol. Psychol.* **56**:1007–1011.

Nicolaidis, S., 1969, Early systemic responses to orogastric stimulation in the regulation of food and water balance: Functional and electrophysiological data. *Ann. N.Y. Acad. Sci.* **157**:1176–1203.

Norgren, R., 1974, Gustatory afferents to ventral forebrain, *Brain Res.* **81**:285–295.

Norgren, R., Grill, H. J., and Pfaffmann, C., 1977, CNS projections of taste to the dorsal pons and limbic system with correlated studies of behavior. *International Symposium on Food Intake and Chemical Senses,* Fukuoka, Japan, 1976, University of Tokyo Press, in press.

Norgren, R., and Pfaffmann, C., 1975, The pontine taste area in the rat, *Brain Res.* **91**:99–117.

Nowlis, G. H., and Kessen, W., 1976, Human newborns differentiate differing concentrations of sucrose and glucose, *Science,* **191**:865–866.

Pfaffmann, C., 1952, Taste preference and aversion following lingual denervation, *J. Comp. Physiol. Psychol.* **45**:393–400.

Pfaffmann, C., 1960, The pleasures of sensation, *Psychol. Rev.* **67**:253–268.

Pfaffmann, C., and Bare, J. K., 1950, Gustatory nerve discharges in normal and adrenalectomized rats, *J. Comp. Physiol. Psychol.* **43**:320–324.

Pfaffmann, C., Fisher, G. L., and Frank, M. K., 1967, The sensory and behavioral factors in taste preferences, in "Olfaction and Taste II" (T. Hayashi ed.), Pergamon Press, Oxford, pp. 361–381.

Pfaffmann, C., Frank, M., Bartoshuk, L. M., and Snell, T. C., 1976, Coding gustatory information in the squirrel monkey chorda tympani, in "Progress in Psychobiology and Physiological Psychology", Vol. 6 (Sprague, J. M. and Epstein. A. N. eds.), Academic Press, New York, pp. 1–27.

Powley, T. L., 1977, The ventromedial hypothalamic syndrome, satiety, and a cephalic phase hypothesis, *Psycho. Rev.* **84**:89–126.

Sato, M., Ogawa, H., and Yamashita, S., 1975, Response properties of macaque monkey chorda tympani fibers, *J. Gen. Physiol.* **66**:781–810.

Steiner, J. E., 1973, The human gustofacial response, in "Oral Sensation and Perception" (J. F. Bosma, ed.), DHEW Publication No. (NIH) 75-546, U.S. Government Printing Office, pp. 254–278.

Strouthes, A., 1977, Saccharin eating in undeprived and hungry rats, *Anim. Learn. Behav.* **5**:42–46.

Teitelbaum, P., and Epstein, A. N., 1962, The lateral hypothalamic syndrome: Recovery of feeding and drinking after lateral hypothalamic lesions, *Psychol. Rev.* **69**:74–90.

Functional Anatomy of the Tongue and Mouth of Mammals

Bruce P. Halpern

1. Introduction

The static and dynamic characteristics of the jaws, tongue, and other structures surrounding the oral cavity of mammals strongly affect the types of
environments in which a particular mammal can successfully function. A
highly specialized mouth and tongue, such as that noted in anteaters by
Sonntag (1925), is an extreme example of an ingestive apparatus that sets
very narrow limitations on usable habitats. With reference to the primary
focus of this chapter, liquid ingestion, less dramatic but equally significant
constraints are found in adult animals unable to suck liquids, presumably
because of their mouth characteristics. (See pp. 62–80, Drinking
Behavior, especially p. 72 and Table 4). A specific liquid intake and
intraoral manipulation mechanism may have great adaptive advantage for a
given ecological niche. However, the possibility of evolutionary modification in the event of major, relatively rapid changes in habitat may be greatly
reduced. Nonspecialized omnivores, having ingestive apparatus usable with
a wide range of foods and liquid sources, avoid these problems, as Hiiemae
and Crompton (1971) and Jolly (1972) have noted. It may be that omnivores may not be able to exploit some habitats as efficiently as specialized
feeders.

Anatomical characteristics of the tongue and jaws are significant limit-

Bruce P. Halpern • Department of Psychology and Section of Neurobiology and Behavior,
Cornell University, Ithaca, New York.

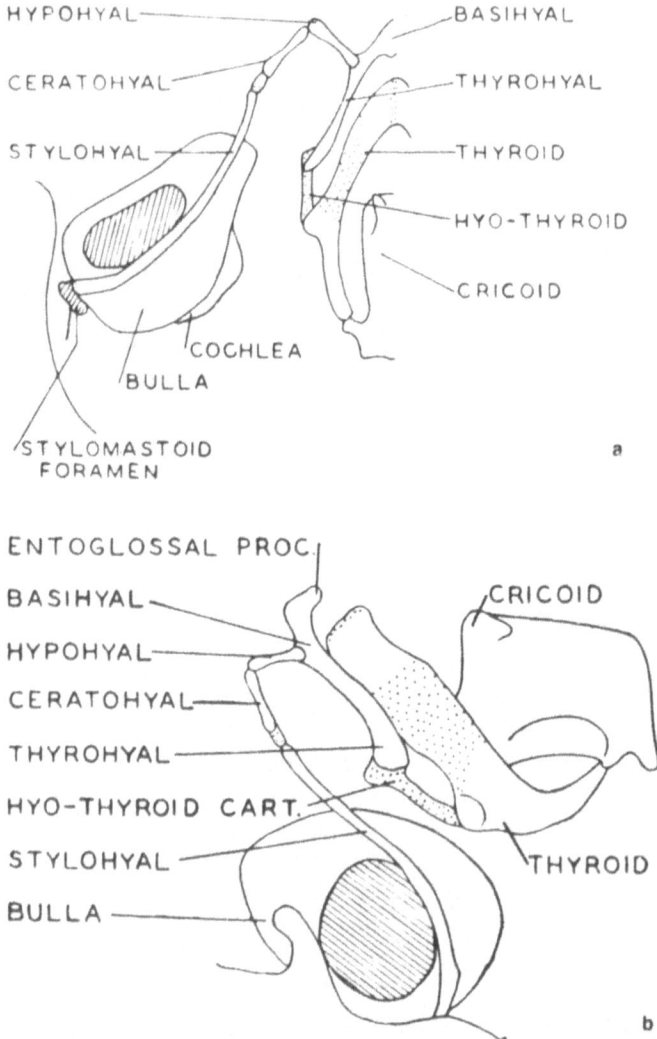

Figure 1.   The hyoid apparatus and deep hyoid muscles of the microchiropteran bat *Tapho-zous nudiventris*: a, ventral view; b, lateral view. This bat has structures similar to those of the family Vespertilionidae. In particular, it is similar to *Eptesicus fuscus* (Tables 2 and 3) and *Choeronycteris mexicana* (Table 4). Stylohyal, ceratohyal, hypohyal, and thyrohyal are hyoid bones. From Sprague (1943).

ing factors in ingestion. The tongue is greatly variable in shape (see Sonn-tag, 1925; Bradley, 1971), and movement patterns in the many orders of therian mammals (e.g., Abd-el-malek, 1955; Herring and Scapino, 1973; Hiiemae and Crompton, 1971). Particular movements are associated with liquid ingestion by licking/lapping and by sucking, while suckling presents a special case. Specialization in form, spacing, and type of occlusion of the teeth sets a limit on jaw closing (e.g., Ardran, Kemp, and Ride, 1958; Crompton and Hiiemae, 1970; Herring and Scapino, 1973; Kallen and Gans, 1972). Comparable constraints are associated with the lower jaw (i.e., the mandible), which varies considerably in possibilities for lateral and/or orbital motion, width of opening (gape), extent of protrusion or retraction, all with reference to the upper jaw (maxilla and premaxilla regions of the skull). Factors in these differences in lower-jaw movement include, in addition to the teeth, the degree of ossification of the symphysis (union) of the mandibles (Figures 1 and 2) and the configuration of the joint between the mandible and the bone(s) at the base of the skull with which the mandible articulates.

For any one animal, characteristics of the tongue, teeth, jaw, and palate are often sufficiently distinctive to provide a basis for grouping that mammal into an appropriate order and family. Taxonomic affinities of jaws and teeth are often the only clues to phylogenetic classification schemes (see Sonntag, 1925; Hildebrand, 1974). These relationships also permit generalizations from living mammals to extinct forms (e.g., Crompton and

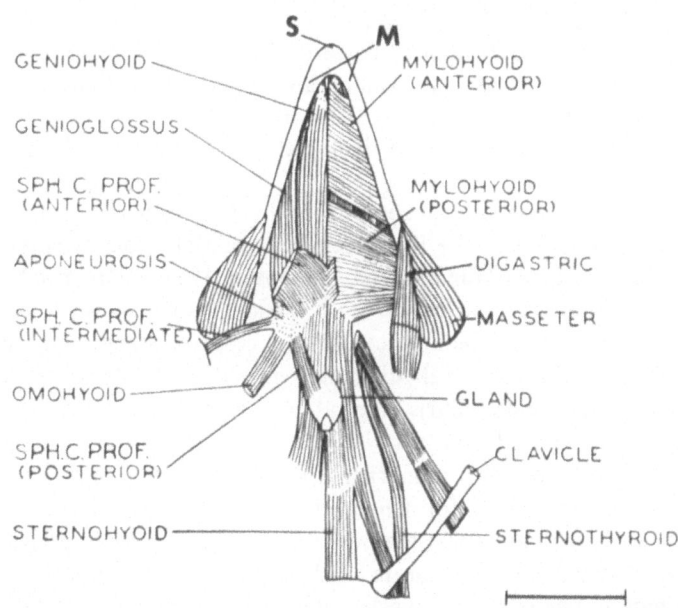

Figure 2. Ventral view of the muscles of the hyoid complex in the microchiropteran bat *Phyllostomus hastatus,* a bat of the family Phyllostomidae. The illustrated structures are similar to those of another member of this family, *Choeronycteris mexicana* (Table 4), and to bats of the family Vespertilionidae, specifically *Eptesicus fuscus* (Table 2) and *Myotis lucifugans* (Table 1). M = mandible. S = mandibular symphysis. Calibration line is 1 cm. Modified from Sprague (1943).

Hiiemae, 1970). A mammal's diet can often be deduced from the oral structures. Relevant information includes the articulations and the wear patterns of hard tissues. These same morphological features grossly set limits on what that animal can ingest.

Structural information on the tongue, jaw, etc., is complemented by functional anatomical studies in living mammals (e.g., Ardran *et al.,* 1958; Hiiemae and Crompton, 1971; Herring and Scapino, 1973; Kallen and Gans, 1972). A combination of functional and structural approaches permits a useful description of the nature of relevant oral-cavity functions:

1. *Ingestion.* Food intake into the oral cavity, that is, gnawing and/or biting of solid food, licking/lapping of liquids and soft foods, and sucking of liquids.
2. *Mastication and transport.* Modification and movement of the food within the oral cavity, that is, chewing, grinding, and manipulation by the tongue.
3. *Storage.*   Accumulation of liquids and masticated solids at the rear of the mouth.
4. *Swallowing.*   Transfer of food from the oral cavity to the digestive structures of the abdomen.

For these functions, a coordinated pattern of tongue, mandible, and cheek motion has been described (e.g., Kallen and Gans, 1972). Factors such as orientation of the head (e.g., of the skull) and use of the forelimbs (or the absence of such use) are properly included, as they are in Hiiemae and Crompton (1971).

Mastication is normally accomplished by the teeth of the lower jaw being moved in relation to the fixed teeth and palatal ridges (rugae) of the upper jaw and skull (see Hiiemae and Crompton, 1971; Hildebrand, 1974). A number of different muscles, typically divided into a group of six, are directly involved in mandibular movement. These six mandibular muscles are the masseter, the temporal, digastric, medial, and lateral pterygoids; and the zygomaticomandibularis (the latter muscle is not recognized as a distinct muscle by all authors).

Two separate muscles comprising the lateral pterygoid have been proposed by Grant (1973) for two anthropoid primates: human (*Homo sapiens*) and rhesus monkey (*Macaca mulata*). The lateral pterygoid is generally considered to have separate heads in these forms (see Møller, 1966). The argument for two separate muscles cites biomechanical and EMG data.

The attachments of the mandibular muscles to the mandible and bones of the skull, and the consequent movements of the mandible, have many similarities in mammals. This is detailed for a single mandibular muscle, the masseter, in Table 1. Some differences can also be noted, especially in specialized feeders such as bats, as Kallen and Gans (1972) have reported.

Major distinctions also occur. Thus, in contrast to the six mandibular muscles described above, a primitive mammal, the American oppossum (*Didelphis marsupialis*), is thought by Hiiemae and Jenkins (1969) to have only four clearly separable jaw muscles: the masseter, the adductor complex, and the medial and lateral pterygoids. A significant factor in mammalian mastication is the shape and location of the teeth (e.g., Ardran *et al.*, 1958; Kallen and Gans, 1972). The ability to grind hard objects, to pierce rigid outer coverings, to shear soft tissues, and to close the jaws in particular orientations depends in part on the teeth. However, since ingestion of liquids is the major concern of this chapter, mastication *per se* will not be specifically considered.

The morphology and functional characteristics of the tongue are of special interest in drinking. Many animals—for example, the rat (Hiiemae and Ardran, 1968), the opossum (Hiiemae and Crompton, 1971), and the bat (Kallen and Gans, 1972)—ingest liquids by either extending the tongue into the volume of an external body of liquid (lapping) or pressing the tongue against surfaces that support a liquid (licking) and then rapidly retracting the tongue. In the laboratory albino rat (*Rattus norvegicus*), tongue-retraction time during licking was found by B. P. Halpern and T. L. Nichols (unpublished observations, 1974) to be about 30 msec. A small volume of liquid is carried into the mouth. This licking/lapping sequence requires considerable synchronization of the intrinsic and extrinsic tongue muscles (Tables 2 and 3). In contrast, other animals ingest liquids by producing a vacuum within the oral cavity (Table 4). No protrusion of the tongue is involved in this suction drinking (Figure 9). Production of the vacuum probably requires a seal produced between the dorsal surface of the tongue and the palate (see Ardran and Kemp, 1955; Herring and Scapino, 1973).

Both licking/lapping and sucking require appropriate positioning of the tongue in space. In the general vertebrate case, precise positioning of a mobile appendage is aided by mechanoreceptors in the joint capsule (Mountcastle, 1968); in tendons, that is, tendon organs (Matthews, 1972); and in muscle, for example, muscle spindles (Granit, 1966; Matthews, 1972). The *jaw* muscles are equipped with these various mechanoreceptors. However, it should be noted that Goodwin and Luschei (1974) conducted a selective denervation—that is, large lesions of the tract of the mesencephalic nucleus of the trigeminal nerve (see Crosby, Humphrey, and Lauer, 1962, for a neuroanatomical description) study on primates (monkeys) that indicated that muscle spindles may be of little significance in control of mastication. For the *tongue,* the intrinsic muscles (Table 3)—that is, wholly contained within the tongue—operate no joints. Both the intrinsic and the extrinsic—that is, passing into the tongue, muscles (Table 2)—have few if any muscle spindles in many species (see papers by Cooper, 1953;

Table 1. Masseter Muscle of the Mandible: Attachments and Function
(Innervation Is from Branches of the Trigeminal Nerve Unless Otherwise Noted)

Animal	Origin	Insertion	Function	Comments
Pig[a] (miniature) Sus scrofa	Zygomatic arch (of temporal bone).	Mandibular angle (junction of ramus and mandible).	Adductor (closes jaw). Active in vertical and lateral phases of mastication.	Very thick.[n]
Rat[b] (laboratory albino) Rattus norvegicus	Lateral surface of maxilla, by a tendon.[b,e]	Caudal part of mandible[d] near pterygoid ridge.[b]	Mastication.[c,e] Power stroke (caudal–rostral mandibular movement over molars).[c]	Superficial. Rostral.
	Maxilla[b] or zygomatic arch.[d,e]	Caudal part of lateral surface of internal aponeurosis of masseteric ridge, which extends between masseteric ridge and deep masseter; or on lateral surface of mandible.[b]	Same as rostral.	Medial (posterior).
	An aponeurosis that develops from the tendon that connects the rostral superficial masseter to the maxilla.	Medial side of mandible.	Mastication[b] during power stroke. Also, beginning of opening movement.	Pars reflexa.[b]
	Zygomatic process[b] (infraorbital fossa)[d,e] of maxilla.	Aponeurosis of masseteric ridge[b] or on lateral mandible.[d,e]	Mastication[b,c,d] especially second half of closing, and power stroke.[b]	Anterior deep.
	Zygomatic arch.	Aponeurois of deep masseter.[b]	Mastication[b,e] especially second half of closing, and power stroke.	Posterior deep.
		Lateral surface of (ramus of[d]) mandible.[b]		Superficial.[b] Deep.

	Zygomatic arch.[f,g]	Coronoid fossa of mandible.[f,g]	Mastication. Adductor.[f,g,h,i]	
Cat *Felis domesticus*	Zygomatic arch.	Coronoid fossa of mandible.[f,g]	Mastication. Adductor.[f,g,h,i]	
Bat[j] (little brown) *Myotis lucifugans*	Zygomatic arch.	Angular process of (posterior) mandible.	Closes jaw. Protrudes lower jaw.	
Rabbit* (European domestic, or gray) *Oryctolagus cuniculus*	Zygomatic arch.	Ramus of mandible.	*Adductor. Mastication* (grinding).	
Opossum[k,l] (American) *Didelphis marsupialis*	Zygomatic process (of maxilla), zygomatic arch. Lower border and medial surface of zygomatic arch.	Mandibular angular process and ramus.	*Adductor. Mastication* (centric occlusion). Retraction (similar to macaque monkey, *Macaca mulata*).[m]	Superficial (muscle spindles present).
	Lower border and medial surface of zygomatic arch.	Mandibular ramus and lateral surface of coronoid process.		Deep.
Human[o,p,q,r] *Homo sapiens*	Zygomatic process of maxilla, and lower border of zygomatic arch.	Angle and ramus of mandible.	*Adductor. Mastication.* centric occlusion.	Superior.
	Lower border and medial surface of zygomatic arch.	Upper half of ramus and lateral surface of coronoid process of mandible.		Deep.

a Herring and Scapino (1973).
b Weijs (1973b).
c Weijs (1973a).
d Chiasson (1969).
e Greene (1968).
f Wischnitzer (1967).

g Walker (1967).
h Kent (1973).
i Kawamura (1964).
j Kallen and Gans (1972).
k Hildebrand (1974).
l Romer (1970).

m Matsunami and Kubota (1972).
n Sisson (1930).
o Basmajian (1974).
p MacConaill and Basmajian (1969).
q Matthews (1972).
r Meader and Muyskens (1962).

Note: *body of the hyoid bone* = basihyal bone. The horns of the hyoid bone (i.e., the cornus) differ much among mammals. The greater horn = the anterior horn in cat and bat (Figure 1), while the posterior (caudal) horn is the lesser horn. In rat, human, and rabbit, the posterior horn is the greater horn, while the anterior (cranial) horn is the lesser horn. The *maxilla* is a portion of the upper jaw, forming part of the skull. An *aponeurosis* is a flattened tendon, that is, a flattened connective-tissue structure bearing muscle fibers (Wischnitzer, 1972; Weijs, 1973b). A *raphe* is a midline connective-tissue ridge joining two bilaterally symmetrical muscles (Wischnitzer, 1972).

Table 2. *Extrinsic Muscles of the Tongue: Attachments and Function*
(Innervation Is from Branches of the Hypoglossal Nerve, Cranial XII)

Muscle	Animal	Origin	Insertion	Function	Comments
Styloglossus	Pig *Sus scrofa*[k]	Greater horn of hyoid bone.	Tongue.	In general, retracts tongue of terrestrial mammals.[h]	
	Rat (laboratory albino) *Rattus norvegicus*[b]	Base of paramastoid process of occipital bone.	Longitudinal intrinsic muscle of tongue, laterally.	*Elevates* tongue. *Moves tip downward* and/or posterolaterally.[f]	Entire Cranial XII, especially lateral division, involved.[f]
	Cat *Felis domesticus*	Lateral surface of greater horn of hyoid bone[c]; mastoid process of occipital bone.[c,d]	Apex of tongue,[c] on anteroventral lamina propria (anterior arch).[f]	*Retracts* tongue.[c] *Elevates*, tip down, moves posterolaterally.[f]	Entire Cranial XII innervates, especially lateral division.[f]
	Bat (big brown)[o] *Eptesicus fuscus*	Stylohyal portion of greater horn of hyoid bone.[g]	Lateral portions of tongue.[g]	In general, retracts tongue.[h]	A member of Vespertilionidae, same family as little brown bat, *Myotis lucifugas*.
	Rabbit (European gray) *Oryctolagus cuniculus*[j]	Jugular process of occipital bone.	Base to tip of tongue.	*Retracts* and *elevates* tongue.[h]	
	Opossum (American) *Didelphis marsupialis*[p]	Paraoccipital process of occipital bone.		Changes shape of tongue.	
	Human *Homo sapiens*[t]	Anterior and lateral surface of styloid process of hyoid (post horn) bone, and proximal part of stylomandibular ligament.	Apex of tongue on connective-tissue framework (anterior arch).	During mastication, functions to place and keep food under molars.	Anterior.
			Lateral portion of tongue in membrane of hyoglossus muscle.		Posterior.

Genioglossus	Pig (miniature) S. scrofa[a]	Genial spine of mandible (region of mandibular symphysis).	Toward tip of tongue.	Active during *drinking*, ingestion of food, contralateral *mastication*, and *swallowing*. Produces "guttered" tongue shape during swallowing.	Anterior.	
			Base of tongue.		Posterior.	
	Rat (laboratory albino) R. norvegicus	Tendon (aponeurosis) of geniohyoid, and mandibular[b,i] symphysis.	Transverse intrinsic muscle of tongue in superficial layer near tongue's ventral surface[b]	In general, *protrudes tongue* of terrestrial mammals.[h,i]	Division into an *anterior portion*, originating at mandible and inserting in middle third of tongue, and a *posterior portion*, originating from geniohyoid and inserting in posterior third of tongue, has been suggested.[i]	
	Cat F. domesticus	Medial surface of mandible at mandibular symphysis.[d,i]	Anterior region of tongue (anterior arch).	In general, protrudes (and depresses) mammalian tongue.[h]	Anterior.	Largest extrinsic muscle.
			Paramedian septum of tongue.[i]		Middle.	
			Hyoglossal membrane (between base of tongue and hyoid bone) and anterolateral part of body of hyoid bone.[i]		Posterior.	
	Bat (big brown) E. fuscus[g]	Mandible in vicinity of mandibular symphysis.[g]	Ventral portion of tongue. Some fibers reach raphe[c] of body of hyoid bone.	Protrudes and depresses.[h]	Member of the same family, Vespertilionidae, as little brown bat, *Myotis lucifugas*.	

(Continued)

Table 2 (Continued)

Muscle	Animal	Origin	Insertion	Function	Comments
Genioglossus (*Continued*)	Rabbit (European gray) *O. cuniculus*[j]	Medial surface of mandible immediately behind symphysis.	Medial portion of base of tongue.	Protrudes and depresses.[h]	Large extrinsic muscle.
	Opossum (American) *D. marsupialis*[p]	Genial depression of mandible behind mandibular symphysis.	Ventral portion of posterior half of tongue.	Important in both tongue position and tongue shape. *Protrudes* tongue. *Depresses* middle and posterior portions of tongue.	
	Human *H. sapiens*	Genial (superior[i]) spine (by musculotendon[i]) of medial surface of mandible at mandibular symphysis.[i,j]	Anterior third of tongue.[i,j] Abd-el-malek[i] finds it in connective-tissue framework of apex of tongue (anterior arch[i]), but others do not find it in anterior sixth of tongue.[i]	*Depresses* tongue.[i] Specifically depresses medial portion while raising lateral margins.[n] *Protrudes* tongue.[i,n] *Lateral shift* of tongue. Presses intermolar portion of tongue against palate.[i] Active during retraction of tongue and swallowing.[e,i]	Anterior.
		Tendon connected to superior genial spine of mandible.[i]	Middle third of tongue.[i]		Middle (not mentioned as a separate part by Abd-el-malek).[i,n]
		Between superior and inferior genial spines of mandible.[i]	Posterior third of tongue[i] and body of hyoid bone.[i,j]		Posterior.
Hyoglossus	Pig *S. scrofa*			In general, retracts and depresses tongue.	
	Rat (laboratory albino) *R. norvegicus*	Rostral surface of body of hyoid bone and greater horn of hyoid bone.	Into longitudinal intrinsic muscle of tongue at lateral borders of tongue.	*Retracts* and *depresses* root of tongue.[i]	

Animal	Bone	Connective tissue / location	Function	Position
Cat *F. domesticus*	Hyoid bone, either lesser horn[d,i] and/or body.[c]	Anterior connective-tissue frame of tongue (anterior arch[j] or dorsum of tongue.[c]	*Retracts*[c,i] and *depresses*[i] tongue.	Anterior.
		Hyoglossal membrane (see "cat—genioglossus") and paramedial septum of tongue connective-tissue framework.		Posterior.
Bat (big brown) *E. fuscus*[g]	Raphe[c] of body of hyoid bone.	In tongue ventral to styloglossal insertion.	Retracts and depresses tongue.[h]	
Rabbit (European gray) *O. cuniculus*	Body of hyoid bone.	Base of tongue medial to styloglossal insertion.	Retracts and depresses tongue.[h]	Anterior.
	Anterior and posterior horns of hyoid bone.			Posterior.
Opossum (American) *D. marsupialis*[p]	Hyoid bone.	Lateral part of body of tongue in posterior two-thirds.	*Retracts* tongue. Changes shape of tongue.	Large muscle.
Human *H. sapiens*[n]	Lateral part of anterior surface of hyoid bone, and greater horn of hyoid.	Anterior connective-tissue framework (anterior arch) of tongue.	*Depresses* medial portion of tongue, and raises lateral margins, during ingestion of liquids.	Anterior.
		Paramedian septum (tongue connective-tissue framework) and tongue muscles.		Posterior.

[a] Herring and Scapino (1973).   [e] Basmajian (1974).   [h] Montagna (1959).   [k] Ellenberber and Baum (1932).   [n] Abd-el-malek (1955).
[b] Weijs (1973b).   [f] Abd-el-malek (1939a).   [i] Doran and Baggett (1972).   [l] Abd-el-malek (1939b).   [o] Barbour and Davis (1969).
[c] Wischnitzer (1967).   [g] Sprague (1943).   [j] Craigie (1960).   [m] Friel (1974).   [p] Hiiemae and Jenkins (1969).
[d] Walker (1967).

Note: *body of the hyoid bone* = basihyal bone. The horns of the hyoid bone (i.e., the cornus) differ much among mammals. The greater horn = the anterior horn in cat and bat (Figure 1), while the posterior (caudal) horn is the lesser horn. In rat, human, and rabbit, the posterior horn is the greater horn, while the anterior (cranial) horn is the lesser horn. The *maxilla* is a portion of the upper jaw, forming part of the skull. An *aponeurosis* is a flattened tendon, that is, a flattened connective-tissue structure bearing muscle fibers (Wischnitzer, 1972; Weijs, 1973b). A *raphe* is a midline connective-tissue ridge joining two bilaterally symmetrical muscles (Wischnitzer, 1972).

Table 3.   Intrinsic Muscles of the Tongue
(Innervation Is from the Hypoglossal Nerve, Cranial XII)

Muscle	Animal	Origin	Insertion	Function	Comments
Superior longitudinal	Cat Felis domesticus[a]	Anteromedian portion of body of hyoid bone, hyoglossal membrane (extends from connective-tissue framework at base of tongue to hyoid bone) and dorsal mucous membrane of tongue.	Connective tissue underlying dorsal mucous membrane of tongue, and anterior portion of connective tissue framework of tongue (anterior arch).	Moves tip of tongue up, deviates tip to contralateral side.	When active together with other intrinsic muscles (electrical stimulation of hypoglossal branches), produces more tongue movement and less anterior–posterior-oriented concavity than extrinsic muscle alone.
	Human Homo sapiens[b]	Hyoglossal membrane (see above).	Anterior portion of connective-tissue framework of tongue (anterior arch).	Mastication. Swallowing. Changes shape of tongue.[c]	
Inferior longitudinal	Cat F. domesticus[a]	Anterolateral part of body and horn of hyoid bone.	Connective tissue underlying mucous membrane of ventral–lateral part of anterior portion of tongue, and anterior arch.	Moves tip of tongue down. Lateral movement of tip.	
	Human H. sapiens[b]	Anterior surface of hyoid bone.	Connective-tissue framework at tip of tongue.		Innervation through chorda tympani nerve.[d]

Transverse	Cat *F. domesticus*[a]	Middle part of median septum of tongue.	Connective tissue under dorsolateral mucous membrane of tongue, longitudinal muscles of tongue, and median septum of tongue.	Protrudes tongue. Narrows tongue.	
	Human *H. sapiens*	Median septum of tongue.	Connective tissue under lateral margins of tongue.		
Vertical	Cat *F. domesticus*[a]	Transverse muscle of tongue.	Connective tissue underlying dorsal and ventral mucous membranes.	Flattens tongue.	
	Human *H. sapiens*	Ventral submucosal connective tissue.	Connective tissue under dorsal mucous membrane.		Long fibers.
		Paramedian connective-tissue septum of tongue.	Septa (connective-tissue internal divisions) of tongue muscles.		Short fibers.

[a] Abd-el-malek (1939a).
[b] Abd-el-malek (1939b).
[c] Friel (1974).
[d] Sogmaes (1954).

Note: *body of the hyoid bone* = basihyal bone. The horns of the hyoid bone (i.e., the cornus) differ much among mammals. The greater horn = the anterior horn in cat and bat (Figure 1), while the posterior (caudal) horn is the lesser horn. In rat, human, and rabbit, the posterior horn is the greater horn, while the anterior (cranial) horn is the lesser horn. The *maxilla* is a portion of the upper jaw, forming part of the skull. An *aponeurosis* is a flattened tendon, that is, a flattened connective-tissue structure bearing muscle fibers (Wischnitzer, 1972; Weijs. 1973b). A *raphe* is a midline connective-tissue ridge joining two bilaterally symmetrical muscles (Wischnitzer, 1972).

Table 4. Liquid-Intake Patterns for Water

Intake mode		Species and genus (common name)	Family	Order	Mean-intake rate		Comments	Literature citations
Lick	Suck				Events sec	Volume (ml) sec		
X		*Didelphis marsupialis* (American opossum)	Didelphidae	Marsupialia	4.6 *licks*/sec.		55 licks/burst. Jaws are opened 3°–10° (mean gape = 6°). About 16 licks/swallow.	Hiiemae and Crompton (1971)
		Didelphis virginiana (Virginia opossum)			3.4 *licks*/sec.		Licking rates from 1.8 to 6.0 licks/sec have been seen in 23-hr deprived animals drinking from 3.25-mm tip i.d. tube, reached through 2-cm diameter hole. From 92 to 150 licks/burst, where burst is ≥ 5 sec licking with ≯ 1 sec nonlicking.	Cone *et al.* (1971) Cone *et al.* (1973) Cone (1974)
X		*Felis domesticus* (domestic cat)	Felidae	Carnivora	3.5–4 *licks*/ sec.			Taylor (1969) Schaeffer and Huff (1965)
X		*Rattus norvegicus* (Laboratory rat)	Muridae	Rodentia	6.5 *licks*/sec.	0.03–0.04 ml/ sec drinking tube; 0.05, open (1 × 1 cm) surface.	Some of drinking-tube volume is lost but none of open-surface. Dorsal surface of tongue is "wiped clean" between licks. Liquid is accumulated in a posterior space (diastema) before swallowing.	Halpern (1975) B. P. Hapern and T. L. Nichols (unpublished observations) Hiiemae and Ardran (1968) Schaeffer and Premack (1961)

X	*Mesocricetus auratus* (Golden hamster)	Cricetidae	Rodentia	5 *licks*/sec.	0.025 ml/sec, drinking tube.	A range from 4.0 to 5.5 licks/sec has been reported.	Faull and Halpern (1972); Schaeffer and Huff (1965)
X	*Choeronycteris mexicana* (Hog-nose bat)	Phyllosto-matidae	Chiroptera			Nectar-eating bat. In captivity, laps aqueous sucrose solution. Bats commonly drink in flight by scooping up water with lower jaw.	Barbour and Davis (1969)
X	*Bos taurus* (Cow)	Bovidae	Artio-dactyla			In calves, sucking occurs at 74 sucks/min. Calves can be weaned from nipple to open liquid surface.	Hafez and Schein (1962)
X	*Cavia por-cellus* (Guinea pig)	Caviidae	Rodentia			Normally obtain water from drinking tube only after lip contact and intraoral insertion. However, can learn to lick with tongue contact only.	Mendelson *et al.* (1973); Alvord (1968)
X	*Sus scrofa* (miniature pig)	Suidae	Artio-dactyla	3.5 *sucks*/sec.		1 swallow/sec. Produces reduced pressure by tongue and soft palate or by closing internal nares plus lowering floor of mouth.	Herring and Scapino (1973)
X	*Homo sapiens* (Human) ADULT	Hominidae	Primates	0.7 sips/sec.	12 ml/sip. Range 1-31 ml/sip.	Each sip is 1000 (893-1051) msec in duration (liquid contact). One sip each 1.5 sec. Swallows start after sip ends (single sip).	Halpern (1975); Schmidt-Nielsen (1964); Halpern and Nichols (1975); B. P. Halpern and T. L. Nichols (unpublished observations)

(Continued)

Table 4   (Continued)

Intake mode		Species and genus (common name)	Mean-intake rate			Comments	Literature citations
Lick	Suck		Order / Family	Events sec	Volume (ml) sec		
		INFANT Nutritive suckling (artificial nipple)		1–1.5 sucks/ sec. (Range = 1.0–2.4 per sec.)	0.2 ml/suck (70 mM sucrose or water).	1.0/sec during first 5 days. 1.5/sec by 7 months. Initial burst of 50–200 sucks, then 12 sucks/burst (range = 6–18) in later bursts. A normal 8-year-old child sucked at 1.6 sucks/sec for apple juice, with 68-suck initial burst, and 16 sucks, later bursts.	Wolff (1972) Kemper (1972)
				0.66 sucks/ sec (on third trial. Lower on first and second trials).	0.03 ml/suck (milk).	Liquid flow rate through nipple is restricted by external calibrated capillary tube used to measure volume per suck.	Kron et al. (1967)
				0.75 sucks/ sec for water; 1.0 sucks/sec for 0.44 M sucrose.	Fixed volume of 0.02 ml/ suck delivered for each criterion suck (automatic pump).	During nonnutritive sucking periods (i.e., when suck-controlled pump was off), suck bursts (≤2 sec between sucks) contained 11–13 sucks, with intersuck interval of 658 msec. During 0.44-M sucrose delivery periods, 24–30 sucks per burst; 856 msec intersuck intervals. During nutritive sucking, swallows occurred	Lipsitt (1977) Crook and Lipsitt (1976)

				Description	Rate	References
				every 3.7–5.4 sec; respirations, every 0.92 sec. In a second study, 20-mM sucrose produced 760-msec intersuck intervals; 440-mM sucrose, 820-msec intervals. The two sucrose concentrations did not produce significant differences in number of sucks or pauses. For both studies, full-term neonates (24–72 hours old) were examined.		Graber (1963)
				Shape of the artificial nipple interacts with intake pattern. Conventional artificial nipples may produce a sequence quite different from natural suckling.		
				Sucking rate increases with age, at least through 7 months. 4–14 sucks/burst. Suck duration = 0.3 sec/suck. 4 sucking bursts/min. with total of 50–60 suck/min.		Kaye (1972) Wolff (1972) Butterfield and Siperstein (1972)
		INFANT Nonnutritive sucking			1–3 sucks/sec.	
X	*Pantroglodytes* (Chimpanzee)	Pongidae	Primates	Sucks from streams (lips); sucks on fingers after dipping into water in tree "bowl." Leaves are rolled or crumpled, and used as "sponge."		Reynolds and Reynolds (1965) Goodall (1965) Jolly (1972) Van Lawick-Goodall (1968)

Hosokawa, 1961; Kawamura, 1964). Cutaneous mechanoreceptors may be of major importance in identifying tongue position, as Yokota, Suzuki, and Nakano (1974) have suggested.

The various superficial and deep receptors of the tongue will be considered in relation to liquid ingestion. All mammalian taste input is through differentiated epithelial structures, the taste buds. On the tongue, the taste buds are usually, though not invariably, located in specialized regions of the mucosa known as *papillae.* Three types of mammalian lingual (i.e., tongue) "gustatory" papillae are commonly recognized: fungiform, foliate, and circumvallate papillae. A wide variation exists in the total number of papillae and taste buds present on the tongue of mammals, as well as in the occurrence of a particular type of papilla in specific groups of mammals. With regard to the number of taste buds, the range is from about 2000 in the domestic cat to approximately 50,000 in the antelope. The adaptive significance of the differences in number of taste buds and types of papillae present is obscure (see Sonntag, 1925; Bradley, 1971).

The sequence of tongue movement in the ingestion of liquids will also be reviewed, as well as the muscles and the efferent neural elements that create the movement patterns. Brief consideration will be given to the already-well-described process of swallowing (deglutition), with emphasis on the ingestion–deglutition sequence.

For all topics covered, the treatment is necessarily selective. Species for which functional studies are available, and which add to the coverage of a wide range of mammalian orders, are included. It will be seen that there are many similarities among the mammals in liquid ingestion, but inter-mammalian differences are also considerable. Broad generalization, even about species of the same order (e.g., a rat, a beaver, and a squirrel of the order Rodentia), should be made only when supported by clear evidence. Still greater caution is required for statements that attempt to include several orders of mammals. Although many mammals can fit a variety of ecological niches, adaptive specializations are common.

2. Receptors

2.1. Superficial Receptors

2.1.1. Chemoreceptors

2.1.1.1. Tongue
(A) Epithelium and Papillae—General Considerations. The mammalian tongue has several types of distinct structures, the taste-bud–bearing papillae, in which taste buds (differentiated chemoreceptor organs) gener-

ally occur. The mammalian taste buds are spherical or ellipsoidal multi-cellular structures, which have been described by many authors, including Sognnaes (1954), Pfaffmann (1959), Beidler (1965, 1969), Scalzi (1966), Graziadei (1969), and Farbman (1971, 1972). The taste bud is composed of 30–60 cells. These cells are long (about 80 μm) and narrow. The superficial ends of taste-bud receptor cells terminate in microvilli, which project into a common space, the taste pit. The taste pit connects with the surface of the tongue through an opening, the taste pore. It is generally assumed that interactions between potential gustatory stimuli and taste-bud receptor cells usually occur at the microvilli (see Beidler, 1965).

Each taste bud is separated from the underlying connective tissue of the papilla or mucosa by its basement membrane, as De Lorenzo (1963) has observed. Beneath the basement membrane, myelinated nerve fibers approaching the taste bud tend to lose their myelin sheath. This region of nerve fibers immediately below the taste bud is called the *subgemmal plexus*. Before entering a papilla, each axon branches many times, averaging about seven such branchings (see Beidler, 1969). Myelinated and unmyelinated axons are present in the papilla beneath the bud, but Beidler (1965, 1969) found that only nonmyelinated axons occur within a taste bud. In addition, these nonmyelinated axons were seen to branch extensively inside the taste bud. Thus, each receptor cell is probably innervated by several different axons from the subgemmal plexus. The branching that is observed beneath a papilla and within a taste bud suggested to Beidler (1965) that a single axon in a nerve innervating taste buds may reach several taste buds. This concept has been confirmed in electrophysiological studies by Miller (1971, 1974, 1975) and other workers. Oakley (1975) has observed that the taste buds innervated by a single axon may be many millimeters apart.

The taste buds of the tongue are generally considered to be of epithelial origin, as Farbman (1971) indicated. The cells of the taste buds continuously differentiate from epithelial cells during life (see Beidler and Smallman, 1965; Beidler, 1965; Conger and Wells, 1969). However, since lingual taste buds are restricted to specific loci (the gustatory papillae or the anteroventral mucosa), either the potential for differentiating into taste buds is present only in certain epithelial cells, as Oakley (1974) has suggested, or perhaps particular environmental factors (e.g., a rich blood supply) must be present. In all cases, a peripheral *taste* nerve *must* be present for mammalian taste buds to appear or remain, as Oakley (1974) has clearly demonstrated. Farbman (1971) indicated that lingual epithelium from which taste buds differentiate probably derives, in turn, from the ectodermal layer in the developing embryo. More specifically, it appears that the oral division of the tongue, which ends just in front of the circum-vallate papilla(e), is covered with ectodermal epithelium, while the pharyn-geal division, which extends back from the circumvallate papillae, is covered with epithelium of endodermal origin (see Arey, 1956). The distinction between an oral, or *anterior,* division of the tongue, and a pharyngeal, or *posterior,* division will be used here many times.

The lingual papillae, specialized regions of the covering tissue of the tongue (the mucous membrane) were described by Sognnaes (1954) and by Ham (1974). The core of a lingual papilla is a cylindrical or conical structure, the primary papilla. At the top, or the top and sides, of the primary papilla, numerous small connective-tissue elevations, the secondary papillae, project outward from the primary papilla. Both the secondary and the primary papillae are extensions of a deep layer of the mucous membrane, the lamina propria. The lamina propria of all lingual papillae is a vascular connective tissue containing fat and nerves. In common with the remainder of the oral cavity, the outermost tissue of the lingual papillae is a sheet of stratified squamous epithelium.

This type of epithelium, as described by Dean (1954), is composed of multiple cell layers. In stratified squamous epithelium, cells migrate upward from a basal layer toward the surface. Cell division occurs only in the deeper layers, while the more superficial cells are flat and degenerate. The most superficial cells are continuously desquamated, or shed. All the cells of this epithelium are closely coupled to each other. The basal-layer cells are embedded in the basement membrane, a fibrous connective tissue sheet that, on the tongue, separates the epithelium from the lamina propria. Blood and lymphatic vessels do not penetrate into the epithelium, but, as Beidler (1965) noted, nerves do. The tight coupling between cells plus the basement membrane results in this epithelium's being a relatively impervious tissue (Mistretta, 1971).

Within the oral cavity, the stratified squamous epithelium of most of the dorsum of the tongue, as well as the hard palate, has a specialized outer layer. In these loci, the surface epithelial layer is, as described by Sognnaes (1954), cornified cells. That is, the superficial cells of most of the epithelium of the dorsum of the tongue contains keratin, a fibrous insoluble protein. Consequently, as Beidler (1965) reported, the epithelium of the dorsum is rather impervious to most substances.

In contrast to most of the lingual dorsum, the epithelial layer of human fungiform and circumvallate papillae, as described by Ham (1974), is not keratinized. Therefore, these papillae appear red in life, because the blood-filled capillaries of the lamina propria can be seen through the nonkeratinized epithelium. Similarly, an intravascular dye will make these lingual gustatory papillae especially visible (Figure 3). It follows that the low permeability of the dorsum of the tongue, to the extent that it is dependent upon keratin in the superficial layer of the epithelium, may be less evident at fungiform and circumvallate papillae. Mistretta (1971) found that fungiform-papillae–bearing lingual epithelium of rat is more permeable than abdominal skin, although the papillae had 5-μm keratin layers.

The lingual gustatory papillae cluster in specific regions of the dorsal (and sometimes ventral) surface and sides of the tongue, as described by many workers (e.g., Sonntag, 1920, 1925; Pfaffmann, 1959; Bradley, 1971; Halpern, 1973). These taste-bud–bearing papillae are usually divided into three classes: fungiform, foliate, and circumvallate papillae. The foliate and circumvallate papillae generally occur in restricted, consistent regions of the tongue for all mammals in which they are found, but they do vary in number and spatial layout.

In contrast, the fungiform papillae show many differences in location and density in mammals. The fungiform and/or the foliate papillae are absent from some groups of mammals, but circumvallate papillae occur in almost all orders of mammals. A further complexity is that, in addition to the well-known taste buds of the lingual papillae, in some mammalian forms taste buds are found embedded in the mucosa of the anteroventral portion of the tongue.

Behavioral investigations of long-term—that is, intake choices made over periods of many hours or days (see Halpern, 1973, p. 345)—taste preferences in domestic rats with lesions of a peripheral nerve which innervates a specific taste-bud–papilla population, have been used to indicate whether grossly different, nonoverlapping behaviors are supported by the input through each gustatory nerve. The answer is clearly a negative one. These lesion studies demonstrate that only small quantitative changes, and often no discernible preference alterations, follow chorda tympani (fungiform papillae) or glossopharyngeal (foliate and circumvallate papillae) nerve transection (see Pfaffmann, 1952; Oakley, 1969). Transection of both nerves generally produces large preference changes as Richter (1956) and Vance (1966) observed. Thus, the specific roles of the several gustatory nerves remains unclear. In addition, the arrangement, number, and type of papillae and related taste buds that are present in the normal, nonlesioned animal does not relate to the diet of that animal in an obvious manner, as Sonntag (1925) has noted.

Neurophysiological studies of responses from each of the three types of lingual gustatory papillae and their associated peripheral nerves indicate broadly tuned, highly sensitive peripheral inputs. The fungiform-papillae-induced responses recorded from the chorda tympani branch of the facial nerve (Cranial VII) have usually been taken as the standard of comparison. An aqueous solution of a monovalent salt such as NaCl is commonly employed as the reference stimulus, to which responses to other solutions are compared. Responses from nerves innervating the several populations of taste-bud–bearing papillae are then reported on a "relative" basis, with the response to a fixed concentration of NaCl taken as the comparison standard. This practice, using the chorda tympani nerve response to one concentration of a single solution as the basis for all response comparisons, can produce misleading conclusions. Consequently, Frank (1968), after a detailed study of foliate and circumvallate papillae inputs in the domestic rat, pointed out "rather than saying that the back of the tongue is deficient in salt receptors because the relative size of the whole nerve response to NaCl (compared with responses to other stimuli) is smaller, it could be that the front of the tongue is deficient in quinine and perhaps sugar receptors" (p. 108). The studies of responses from the circumvallate and foliate papillae do suggest that they provide a sensitive, broadly tuned yet dis-criminating, input.

Investigations on the functional anatomy of the tongue and mouth of mammals, such as those referred to in the preceding paragraphs, employ a wide range of techniques. Many of these techniques are the standard approach of electrophysiology and neuroanatomy. I will briefly describe a few procedures that are especially pertinent to the following portions of this chapter. A more technically detailed presentation of these and other topics is available in Halpern (1973) and the references cited therein.

(B) Analytical Techniques

(1) Neurophysiological Recording. Electrophysiological analyses are most commonly made by bringing a nontoxic, stable, voltage sensing probe—a bioelectrode—to the vicinity of one or more functional neurons. The general concerns have been reviewed by Geddes (1972) and Brown, Maxfield, and Moraff (1973). Two classes of recording are done: single-unit recording and multiunit recording.

In *multiunit* neurophysiological studies, the bioelectrode, usually refer-red to as the electrode, records simultaneously from a large number of neurons. One typical case would be an intact peripheral nerve lifted out of contact with body fluids by a hook-shaped electrode, or the cut end of such a nerve sucked into a tube containing an electrode. Another frequent application of multiunit recording would be a population of neurons in the central nervous system studied by the insertion of an electrode with a rela-- tively large uninsulated tip (say, 25–100 μm in diameter) into the mass of neural tissue. Thus, multiunit recording measures the voltages produced by a population of neurons. All bioelectric activity in the general vicinity of the electrode affects the measured response, but some elements in the popula-tion contribute more than others. The problems associated with quantifying multiunit responses have been reviewed by Halpern (1973). An analogue summator, commonly called an *integrator* in the chemical-senses research literature, is often used to provide a readily quantified transformation of multiunit responses. The integrator produces a running average of the popu-lation response. As typically employed, this device yields a rather distorted picture of the temporal pattern of the bioelectric activity.

Single-unit recording differs from multiunit recording in that the bioelectric activity of an individual neuron—or a few, independently identi-fiable neurons—is measured. A functional analysis of individual neural or receptor elements is thus available. For peripheral nerves, this may be done by a surgical dissection of the nerve until only one or a few functioning axons remain. In effect, a nerve branch is damaged so that most of the axons no longer propagate action potentials. This technique makes isolation of the small-diameter axons, which are especially sensitive to trauma, very difficult. Within the central nervous system, as well as the ganglia of cranial nerves in which their cell bodies are located, single-unit recording is done by the use of a specialized bioelectrode with a very small, specifically prepared tip region, a microelectrode. The microelectrode technique has its own limitations. Most notably, only one or a few neurons are recorded from, although the electrode is within, and is moved through, a mass of neurons. The reasons for this selectivity, and the possible biases it introduces, are unknown. Microelectrodes may also be used to record from peripheral nerve axons and from nonneural receptor elements. Taste-bud receptor cells have been studied by using microelectrodes. The interpretations of these experiments are quite difficult, as Halpern (1973) has noted. With reference to peripheral nerves, attempts to record gustatory afferent responses with a microelectrode in mammals have had little success, perhaps because of the small diameter of the axons. A general discussion of microelectrodes is available in Kay (1964).

(2) Nerve Transection and Regeneration. When peripheral nerves are cut at a distance from the cell body that is sufficiently great to permit the cell body to survive, the nerve will begin to grow again; that is, it will regenerate. Often this regeneration eventually reestablishes functional connections with motor or sensory structures comparable to those existing before the nerve was cut. The growth path of the regenerating nerve typically follows the course of the distal segment of the axon, which degenerates after its separation from the cell body. The process of peripheral-nerve degeneration and regeneration is described in many sources (e.g., Ochs, 1965; Ham, 1974). Experimental manipulations can cause the regenerating nerve, growing out toward the periphery, to proceed along the path normally followed by some other nerve. This process, in which the proximal portion of a nerve regenerates along the degenerated distal segment of another nerve, is called *cross-regeneration.* Cross-regeneration permits tests of the relative roles of the nerve and of the tissues reached by a nerve on the functional properties of the system (e.g., see Oakley, 1974).

(C) Lingual Gustatory Papillae

(1) Fungiform papillae. On the anterior division of the tongue of many mammals, cylindrical, typically elevated structures—the fungiform papillae—are located (Figure 3). These papillae sometimes also occur in the region of the circumvallate papillae (i.e., the "vallate region"), which is located on the posterior–anterior divisions border. Each fungiform papilla contains from one to many taste buds on its top surface. A single taste bud

occurs in fungiform papillae of the dormouse, *Glirus japonicus* (Kubota and Togawa, 1966), the laboratory albino rat, *R. norvegicus* (Beidler, 1965; Beidler and Smallman, 1965; Bernard and Halpern, 1968; Miller, 1974), and most other myomorphic rodents (see Sonntag, 1925; Bradley, 1971). Members of this suborder, Myomorpha, include mice, rats, hamsters, and voles. In contrast, three to four taste buds occur in a human's (see Pfaffmann, 1959; Henkin, 1967; Bradley, 1972) fungiform papilla or that of a domestic cat, *Felis domesticus* (e.g., Oakley, 1975), while many taste buds were noted by Kubota and Hayama (1964) in each fungiform papilla of some mammals, such as the marmoset.

The fungiform papillae are scattered over the tongue of some animals but distributed nonuniformly over the surface of the tongue of others. An increase in apical and lateral clustering in "higher" orders of mammals has been suggested by Sonntag (1925). Fungiform papillae are clustered near the tip and/or the anterolateral margins, as Sonntag (1925) has noted, in anthropoid and some lemurioform primates, including humans (Arey, 1956), as well as in hare and rabbit (Leporidae), some families of the order Carnivora (especially Felidae, Procyonidae, and Ursidae), the aardvark (Tubulidentatia), suiformes (pig, hog, hippopotamus), and the laboratory albino rat (e.g., Miller, 1974; Mistretta, 1972). In the marmoset, Kubota and Hayama (1964) observed a large fungiform papilla at the tip of the tongue.

In mammals that have an anterolateral clustering of fungiform papillae, fungiform papillae are less densely distributed on the medial portions of the anterior surface and are usually absent from the midline region of the anterior dorsum. A quantitative study on the rat reported by Miller (1974) and Miller and Preslar (1975) provides a specific example of a nonuniform, anterolateral-concentrated, fungiform-papillae distribution. This is illustrated in Figure 3. In addition to the "dorsum-only" location, Kubota and Iwamoto (1967) and Bradley (1971) described these papillae on the ventral surface of the tongue of the slow loris, *Nycticebus coucang,* a lorisioform Asian primate, and members of the suborders Tylopoda (camels and llamas) and Ruminantia. In one lagomorph, the pika, *Ochotona hyperborea yezonensis,* Kubota (1966) found no fungiform papillae, but the ventrum of the tip had many taste buds in the epithelium. As well as in the pika, fungiform papillae are absent in the duck-billed platypus, Cetacea (whales), the gray seal, and Myrmecophageae (giant and lesser anteater; see Bradley, 1971).

Fungiform papillae are not restricted to anterior tongue loci. These papillae also occur in the vallate region (i.e., the region of the circumvallate papillae) of some primates (Sonntag, 1921; Ishiko, 1974) and ungulates (Sonntag, 1922). For one ungulate, the cow (*Bos taurus*), chorda tympani nerve responses were recorded by Bernard (1962, 1964) upon posterior

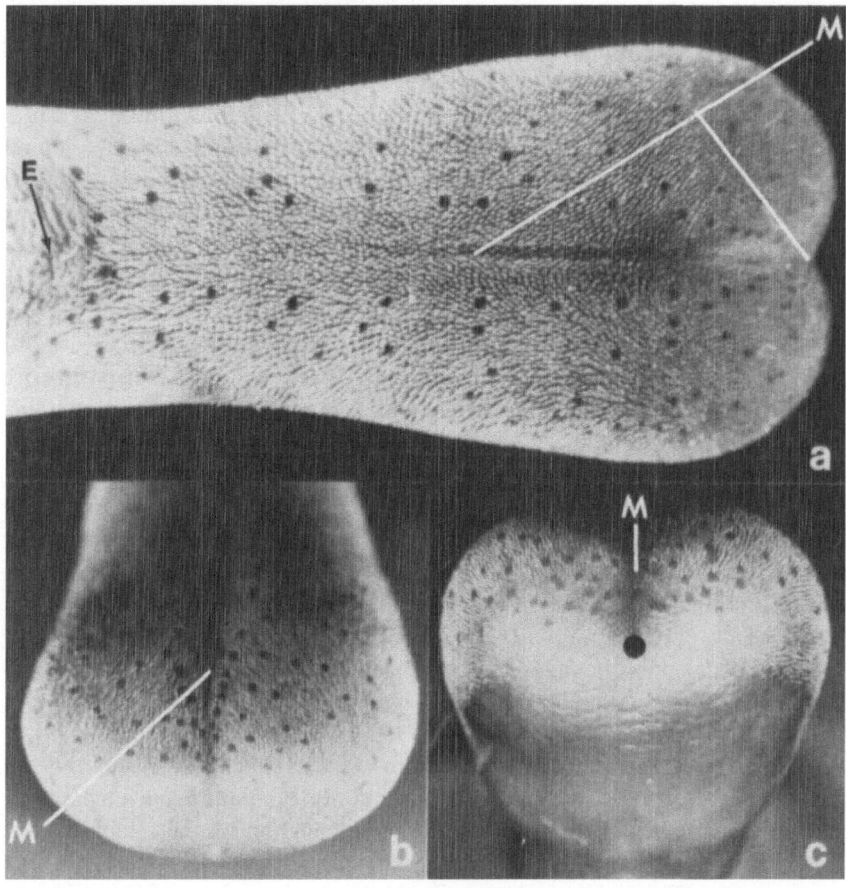

Figure 3.   Photomacrograph of exterior surface of the anterior (oral) portion of the tongue of a laboratory albino rat (*Rattus norvegicus*). The fungiform papillae, which appear as dark spots, were made readily visible by vital staining with an intracardiac perfusion with 0.1% methylene blue in 5% formalin. The filiform papillae, which do not bear taste buds, give a rough, gray appearance. M is the median dorsal sulcus; E, the intermolar eminence. Calibration line in lower right is 2 mm. Fungiform papillae are most common near the tip of the tongue, with 52% of all fungiform papillae within 4 mm of the tip. At the tip, the density is 3.4 papillae/mm^2. In the medial-lateral direction, the maximum density is between 0.2 and 1.0 mm from the midline (M). a, Dorsum of anterior portion of the tongue. There is an average of 187 $\pm$ 9 (mean $\pm$ SD) fungiform papillae on the rat's tongue. Average counts of 179-149 fungiform papillae per rat's tongue were reported in a study by Beidler and Smallman (1965). b, tip of tongue, top view. The most rostral papillae are out of focus. c, ventral view of tip of tongue. High density at the tip and absence of ventral papillae are clear. The black circle is a reference mark used to plot the location of each papilla. Modified from Miller and Preslar (1975).

tongue-region gustatory stimulation. Similar evidence of functional overlap has been obtained in macaque monkeys (*Macaca irus*) by Ishiko (1974) and in the domestic cat by Oakley (1972, 1975) and Ishiko (1974). The apparent anatomical absence of fungiform papillae in the vallate region of the rat is confirmed by Frank's (1968) observation of a lack of rat chorda tympani nerve responses to posterior tongue-region stimulation. A similar lack of chorda tympani responsiveness to gustatory stimulation of the posterior region of the tongue has been described in sheep by Iggo and Leek (1967).

Anatomical studies (e.g., Elliott, 1963; Farbman and Allgood, 1971; Rollin, 1973, Sato, 1973; Williams and Warwick, 1975) indicate that the taste buds of the fungiform papillae are innervated by special visceral afferent fibers of the facial nerve, that is, the seventh cranial nerve. More specifically, those facial-nerve axons which innervate the fungiform papillae travel primarily as the chorda tympani nerve, with cell bodies in the geniculate ganglion (see Boudreau, Bradley, Bierer, Kruger, and Tsuchitani, 1971). Before reaching the tongue, the chorda tympani nerve joins with the lingual branch of the trigeminal nerve. The gustatory afferent axons of the chorda tympani nerve thus are part of the facial nerve but reach the tongue and its papillae in branches of the lingual nerve. This arrangement has been described by Pfaffmann (1959) and studied electrophysiologically by Ishiko (1974). Occasionally, a suggestion of trigeminal-nerve (Cranial V) innervation of fungiform-papillae taste buds is made. However, tests of this concept have routinely provided negative results (e.g., Beidler, 1969; Rollin, 1973).

Nonetheless, as already discussed, an identifiable chorda tympani nerve does not reach the tongue. Instead, the chorda tympani nerve innervation reaches the tongue in branches of the distal portion of the lingual nerve. These trigeminal branches have been reported in detail by Ishiko (1974) for some species. Thus, three relevant lingual-nerve branches are described in the domestic cat, *F. domesticus* (anterior, medial, and posterior branches), while in the "crab-eating" macaque monkey, *M. irus,* five branches are recognized. In the domestic cat, the medial branch of the distal lingual nerve, which innervates 38% of the total length of the anterior division of the tongue, carries the major taste input. This conclusion is based upon multiunit response magnitudes of the chorda tympani nerve to localized chemical stimulation before and after selective transection of the distal branches, as well as direct recording from the distal branches. The three distal lingual-nerve branches to the tongue also differ among themselves in their threshold and relative response magnitude to quinine hydrochloride and hydrochloric acid. Overall, these observations indicate that in the domestic cat the tip of the tongue, innervated through the anterior branch of the distal linqual nerve, is *not* highly responsive to taste stimuli. This tip region, which constitutes 29% of the length of the anterior division of the tongue (the first 16 mm of a 54-mm anterior-division length), appears less sensitive than either the medial or the posterior regions of the anterior division of the tongue. Although these latter two regions are somewhat larger than the anterior lingual-branch region (38% and 33% of the length of the anterior division of the tongue), their responses exceed those of the anterior region by 152% and 64%, respectively.

In contrast, results from the laboratory albino rat (*R. norvegicus*) do indicate that taste input from the tip of the tongue is of major importance. In this rat species, a very pronounced concentration of lingual gustatory papillae is present at the tip of the tongue. The fungiform papillae of the most anterior 4 mm of the tongue (20% of the total length of the anterior division of the tongue) are approximately equal in *number* to the fungiform papillae of the "mid-region" of the tongue, which starts 5 mm from the tip of the tongue and extends to 20 mm from the tip (see Miller and Preslar, 1975; Miller, 1975). That is, the rear 15 mm of the anterior division of the tongue, extending to the intermolar eminence (Figure 3), constitutes the mid-region of the tongue and contains one-half of the total number of fungiform papillae, each with one taste bud. However, multiunit response magnitudes of the chorda tympani nerve to stimulation of the anterior 4 mm of the tongue are *not* equal to response magnitudes from stimulation of the mid-region. Instead, the chorda tympani nerve responses to chemical stimulation of the anterior 4 mm of the rat tongue (separated from the mid-region with a diaphragm) with 300 mM NaCl, 10 mM HCl, or 300 mM sucrose were 30% to 40% larger than responses to stimulation of the mid-region of the tongue. An explanation for this substantial difference, in terms of possible greater overlapping innervation of taste buds and more lateral enhancement in the anterior region, was proposed by Miller (1975).

The results in cat and rat are clearly different. It may be that there are few taste buds in the tongue region innervated by the cat's anterior branch of the distal lingual nerve, although anatomical reports suggest a clustering of fungiform papillae near the tip of the tongue in Felidae. Since several taste buds are present in each domestic-cat fungiform papilla, branching and lateral interactions may be less directly related to the number of fungiform papillae *per se.* Whatever the reason, it seems likely that substances may have to reach the middle of the tongue of the domestic cat before substantial taste input is available, while appreciable taste input is provided by the tip of the tongue in the laboratory albino rat.

During liquid ingestion (see pp. 62–79), the many chemoreceptors of the anterior and/or anterolateral regions of the tongue make early contact with the liquid, perhaps facilitating rapid decision on continued ingestion and swallowing. For licking/lapping animals that are equipped with a concentration of taste-bud–bearing papillae at the tip of the tongue, this anterior chemoreceptor array initially contacts substances in an extraoral phase of ingestion (Figure 3). For one licking/lapping mammal, the laboratory albino rat (R. norvegicus), decisions to stop further ingestive licking that are based upon a single initial lick have been observed by Halpern and ·Tapper (1971) and by Scott (1974). The elapsed-time period for this decision must be less than the duration of a licking cycle, plus the duration of a noncontact (i.e., aborted) lick and minus the gustatory neural-response latency. At the beginning of a period of licking, a single-lick cycle from the first contact to the beginning of the second contact was found by Halpern (1975) to be 131 msec. The neural-response latency measured by Halpern and Marowitz (1973) was 30 msec, while tongue-extension time for an aborted, noncontact lick was found by B. P. Halpern and T. L. Nichols (unpublished observations, 1974) to be 20 msec. The resultant (131 msec − 30 msec + 20 msec) 121 msec is the time period available for a single-lick rejection decision. This rapid cessation of ingestion limits intake to about 6 μl (Table 4) (see Halpern, 1975). Perhaps even less intake is possible. Thus, it may be that this type of single-lick rejection decision can reduce or even prevent the small but potentially significant ingestion that would normally be associated with a single lick (i.e., 6 μl). This zero-volume ingestion would require sufficient gustatory information to be available during the extraoral stages of ingestion. To estimate the possibility of this, both behavioral and neural data are needed. Relevant data do exist: behavioral tongue contact with a drinking tube during the *first* lick of a liquid that has novel, not previously experienced, taste properties has a median duration of 48 msec in the rat (Halpern and Nichols, unpublished observations, 1974; Halpern, 1975). After contact ends, about 15 msec pass before the tongue is fully retracted into the mouth. Consequently, an extraoral ingestion interval of 63 msec (48-msec contact plus 15-msec tongue-retraction time) precedes the intraoral ingestion phase.

Electrophysiological studies can indicate the duration of neural information available to the domestic rat during this 63-msec extraoral ingestion period. Since the taste buds of the fungiform papillae are innervated by the chorda tympani branch of Cranial VII, as previously discussed, this nerve is recorded from. Chorda tympani gustatory-response latency for the anterior chemoreceptor array is about 30 msec (see Beidler, 1965; Faull and Halpern, 1972; Halpern and Marowitz, 1973). Therefore, some information would be available on taste properties of the first lick beginning with the final 18 msec (48-msec contact minus 30-msec latency) of contact. When gustatory stimulation with 200 mM NaCl is used, the chorda tympani response of the domestic rat to an "artificial lick" (5-μl volume, 55-msec flow duration, 6 mm² of anterolateral tongue wet) begins with a rapidly rising, linear, phasic component that reaches maximum magnitude 107 msec after liquid-contact start, as Halpern and Marowitz (1973) have reported. This phasic response would exceed 25% of its maximum amplitude during the final 18 msec of tongue contact with the drinking tube; 50% of maximum amplitude during tongue retraction. It follows that neural input upon which a zero-volume ingestion decision could be based is available. Appropriate behavioral tests, in which behavior is measured with high-speed motion-picture techniques, will be required to assess this possibility.

If the ability to make the rapid, low-volume (or zero-volume) decisions described in a licking/lapping animal is to have adaptive significance, the relevant chemoreceptors must be broadly sensitive. This seems to be correct because, from the point of view of avoidance behavior, a very narrow tuning of this chemoreceptor *population* might lead to frequent missing of significant events. A restriction to a narrow ecological niche might also be a consequence of narrow tuning of the population. However, an alternative argument is possible. Thus, defective avoidance response would be less likely if the basic strategy were to reject anything that did not fit the criteria of a limited set of "acceptable substances." Nonetheless, the complex nature of taste substances, as Meiselman and Halpern (1970) have described it, and the wide diversity of the foods of many mammals suggest to me that a narrowly tuned fungiform-papillae taste-bud population would not be a prudent approach.

Taste responses to aqueous solutions have been studied electrophysiologically in a number of therian mammals. Mammals studied include marsupials, bats, primates, lagomorphs, rodents, carnivores, and artiodactyls. Neural responses to taste stimulation of the anterior portion of the tongue, recorded from the chorda tympani nerve, are produced by solutions containing a wide range of solutes. These effective taste-stimulus liquids include solutions of inorganic salts, sugars, inorganic and organic acids, and alkaloids (e.g., see Erickson, Doetsch, and Marshall, 1965; Bartoshuk, 1965; Iggo and Leek, 1967; Oakley, 1967a,b; Mistretta, 1970; Frank, 1972; Sato, 1973; Pfaffmann, 1975), as well as benzosulfimides, for example, saccharin (Frank, 1972; Sato, Yamashita, and Ogawa, 1969; Sato, 1975) and amino acids (Halpern, Bernard, and Kare, 1962; Mistretta, 1970). This broad range of effective-stimulus molecules seems appropriate for the leading edge of the tongue. Recordings from individual chorda tympani neurons similarly show a broad range of responsiveness in many cases. A differential response to various stimuli rather than equal sensitivity to all stimuli characterizes the degree of responsiveness across single peripheral gustatory neurons, as Sato (1973) and Ogawa, Sato, and Yamashita (1973) noted. Chorda tympani neurons with quite narrow response spectra do also occur (e.g., see Frank, 1972). These instances of narrow tuning of individual chorda tympani neurons might be interpreted to show that single fungiform papillae, and perhaps clusters of such papillae, could be unresponsive to many potential stimuli. However, the taste bud(s) of each fungiform papilla is innervated by several chorda tympani axons, as previously discussed. This probably ensures the ability of every part of the anterior chemoreceptor array to provide information on any solution that contacts it, thus conforming to a nontopographic modality in the Erickson Model sense (see Erickson, 1968), even if the individual "input channels" (neurons) are narrowly tuned.

Very rapid, low-volume intake decisions may not occur in sucking ingestors, since suction-drinking behavior is very much slower than licking. For example, Halpern (1975) and B. P. Halpern and T. L. Nichols (unpublished observations) have reported that in humans a single sip has a duration of 900–1000 msec, a volume of 10–30 ml, and a contact-cycle length of 2–3 sec (Section 4). Nonetheless, behavioral studies of responses to stimulation of human single fungiform papillae, with psychophysical judgments as the measure, have generally found a wide range of responsiveness (see Harper, Jay, and Erickson, 1966; McCutcheon and Saunders, 1972; Bealer and Smith, 1975), even in this suction-drinking species (Table 4). The combination of broadly tuned sensory input and slow, large-volume liquid ingestion suggests that recognition of undesired liquids, and rejection, can occur prior to swallowing. With single-sip drinking, Halpern (1975) and Halpern and Nichols (1975) have found that swallows do *not* occur during the ingestion of liquid from a drinking glass. If a strong rejection response does occur, it is reasonable to expect that the inside of the mouth will subsequently be actively washed and wiped, with saliva and the fluid used spit out.

Neurophysiological studies of the individual fungiform papilla in non-human species by Miller (1971, 1975), Oakley (1972), and others (see review by Halpern, 1973, especially p. 315) provide relevant data on the basis for the commonly observed neural and behavioral indications of broadly tuned gustatory input. In these studies, taste solutions were applied to a single papilla, while responses were recorded from the axon or cell body of an innervating neuron. Again, many different solutions were effective stimuli. As was noted earlier, each individual chorda tympani axon branches, both below and within the fungiform papillae, and reaches several papillae. The result is that each chorda tympani axon carries input from many receptor cells in an array of papillae. Thus, Beidler (1965, 1969) and Boudreau and Tsuchitani (1973) found five to six fungiform papillae reached by a single chorda tympani nerve axon in the laboratory albino rat. Similar results are obtained with carefully controlled nongustatory stimulation of fungiform papillae. For example, Wang and Frank (1970—see their figure 2 and their pages 35–37) used electrical stimulation, 0.1 msec anodal pulses at 1 hz, which evoked one or two single-unit spikes with 20 msec latency when a fungiform papilla was stimulated, but was ineffective when their 0.3 mm diameter gold electrode was placed between papillae. Under these conditions, Wang and Frank found in the laboratory albino rat that the 6 single chorda tympani nerve axons which they studied reached 1-9 papillae, with a mean and median of 4.5 papillae per axon. For four of the papillae which excited their axons, two axons innervated the same papilla, while in one case, three axons reached one papilla. In this latter instance, each of the three innervating axons also reached at least one other papilla, with spatial separations greater than 2 mm occurring between the innervated papillae. These papilla-axon relationships, together with those reported by Beidler and by Boudreau and Tsuchitani, confirm that, for the laboratory albino rat, the axon branching observed outside a papilla by Beidler (1969) does have the functional meaning one would assume: one fungiform papilla taste bud can excite several chorda tympani nerve axons, and each of these axons is likely to be excited by several spatially separated fungiform papillae taste buds.

For the domestic cat (*F. domesticus*) Oakley (1972, 1975) reported that one to seven (mean = 2.7) fungiform papillae are reached by the same chorda tympani axon. The innervation of several papillae by the branches of each axon means that many taste buds, perhaps as many as 28 in the domestic cat, and consequently a sizable array of receptor cells (perhaps over 100) are reached by a single peripheral nerve axon. This could lead to a broadly tuned primary-neuron input derived from random connections of branches to narrowly tuned receptor cells. However, Oakley (1975) found that the sensitivity profile of each pair of papillae innervated by a single axon is quite similar, suggesting that breadth of tuning is not dependent upon random relations between taste-bud receptor cells and innervating axon branches. The overall pattern, from both neural and behavioral studies, is high-sensitivity, rapid, wide-spectrum chorda tympani responsiveness to stimulation of the anterior portion of the tongue.

The broadly tuned response spectra of mammalian fungiform taste buds discussed in the immediately preceding paragraphs make it unlikely that specific behavioral predictions for free-living conditions can be made from chorda tympani responses. It is often assumed that taste responses *per se* are directly related to the adaptive food- and water-selection patterns of mammals. An electrophysiological study of this in four Gerbillidae has been done by Jakinovich and Oakley (1975). Chemical stimulation of the anterior portion of the tongue and multiunit recordings of the chorda tympani were used. Species that survive on dry diets, *Meriones shawi* and *M. libycus,* have a much greater response to sucrose then do those gerbils that eat succulents or greens (e.g., *M. unguiculatus* and *Psammomys obesus*). However, water responses in these four animals did not correlate with habitat.

(2) Foliate Papillae. On the sides of the tongue of some mammals, taste-bud-bearing invaginations, the foliate papillae, are located. These foldlike papillae (see Sonntag, 1920; Bradley, 1971; Oakley, 1974), which

are sometimes called the *lateral organs,* generally occur near the base of the tongue, near the division between the oral (i.e., anterior) and pharyngeal (i.e., posterior) divisions of the tongue. A detailed description of rabbit (genus *Oryctolagus*) foliate papillae was provided by Scalzi (1966): "On the posterio-lateral portion of the tongue of rabbits and immediately anterior to the glossopalatine arch . . . are two discrete corrugated patches . . . each of which is approximately 5 mm long and 3 mm wide. These patches consist of numerous foliate papillae, the contours of which are sinusoidal in cross section" (p. 11). In the rabbit, taste buds occur only on the walls of the foliate papillae, not on the crest, that is, not on the top of the "body" (see Scalzi, 1966, pp. 44–45, for a gross photograph of the rabbit tongue and a low power photomicrograph of a rabbit foliate papilla). Yamada (1967)—pp. 104–106, and his figures 9, B and C; 10, b—reported that the shallowest taste buds of rabbit foliate papillae were 13 μm below the surface, while nearly half the foliate taste buds were within 100 μm of the surface. Laboratory albino rats have foliate papillae in a locus similar to that of rabbits—see Hamosh and Scow (1973), especially figure 20-1, for a gross photograph of rat foliate taste buds *in situ.* Human foliate papillae, in common with rabbit and laboratory albino rat, have a folded appearance, with taste buds on the walls of the papillae (Sognnaes, 1954, figure 326). Innervation of foliate papillae is primary by the glossopharyngeal nerve, that is, cranial IX, as Frank (1968) and Yamada (1967) have demonstrated, but some chorda tympani innervation also occurs (see Yamamoto and Kawamura, 1975).

Foliate papillae are relatively uncommon across mammals. In some cases, it may be difficult to distinguish between modified fungiform papillae and foliate papillae. Mammals lacking foliate papillae include several genera of microchiropteran bats, i.e., *Vespertilio* and *Artibues;* the common vampire bat of the genus *Desmodus;* primitive primates related to the slow loris, i.e., the lorisioforms; some members of the orders Edentata (sloths) and Pholidota (the scaly anteater); many of the carnivores of the suborder Fissipeda, including most *Vulpes* (foxes) of the family Canidae, the family Mustelidae (otter, skunks, badgers), and most of the family Procyonidae (coati, kinkajou, panda); all members of the order Cetacea (whale, dolphins, porpoises); some of the order Pinnipedia (seals and sea lions); the aardvark (Tubulidentata); a sirenid (the dugong); most members of the order Perissodactyla (odd-toed hoofed mammals such as the horse, zebra, rhinoceros, and donkey); and many animals in the order Artiodactyla, including most ruminants of the suborders Ruminantia (common deer, moose, cattle, sheep, goat, bison, buffalo, etc.); and Tylopoda (llamas and camels). However, within the same order, Artiodactyla, most of the suborder Suiformes (pig, hog, hippopotamus) do have foliate papillae. Other exceptions are a few members of the suborder Ruminantia, specifically the genus *Giraffa* (the giraffes) and the family Tragulidae (deerlike species of southeast Asia and western Africa), which do have foliate papillae (see Bradley, 1971; Brightman, 1976; Sonntag, 1925, pp. 714–717). In addition, although foliate papillae are present in the order Proboscidea (elephants), no taste buds are found on them (Bradley, 1971). In the slow loris (*N. coucang*), fungiformlike papillae were seen in the usual foliate location by Kubota and Iwamoto (1967). In many other animals, rows of circumvallate papillae occupy the "usual" location of the foliate papillae (see Sonntag, 1922, p. 652).

In domestic cats (*F. domesticus*), Sonntag (1923) found no foliate papillae. However, he described a group of club-shaped papillae that occur in a posterolateral region of the dorsum typically associated with foliate papillae. He considered these papillae, in which he observed taste buds, to be a type of fungiform papillae and designated them "clavate papillae." Such papillae also occur in a few other cats of the genus *Felis* (Bradley, 1971). Foliate papillae have been reported in the domestic cat by Yamada (1967). However, his published photomicrograph of the "Foliate papilla of a cat" bears little resemblance to typical foliate papillae, for example, as seen in the rabbit (Yamada, 1967, Figures 10b and 10d). Moreover, the structure identified as foliate papillae in the cat by Yamada may have no taste functions, since "extirpation of the foliate papillae did not modify the glossopharyngeal nerve response to taste" (p. 102).

Instead of depicting foliate papillae in the cat, Yamada's illustration shows a fingerlike papilla that probably corresponds to Sonntag's (1925) clavate papilla. If the clavate papillae are indeed a form of fungiform papilla, innervation solely from the chorda tympani nerve would be expected. However, Ishiko (1974) reported that responses to 1 M NaCl stimulation of the posterior region of the anterior division of the tongue can be recorded from both the glossopharyngeal nerve and the posterior branch of the distal portion of the lingual nerve (carrying axons that reached the proximal portion of the lingual nerve from the chorda tympani nerve). The glossopharyngeal nerve responses to stimulation of the anterior division of the tongue rise to maximum magnitude much more rapidly than do those to stimulation of the posterior division of the tongue. This suggests that the taste buds innervated by the glossopharyngeal nerve on the anterior division of the tongue are more accessible to stimulus solutions than those on the posterior division. This hypothesis would fit the structural differences between the clavate and the circumvallate papillae. The taste-bud–bearing papillae responsible for these responses were not directly identified. However, assuming that the clavate papillae are the source, the glossopharyngeal innervation, apparently in combination with axons of chorda tympani origin, is very similar to the typical foliate-papilla pattern. A "chorda tympani" innervation of some clavate papillae, at least, is demonstrated for a geniculate ganglion unit that responded to stimulation (electric) of both a fungiform and a clavate papilla by Boudreau *et al.*

(1971). In this context, both clavate papillae and foliate papillae were reported in one study, though only clavate papillae are illustrated (Boudreau and Tsuchitani, 1973; see especially his figure 12-6 and pp. 405–406). However, it is stated that the cat has a very small structure that may be identified as a foliate papilla "which may however be absent from one or both sides" (Boudreau et al., 1971, p. 473).

Many taste buds are present in the foliate papillae (see Sonntag, 1925; Yamada, 1967; Bradley, 1971). An average of 36 taste buds per foliate fold occurs in the laboratory albino rat (Oakley, 1974), while in humans, 50–100 taste buds per foliate papilla typically are present (Moncrieff, 1967).

In the preceding section of this chapter, "Fungiform Papillae," it was argued that the chorda tympani nerve, the carrier of the axons that innervate the taste buds of the fungiform papillae, is broadly and sensitively responsive to a wide range of molecules in solution. In the adult mammal, ingesta would typically first contact the anterior division of the tongue, on which most fungiform papillae occur. For example, in the domestic cat (F. domesticus), the chorda tympani nerve responds to chemical stimulation of 86% of the total length of the tongue (Ishiko, 1974). Given this extensive region of lingual gustatory innervation provided by the broadly sensitive chemoreceptors of the fungiform papillae, what roles might be served by the taste buds of the posterior division papillae, the foliate (if present) and circumvallate papillae? One possibility, of course, is a back-up system for the fungiform papillae. I consider this highly unlikely, since it seems a very uneconomical use of a complex and unique system. More importantly, the gustatory neural responses recorded from the glossopharyngeal nerve to chemical stimulation of the posterior division of the tongue will be seen to differ in relative, and sometimes absolute, sensitivity from those responses recorded from the chorda tympani nerve to comparable chemical stimulation of the anterior division of the tongue. These differences indicate that gustatory input from the foliate and circumvallate papillae probably serves functions somewhat divergent from those of the fungiform papillae input. One specific function for posterior-division taste buds may exist in nursing terrestrial mammals. In such preweaning mammals, suckling typically involves the securing of milk through a nipple that is inserted deeply into the mouth (see Bosma, 1967; Sameroff, 1973). The milk may largely bypass the fungiform papillae, since the front of the tongue often is in close contact with the sides of the nipple, providing intermittent compression of the nipple. However, the foliate and/or circumvallate papillae are certainly reached by the milk. The tongue motions during suckling should be effective in promoting access to the taste buds of the circumvallate papillae. Thus, a major role of the posterior division of the tongue in taste in suckling mammals is possible.

A different function can be outlined for adult mammals. In adults, substances that have been ingested and, if necessary, chewed are often accumulated as a bolus on the posterior dorsum of the tongue for a number of ingestion or mastication cycles before swallowing begins. This accumulation of ingesta on the posterior division of the tongue has been noted in many animals, including the European domestic rabbit (Ardran *et al.,* 1958), the American opossum (Hiiemae and Crompton, 1971), the laboratory albino rat (Hiiemae and Ardran, 1968), and the pig (Herring and Scapino, 1973; see Figure 10). This accumulation may provide sustained preswallowing gustatory stimulation of the taste buds of the posterior division of the tongue at a time when stimulation of the anterior division may be rather brief. In addition, a substantially larger stimulus volume separated into small pieces, with components partially suspended or dissolved in saliva in the case of masticated ingesta, would be available for the taste buds of the posterior division of the tongue. This sustained and extensive stimulation may mean that the foliate and/or circumvallate papillae can provide a more complete and detailed response to ingesta that the fungiform papillae. To accomplish this, the posterior division taste buds must be, in general, very broadly responsive to taste stimuli. However, the relatively large volume, applied to spatially concentrated taste buds, could permit individual taste buds to be more specialized than fungiform-papilla taste buds.

Electrophysiological studies of responses from peripheral nerves innervating foliate papillae have been made. While stimulus liquids were slowly flowed (23 μl/sec) through a foliate fold of the laboratory albino rat's tongue, responses to 300 mM NaCl, 10 mM HCl, 300 mM sucrose, and occasionally 1 mM quinine hydrochloride were recorded from single functional units of the glossopharyngeal nerve by Frank (1968). All the taste-sensitive units had an ongoing, "spontaneous" neural activity in the absence of any known taste stimuli. In most cases, the rate of spontaneous activity was quite low, averaging about 0.5 impulses/sec. Only increases in action potential rate that *exceeded* the spontaneous rate by 50% or more were accepted as responses. These two criteria—that is, $\geq$ 50% difference from spontaneous rate and increases only—establish a signal-to-noise relation and exclude decreases below the spontaneous rate as possible responses.

Responses to only one of the first three stimuli occurred in about one-third of Frank's (1968) units, while the rest showed an increase in firing rate for two or three of the four solutions. Median thresholds were 100 mM NaCl, 1 mM HCl, and 30 mM sucrose. One unit responsive to 1 mM quinine hydrochloride was observed. With a wider range of stimuli, units were observed that responded to:

1. 300 mM NaCl, 300 mM KCl, 300 mM NH$_4$Cl, and 10 mM HCl.
2. KCl, NH$_4$Cl, 10 mM citric acid.
3. 300 mM sucrose or 0.1 mM sucrose octa-acetate, 3 mM sodium saccharin (M. Frank, personal communication), citric acid, HCl, NH$_4$Cl, and KCl.

It is clear that a broad range of responsiveness and reasonable sensitivity occur in glossopharyngeal-nerve–foliate-papillae units in the laboratory albino rat.

For the concentrations used above, stimulus-solution flow through the foliate invaginations (from a 0.1–0.2 mm i.d. pipette) produced no chorda tympani nerve responses (Frank, 1968). This finding appears to confirm a previous report by Pfaffmann (1952) that all foliate-papillae taste buds disappeared in laboratory albino rat following bilateral transection of the glossopharyngeal nerve. However, by using an external flow chamber and much higher concentrations ($\geq$ 500 mM NaCl, 1 M sucrose, 100 mM HCl, 100 mM quinine hydrochloride, 1 M KCl), Yamamoto and Kawamura (1975) could record chorda tympani multiunit responses during foliate papillae stimulation in the laboratory albino rat. Unfortunately, the high concentrations used in this study preclude behavioral interpretations of the chorda tympani responses and put their relevance to gustatory input in question.

In rabbits, a combined foliate- and circumvallate-papillae multiunit electrophysiological response from the glossopharyngeal nerve was obtained by Yamada (1967) to 500 mM NaCl, KCl, and sucrose; 20 mM quinine hydrochloride; and 5 mM HCl. These responses were thought to be primarily of foliate-papillae origin because they were observed when the stimulus solutions flowed over the foliate papillae and the circumvallate papillae but did not occur when stimulus flow was restricted to the circumvallate papillae. The sucrose was the most effective stimulus, with the KCl, HCl, and quinine hydrochloride following, in that order.

Serous glands connect with the bottom of the folds (grooves) of the foliate papillae (see Sognnaes, 1954; Yamada, 1967; Hamosh and Scow, 1973). A lipase may be present that could be of some significance in the digestion of liquids (Hamosh and Scow, 1967).

Electrophysiological responses from the glossopharyngeal nerve of the domestic cat were *not* produced upon stimulation of the anterior 65% (41 mm) of the tongue, even with a stimulus solution as strong as 1 M NaCl (Ishiko, 1974). The report of this unresponsive region, which approximately corresponds to the anterior (oral) division of the tongue, supports the anatomically based conclusion that the glossopharyngeal nerve does not innervate the anterior division of the mammalian tongue. However, it does not follow from this that there is no overlap of chorda tympani and glossopharyngeal innervations. Indeed, approximately half of the tongue region from which responses are produced in the posterior branch of the distal lingual nerve (arising proximally from the chorda tympani nerve) overlaps with the area of glossopharyngeal nerve innervation. As Ishiko (1974; Figure 1, top) indicated, the chorda tympani–glossopharyngeal-nerve overlap area corresponds to the locus of lateral organs on the tongue. Innervation of the lateral taste organs by both the chorda tympani nerve and the glossopharyngeal nerve has been suggested for the cat and the rat by Yamamoto and Kawamura (1975), although not confirmed in all studies of the rat (e.g., Frank, 1968). It may be that in the cat, fungiform papillae of the posterior division of the tongue are innervated by the posterior branch of the distal lingual nerve, through which neurons arising proximally from the chorda tympani nerve travel.

Boudreau and Tsuchitani (1973) described a "large agglomeration of fungiform papillae in the center rear of the tongue in front of the vallate papillae" (p. 407) in the domestic cat. These fungiform papillae of the vallate area were previously illustrated in a plot of the location of lingual papillae in the cat, which when stimulated electrically produce neural responses in single units of the geniculate ganglion (cell bodies of the chorda tympani nerve) (Boudreau *et al.*, 1971; see especially Figure 13). Fungiform papillae in the vallate area of the domestic cat, innervated by chorda tympani nerve units that are responsive to gustatory stimuli, have also been described by Oakley (1975).

(3) Circumvallate Papillae

(a) Anatomy. On the posteriodorsal portion of the tongue of most mammals, one or more relatively large, rounded, typically elevated, taste-bud-bearing structures occur. These rounded structures—together with the surrounding narrow, concentric toroidal depressions—are the circumvallate papillae. These papillae are found only in mammals. The most common architecture is for the central structure (the "body") to be elevated above the surrounding surface of the tongue (Figure 4). Secretory glands always open into the bottom of the torus (Sonntag, 1925), as was previously noted for the foliate papillae. In the laboratory albino rat (*R. norvegicus*), the serous glands that empty into the circumvallate torus contain a lipase that may be important in lipid digestion (Hamosh and Scow, 1973). In some metabolic disorders (e.g., vitamin A deficiency), the ducts of the serous glands and the circumvallate torus become clogged with debris, which may cause the associated taste losses (Bernard and Halpern, 1968).

More specifically, the circumvallate papillae are located at the intersection between the oral and the pharyngeal divisions of the tongue (Sonntag, 1925; Bradley, 1971), in a medial position on the tongue. The glossopharyngeal nerve (Cranial IX) innervates the taste buds of the circumvallate papillae (Pfaffmann, 1952; Guth, 1957; Frank, 1968; Yamada, 1967). In humans some circumvallate-papillae innervation by the vagus nerve (Cranial X) may occur as Benjamin, Halpern, Moulton, and Mozell (1965, p. 383) suggested. Many taste buds are found on the walls of the toroid (Bernard and Halpern, 1968), but taste buds are rare on the elevated central portion (Figures 4 and 5). (See Sonntag, 1925, p. 711, for exceptions). However, a taste bud near the top of a circumvallate-papilla body is illustrated for a cat papilla by Yamada (1967).

The toroidal design, with the taste buds located in a narrow depression,

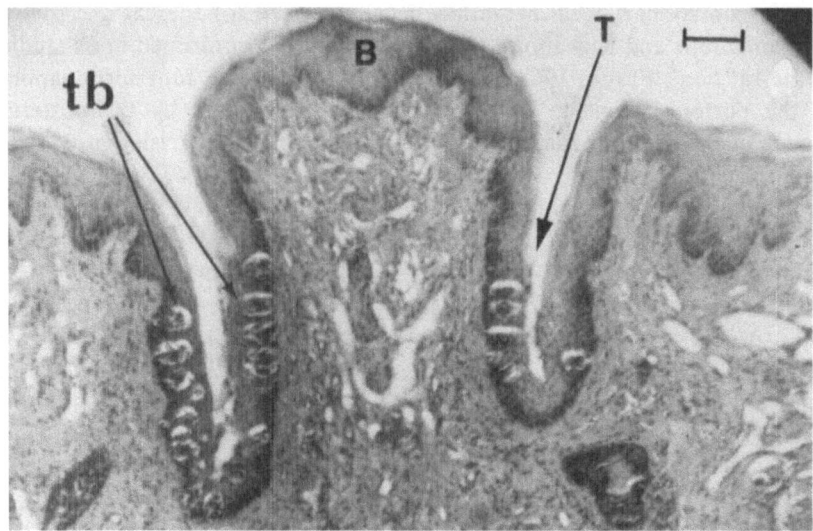

Figure 4. Transverse section through the circumvallate papilla of a laboratory albino rat, *Rattus norvegicus*. The body of the papilla (B) is well elevated above the surrounding mucosa. In the surrounding toroidal depression (T) the taste buds (tb) are visible. Hematoxylin and eosin stained, 5 μm paraffin-embedded section. Calibration line in upper right is 100 μm. The circumvallate papilla of a mouse (Jackson Laboratory LAF$_1$) is very similar (see Conger and Wells, 1969, especially the top left of Figure 1). The possibility that the toroidal depression is a histological artifact has been raised by V. J. Brightman (unpublished observations, 1976).

may decrease access to the receptors. Indeed, laboratory recording studies have often reported difficulty in the stimulation of circumvallate papillae with a surface flow of liquids, as Frank (1968) indicated (see discussion in Halpern, 1973, p. 315). However, during normal ingestion, the circumvallate-papillae region of the tongue undergoes mechanical deformation, which may introduce liquids into the torus under some pressure (Sonntag, 1925). In the cat and goat, glossopharyngeal (Cranial IX) nerve responses to posterior tongue-surface stimulation with solutions are greatly facilitated by a stroking of the papillae (Kitchell, 1961). Indeed, in these two animals, responses to chemicals were almost undetectable by Kitchell without mechanical deformation of the circumvallate area. This finding supports the proposal that access to the circumvallate-papilla taste buds is difficult in the nonmoving, noningesting mouth. Changes in access to taste-bud receptor cells as a function of tongue movement were suggested by Beidler (1969) for not only the circumvallate papillae but also the fungiform papillae.

In the calf (*B. taurus*), Bernard (1962, 1964) reported that multiunit glossopharyngeal-nerve responses were regularly produced by surface circumvallate-papillae stimulation. The magnitude of the responses was relatively small, although the rise rate was not slower than that observed for chorda tympani responses. Perhaps improved access to taste buds in the torus would have permitted stimulation of a larger population of receptors and, consequently, a larger, more "normal" response. Long-latency, slowly rising glossopharyngeal-nerve responses have been reported for stimulation of the rabbit (Yamada, 1967) and the cat (Ishiko, 1974) circumvallate papillae.

The number of taste buds per circumvallate papilla varies widely among mammalian species, ranging from 100 per papilla in the cat (*F. domesticus*) and 181–375 in the laboratory albino rat's single papilla (see Bernard and Halpern, 1968; Bradley, 1971), through 1200 per papilla in the rabbit (*Oryctolagus cuniculus*) and some fugivorous bats and 1467 in the calf, to 5380 per papilla in the pig (*Sus scrofa*). There are 600–700 taste buds per circumvallate papilla in rhesus monkeys, but 300–700 in humans, both of which are primates of the suborder Anthropoidea. Sonntag (1925) and Bradley (1971, 1972) summarized the data on taste-bud counts. As was

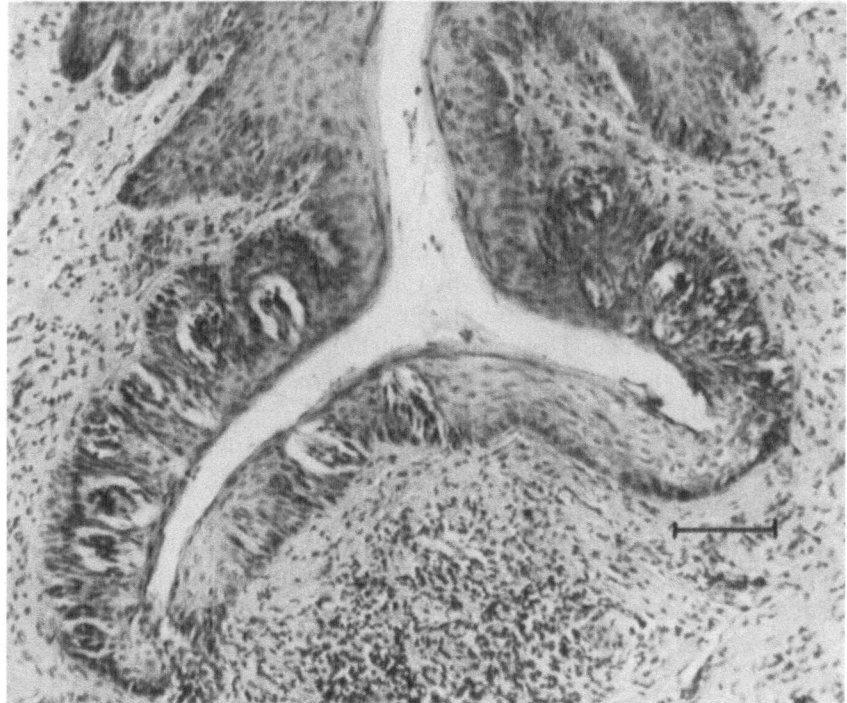

Figure 5. Photomicrograph of an oblique section through the torus of the circumvallate papilla of a laboratory albino rat. Taste buds (with shrunken and distorted receptor cells) can be seen on the walls. Hematoxylin and eosin stain. Calibration line is 100 μm.

previously suggested, the large number of circumvallate-papillae taste buds, located in an oral region in which preswallowing accumulation occurs, may constitute the most sensitive lingual gustatory population.

Although circumvallate papillae are found on the tongue of most mammals, there are two complete orders, and a few additional species, without them. Mammals lacking circumvallate papillae include some specialized microchiropteran bats, for example, the nectar-eating bat *Choeronycteris* (see Table 4) and the common vampire bat, *Desmodus rotondus;* the whales, that is, the order Cetacea; and the Hyracoidae (this order, which is a part of the cohort Ferungulata, includes the coney, a small guinea-pig-like herbivore). It should be noted that circumvallate papillae are present in some bats of both the mega- and the microchiropteran suborders. Bats with circumvallate papillae include species as diverse as the microchiropteran Vespertilionidoia *Vespertilio subulaties* and the very large, frugivorous megachiropteran bat, the "flying fox" *Pteropus pselophon* (Sonntag, 1925; Bradley, 1971; Brightman, 1976).

The total number of circumvallate papillae present and the geometry of their arrangement on the posterior (i.e., pharyngeal) division of the tongue show wide variation among mammalian species. Within one order, the total number of circumvallate papillae may range within a factor of five, but a somewhat narrower range occurs within a single family. Generally, only a single circumvallate papilla occurs on the tongue of myomorphic rodents (e.g., hamsters, mice, rats, gerbils), although more are found in the members of *Cricetomys,* the West African "hamster rats." These single papillae of rodents in the suborder Myomorpha are in a median position. In other rodents, for example, those of the suborders Sciuromorpha (squirrels, chipmunks, beavers) and Hystricomorpha (e.g., guinea pig, porcupine, chinchilla), two to five papillae are present. Many mammals have two to five circumvallate papillae. These animals include some bats, for example, the microchiropteran *V. subulatus;* rabbits (*Oryctolagus;* see Yamada, 1967); pigs *S. scrofa*); nonmyomorphic rodents (see above); some carnivores, especially the genera *Vulpes* (foxes), *Mustela* (e.g., the ferret), and *Mephitis* (skunk); opossums (genus *Didelphia*); and most primates. Primates with two to four circumvallate papillae include the suborders Prosimiea (lemuriformes and lorisiformes) and Tarsioidea (tarsiers); the genus *Hylobata* (gibbons); and many of the Old World monkeys of the family Cercopithecidae (e.g., langur monkeys, such as *Presbytis entellus,* and rhesus and cynomolgus monkeys, *Macacus rhesus* and *M. cynomolgus*) (see Sonntag, 1921, 1924, 1925, and Bradley, 1971; for overall circumvallate counts).

From 6 to 12 circumvallate papillae are present in fissiped carnivores such as the dog, *Canis familiaris;* the otter, *Lutra canadensis;* the raccoon, *Procyon;* and the cat, *F. domesticus* (Yamada, 1967); and in the hominid primate human, *H. sapiens.* Circumvallate papillae, and total taste buds, are most numerous in mammals of the suborder Ruminantia. About 12 circumvallate papillae are found in the goat, *Capra hircus,* while the sheep, *Orvis aries,* and cattle, *Bos,* have two dozen (or more; see Mistretta, 1972), and the American pronghorn antelope, *Antilocapra americana,* more than 50 (Sonntag, 1925; Bradley, 1971). It is not obvious why many mammals, including omnivores such as the rat, are adequately served by a single circumvallate papilla and numerous others by only a few, while dozens occur in most ruminants. Monitoring of the progress of rumination may be a justification, as Bernard (1962) has proposed.

The geometrical arrangements of the circumvallate papillae of mammals are even more diverse than the number of papillae. Rows, triangles, inverted triangles, W-arrangements, and clusters all appear. Of these, three papillae in a triangle, with the vertex toward the back of the mouth, is the most common and perhaps, as Sonntag (1925) suggests, the most "primitive" arrangement. In general, the arrangement of circumvallate papillae and the total number do not correlate with diet in any obvious manner. Within any single order of mammals, the circumvallate papillae usually occur in only two of all the possible arrangements found in mammals. Within specific suborders or families, just a single layout of papillae is common. However, for the several families of the order Carnivora—and, even more so, for most of the order Primates—two to five arrangements are found. Detailed reviews of the circumvallate-papillae layouts are provided in Sonntag (1925) and Bradley (1971).

Despite the diversity of circumvallate-papilla arrangements among mammals, several commonly studied species have nonvarying geometries. Thus, the laboratory albino rat, *R. norvegicus,* and the laboratory golden hamster, *Mesocricetus auratus,* invariable have a single median papilla. The rabbit, *Oryctolagus,* regularly has a pair of papillae on either side of the midline. Both the domestic cat, *F. domesticus,* and the human consistently have a V-shaped arrangement of papillae, with the apex toward the back of the mouth. The "V" contains 4–7 papillae in the cat and 8–12 in humans.

(b) Electrophysiological Studies. The functional properties of the circumvallate papillae have been examined in a number of neurophysiological studies of glossopharyngeal-nerve responses to vallate-area stimulation, though less extensively than chorda tympani nerve responses to stimulation of the fungiform papillae (typically, on the anterior division of the tongue; see preceding section). In the rabbit glossopharyngeal nerve, Yamada (1967) found that responses to surface liquid flow over the circumvallate papillae occurred only to very concentrated aqueous solutions such as $2M$ NaCl and $1M$ KCl. Solutions of these concentrations also produced responses in the lingual nerve (nongustatory branch of Cranial V) of the rat when applied to the tongue by Kawamura, Okamoto, and Funakoshi

(1968). Therefore, the described rabbit responses are not necessarily "taste" *per se*. The unclear nature of the "circumvallate" responses in the rabbit is reinforced by Yamada's (1967) observation of an absence of responses to other, less strong solutions, such as 20 mM quinine hydrochloride and 500 mM sucrose. Of course, access to the taste buds in the torus may be the limiting factor. All rabbit circumvallate-papillae taste buds were located 150 μm or more below the surface of the tongue. This taste bud location strongly contrasts with the relatively superficial location of some cat circumvallate-papillae taste buds, discussed below.

Cat glossopharyngeal-nerve responses to circumvallate-papillae surface liquid flow were found by Yamada (1967) to show a relatively short-latency, phasic–tonic response pattern similar to that of chorda tympani-nerve response to fungiform-papillae stimulation. Moderate strength solutions, such as 10 mM HCl and 20 mM quinine hydrochloride, were effective stimuli. The cat circumvallate-papillae taste buds are located along the full depth of the torus from surface to bottom, and they sometimes occur on the elevated portion of the body. The effectiveness of surface liquid stimulation in producing neural responses from the cat circumvallate area, without mechanical deformation of the papillae, may be related to the relatively superficial location of some of the taste buds. However, Kitchell (1961) reported that glossopharyngeal responses in the cat may be enhanced by the stroking of a circumvallate papilla.

Glossopharyngeal-nerve responses to surface stimulation of cat circumvallate papillae were also observed by Ishiko (1974). In this case, however, the neural responses rose slowly after the liquid was applied. Relatively low-concentration stimuli were effective. With the neural response to 1 M NaCl taken as 100, the response to 7 mM quinine hydrochloride was 162 and to 3.5 mM hydrochloric acid, 80. Perhaps differences in flow *rate,* or in spread of the tongue, would account for the slow rise-rate that Ishiko (1974) observed, for the enhancement due to papilla manipulation described by Kitchell (1961), and for the prompt, rapidly rising cat glossopharyngeal-nerve responses reported by Yamada (1967).

In sheep (genus *Ovis*), Iggo and Leek (1967) reported that glossopharyngeal-nerve responses to gustatory stimulation of the vallate region were small or absent unless a papilla was mechanically moved by a probe, as Kitchell (1961) had previously noted for the goat. Response latencies were about 200 msec (the time of stimulus arrival at the tongue surface was not precisely known), whether or not a papilla was moved. In common with single-unit responses from the chorda tympani nerve, the majority of glossopharyngeal-nerve units showed pronounced phasic–tonic responses to 500 mM NaCl and 200 mM citric acid and little response to 500 mM sucrose or 10 mM quinine hydrochloride. However, explicit functional differences between chorda tympani and glossopharyngeal-nerve gustatory responses were also observed. Thus, units highly responsive to 500 mM sucrose (8 units), to 10 mM quinine hydrochloride, or to both were observed in the *glossopharyngeal* nerve, but these solutions produced little or no response from chorda tympani units. Thus, input from the sheep fungiform and circumvallate taste-bud populations differs, but a simplistic dichotomy into anterior division = "good" substances, posterior division = "bad" substances, does not appear to fit the data.

Multiunit chorda tympani and glossopharyngeal-nerve responses of the laboratory albino rat to surface stimulation of the tongue with solutions have been quantitatively compared by Oakley (1967a,b). No mechanical manipulation of the circumvallate papilla was done. Both absolute and relative differences (300 mM NH$_4$Cl was used as the standard) were observed. In absolute terms, 300 mM KCl consistently produced responses in the glossopharyngeal nerve that were larger than responses to 300 mM NaCl or to mammalian Ringer's solution, while responses to the latter two solutions were consistently larger than that to KCl for the chorda tympani nerve. Absolute differences in responses to pH 2.05 citric acid also occurred. On a relative basis, the glossopharyngeal nerve was much more responsive to 50 mM sodium saccharin, to pH 2.55 acetic acid, and to 1.0 M sucrose than the chorda tympani nerve. A greater relative glossopharyngeal-nerve response to 50 mM quinine hydrochloride also seemed to occur, but interanimal variability made this finding less clear. As the absolute differences would suggest, the relative responses of the chorda tympani nerve to NaCl and Ringer's were much larger than the glossopharyngeal responses. Once again, chorda tympani–glossopharyngeal response-differences, intriguing but challenging, are evident.

An earlier study of multiunit glossopharyngeal-nerve gustatory responses in rat by Yamada (1966), in which a circumvallate-papilla flow chamber was used, reported rather different results. Only slight responses occurred to 500 mM sucrose, but large responses occurred to concentrated NaCl (500 through 2000 mM). KCl was more effective than equimolar NaCl. The relative ineffectiveness of sucrose in this experiment, compared to the response reported subsequently (Oakley, 1967a,b), may be related to the details of stimulus application or to selective damage to one subset of glossopharyngeal-nerve axons during the protease treatment used in the Yamada (1966) study to aid in the removal of connective tissue.

For the rat circumvallate papilla, Frank (1968) overcame the problem of receptor access by introducing liquids through a glass pipette inserted into the torus of the papilla. A very low flow rate (24 μl/sec) minimized responses to liquid flow *per se* but may have attenuated phasic responses. That is, the initial, phasic component of gustatory neural responses is stimulus flow-rate dependent, as Halpern (1973, pp. 302–303) has noted. Frank's (1968) multiunit recordings from the glossopharyngeal nerve showed that 1 mM quinine hydrochloride, 300 mM NaCl, 300 mM sucrose, and 10 mM HCl gave neural responses, with the response magnitudes decreasing in that order. Single glossopharyngeal-nerve units responded to various sets of gustatory stimuli flowed through the torus of a circumvallate papilla. About 20% of the units responded to only one of the above four stimuli, 50% responded to two, about 20% to three, and 10% to all four. However, average response rates to one stimulus were generally appreciably higher than responses to other stimuli when more than one was effective. In terms of "broad-spectrum" sensitivity, single units responsive to KCl, NH$_4$Cl, and sucrose are found.

Single-unit responses of rat glossopharyngeal nerve to circumvallate-papilla stimulation (via pipette in the torus) have been directly compared with single-unit responses of rat chorda tympani nerve to fungiform-papillae surface stimulation by Frank (1975). On an absolute response-magnitude basis, with the phasic (see Halpern and Marowitz, 1973) component deemphasized by an averaging of the total number of action potentials (i.e., "impulses" or "spikes") over 5 sec, NaCl (300 mM) generally produced a small response (less than 100 impulses/5 sec). A "best-response profile" analysis, as described by Frank (1974, 1975), was used. Of 32 circumvallate-glossopharyngeal units, only 1 was most responsive to 300 mM NaCl, while 9 or more were most responsive ($\geq$ 100 impulses/5sec) to 300 mM sucrose, 3 mM HCl, or 1 mM quinine hydrochloride. In contrast, of 39 fungiform–chorda tympani units, 25 were most responsive to 100 mM NaCl, and 11 were most responsive to 10 mM HCl, while only 1 or 2 were most responsive to 500 mM sucrose or 2 mM quinine hydrochloride. It is clear that the rat circumvallate-glossopharyngeal input is quite different from the fungiform–chorda tympani input, although both populations are relatively broadly tuned and sensitive. There are potentially specialized roles of the circumvallate (and foliate) papillae in gustation. Two such roles are during nursing and in responding to the ingesta bolus at the rear of the tongue prior to a swallow. These potential roles (see discussion on Foliate Papillae) together with the differences between responses recorded from the chorda tympani and the glossopharyngeal nerves, suggest that the gustatory receptors of the anterior and posterior division may serve largely separate functions.

2.1.1.2. Extralingual Loci of Taste Buds. In addition to the generally recognized taste buds of the lingual "gustatory" papillae (and, sometimes, of the anteroventral mucosa of the tongue), taste buds have been described in a number of nonlingual oral regions of adult mammals by several workers, including Pfaffmann (1952), Mistretta (1972), and Miller (1977). In adult rats, taste buds have been found in the mucosa of the pharynx and the nasoincisor ducts (ducts that go from the mouth to the nasal passages) and in the epiglottis. For various mammals, including lagomorphs, rodents, insectivores, carnivores, and primates, taste buds have been reported in the mucosa of the palate, the epiglottis, the pillars of the fauces (the lateral boundaries of the passage from the mouth to the pharynx), the pharynx, the larynx (a puzzling location, since ingesta would reach these laryngeal taste buds on the way to the lungs) and the esophagus by Storey (1967), Bradley (1971), Farbman and Allgood (1971), and Mistretta (1972).

The nonlingual taste buds have been thought by Farbman and Allgood (1971) to be innervated by several different nerves. The greater petrosal nerve, carrying facial nerve (Cranial VII) fibers, appears to reach the taste buds of the palate, which would thus be included in the "functional" anterior division of the tongue. That is, these palatal taste buds would be innervated by the same nerve that reaches the taste buds of the fungiform papillae. The glossopharyngeal nerve is said to innervate taste buds on the pillars of the fauces (posterior division "functional group"), while the vagus nerve (Cranial X), which is not known to innervate the taste buds of any "lingual gustatory papillae," is thought to innervate the taste buds in the pharynx and the larynx.

A review of previous studies of nonlingual taste-bud location and innervation in mammals, and a detailed experimental analysis of nonlingual taste-bud loci in the laboratory albino rat, has been presented by Miller (1977). Overall 221 $\pm$ 16 palatal taste buds were counted in the laboratory rat. This represents 17% of all oral taste buds in this animal. The largest number of palatal taste buds, about 88 of the 221, occurred along the midline on the posterior soft palate. Smaller numbers of taste buds, about 66 of 221, were found at the soft palate–hard palate interface, (the "Geschmacksstreifen" of Kaplick), while 67 of the 221 palatal taste buds were located in the region of the nasoincisor duct. A series of lesion-degeneration experiments demonstrated that the nasoincisor duct taste buds are innervated by the palatine nerve, through which they presumably receive facial nerve (Cranial VII) fibers. The Geschmacksstreifen taste buds were partially innervated by the palatine nerve (about 32% of them received this innervation), but the posterior soft palate taste were mostly (about 60%) innervated by overlapping fields of the greater superficial petrosal nerve branch of the facial nerve, and by the glossopharyngeal nerve.

The specific functions of the nonlingual taste buds are not known. Reasonable, though not necessarily correct, hypotheses could be offered relating palatal taste buds to gustatory input during mastication or bolus movement to the base of the tongue, faucial and epiglottal taste buds to preswallowing input, and so on. Some of these nonlingual taste buds, especially those not innervated by the chorda tympani or the glossopharyngeal nerves, presumably supply the gustatory input that has been observed by Richter (1956) and Vance (1966) to permit taste-preference behavior after these nerves have been cut.

2.1.1.3. Responses to "Nongustatory" Stimuli. Temperature and/or mechanical deformation changes on the surface of the tongue, localized to the gustatory papillae, often produce responses in chorda-tympani-nerve or geniculate-ganglion single units. In many instances, these same units respond sensitively to gustatory stimulation of the tongue (see, e.g., Pfaffmann, 1941; Sato *et al.,* 1969; Boudreau *et al.,* 1975). Such chemo-thermoreceptors and/or chemo-mechanoreceptors seem relatively uncommon in the glossopharyngeal nerve, as Iggo and Leek (1967) and Frank (1968) have reported. It may be that their fiber diameters in the glossopharyngeal nerve make single-unit isolation unlikely.

The chemoreceptor neurons discussed above may not be as responsive to mechanical and thermal events as other neurons of the chorda tympani, the glossopharyngeal, and the trigeminal nerves that are primarily thermo- and/or mechanoreceptors. However, this does not preclude the chemo-thermo-mechanoreceptors as useful sources of nonchemical input. Large temperature differences between the tongue surface and ingested substances are common, while lingual manipulation of foods during ingestion and mastication represents significant mechanical deformation. Therefore, it may be reasonable to describe the taste "chemoreceptors" of the tongue as chemo-mechano-thermoreceptors, involved in the initial evaluation and oral processing of foods. This concept—that variables such as the surface texture, consistency, relative temperature, etc., of ingesta, as well as the sapid components that appear in the mouth, are all part of the available information in tasting—was proposed by Gibson (1967) as part of a general ecological analysis of perception.

2.1.2. Oral Thermoreceptors and Mechanoreceptors

Neural responses to tongue temperature, both steady-state and changes, and to mechanical deformation of the tongue can be recorded from all afferent nerves that innervate the tongue. Such responses have been seen in the *chorda tympani nerve* by Zotterman (1935), Pfaffmann (1941), Nejad (1962), Iggo and Leek (1967), Sato *et al.* (1969), Ishiko (1970, 1974), Sato (1973), and Ogawa *et al.* (1973); in the *geniculate ganglion,* which contains the cell bodies of sensory axons of the chorda tympani nerve, by Boudreau *et al.* (1971); in the *glossopharyngeal nerve* by Zotterman (1935), Bernard (1962), Yamada (1966), Iggo and Leek (1967), Frank (1968), and Ishiko (1970, 1974); in the proximal portion of the lingual branch of the *trigeminal nerve* by Kawamura *et al.* (1968) and Iggo and Leek (1967); and in the *gasserian ganglion* (i.e., ganglion trigeminal), which contains the cell bodies of the sensory axons of the trigeminal nerve, by Beaudreau (1968) and Poulos and Lende (1970). In addition to the "somesthetic" responses that can be recorded from stimulation of the mucosa of the tongue, Jerge

(1967) and Poulos and Lende (1970) reported that mechanical deformation of the mucosa in other portions of the oral cavity also elicits trigeminal responses, while Storey (1967) has observed nontrigeminal responses (of the glossopharyngeal nerve and the vagus nerve).

The substantial trigeminal-nerve response to thermal and mechanical stimulation of the tongue certainly includes input from the lingual papillae. For the fungiform papillae of the laboratory albino rat, Beidler (1969) has demonstrated that many trigeminal-nerve axons are present. Specifically, axons of trigeminal-nerve origin, traveling to the tongue in the lingual nerve, represent the majority (> 75%) of the nerve fibers in a fungiform papilla. Under these circumstances, a possible relationship between these plentiful trigeminal axons and the taste bud of the papilla could be imagined. However, destruction of these nerve fibers by an appropriate lingual-nerve transection does *not* cause degeneration of taste buds. On the other hand, the various organized somesthetic endings reported by Beidler (1965) for the fungiform papillae of some mammals are presumably related to the axons of trigeminal origin in these papillae. These trigeminal axons may also be the source of the nerve fibers that terminate as free nerve endings in the epithelium of the papillae.

Some of the somesthetic responses to lingual temperature or deformation that have been recorded from cranial nerves or their ganglia are from single units that do not respond to gustatory stimuli. I intend *gustatory stimuli* to mean solutions that do produce a neural response when applied to taste-bud–containing lingual papillae preadapted to the temperature of the solution but do not produce a neural response when applied to taste-bud–free mucosa. Unfortunately, this definition of *gustatory stimuli* is defective. It is both circular and nonsymmetrical. The circularity is obvious, since only those stimuli that affect arbitrarily defined receptor structures are accepted. A more serious problem is the asymmetry. Thus, the definition is useful soley at the low-concentration end of the stimulus spectrum. Many solutions evoke neural response when applied to any mucosal region if the solution is sufficiently concentrated, as Kawamura *et al.* (1968) have demonstrated. A mixture representing suprathreshold but "low" concentrations of a range of gustatory-stimulus molecules may represent an effective test liquid. Halpern (1973) has pointed out that the most appropriate stimuli would be the molecules and ions of species-typical ingesta at the concentration at which they occur. However, knowledge of such ecologically valid stimuli is often limited or absent, although recent studies (e.g., Boudreau, Anderson, and Oravec, 1975) are attempting to approach natural foods.

Somesthetic units as "defined" above (i.e., purely somesthetic single units) have been reported in the *chorda tympani nerve* by Iggo and Leek (1967) and Boudreau *et al.* (1971) in its ganglion, the *geniculate ganglion,* by Iggo and Leek (1967), and by Frank (1968) in the *glossopharyngeal nerve,* as well as by Kawamura *et al.* (1968) in the proximal portion of the lingual branch of the *trigeminal nerve.* The receptive fields of the units (i.e., regions of the mucosa from which responses are elicited) are often quite restricted. In distal lingual-nerve branches in the domestic cat, fields receptive to somesthetic response, are all on the dorsal surface of the tongue, with areas ≯ 3 mm².

The receptor structures for the somesthetic responses recorded from the chorda tympani, the glossopharyngeal, the vagus, and the trigeminal nerves or their ganglia upon appropriate stimulation of the oral mucosa must be located in or near the mucosa. For those units that do not also respond to gustatory stimuli, and perhaps for some of the units that do respond to gustatory stimuli, the taste buds are not the somesthetic receptor structures. Normal somesthetic responses upon stimulation of gustatory papillae, specifically foliate and circumvallate papillae, were recorded by Oakley (1974) after cross-regeneration by nerves that neither support nor induce taste buds in these papillae. There is no lack of reported mucosal-receptor structures. A plethora of "endings" has been described. They include encapsulated and nonencapsulated organized end-organs (e.g., Winkelmann, 1962; Marlow, Winkelmann, and Gibilisco, 1965), free-nerve endings (Marlow *et al.,* 1965; Grossman and Hattis, 1967, and essentially all intermediate types of cutaneous endings, as described by the preceding workers and by Farbman and Allgood (1971) and Seto (1972). From the observations of Oakley (1974), it appears that any nerve with sensory properties can, if persuaded to grow into lingual papillae, elaborate functional somesthetic endings.

In general, all parts of the oral mucosa are supplied with somesthetic receptors, a particularly extensive allotment being present on the tongue. Specific energy-capture functions (i.e., temperature reception) cannot be readily assigned to mucosal receptor structures of a given structural type. The input from these variegated receptors moves along the classical taste nerves (chorda tympani and glossopharyngeal nerves), as well as the vagus nerve and the trigeminal nerve. Thus, a rich flow of afferent information is supplied by the superficial receptors of the oral cavity.

2.2. Deep Receptors

2.2.1. Tongue

The evidence for anatomically or functionally recognizable muscle spindles in the intrinsic or extrinsic muscles of the tongue is somewhat contradictory. Many studies (e.g., Carleton, 1938; Beaudreau and Jerge

(1968) and reviews (e.g., Storey, 1967) conclude that few if any lingual muscle spindles have been demonstrated. However, a review by Kawamura and Morimoto (1973) and anatomical investigations by Kubota and Togawa (1966) and Kubota and Iwamoto (1967) do recognize the presence of muscle spindles in the tongue.

Overall, there appear to be great species differences in the presence or absence of muscle spindles in the tongue. Cooper (1953) recognized muscle spindles in the intrinsic tongue muscles (see pp. 54–62 and Table 3) of humans, but limited to the posterior two-thirds of the tongue. She also found spindles in the extralingual portion of the genioglossus, an extrinsic tongue muscle (see Table 2). Spindles were relatively common in the rhesus monkey, a member of genus *Macaca* of the superfamily Cercopithecoidea, but were absent in the kitten, *F. domesticus,* and rare in the newborn lamb. In a later review, Hosokawa (1961) reported spindles in some human hypobranchial muscles of lingual significance (see pp. 61–62) but not in others. No spindles were recognized in the tongue muscles of the cat, the rabbit, or the rat, a finding that confirms and extends the previous negative observations by Cooper (1953).

Functionally, muscle-spindle afferents would be expected to have a "spontaneous" activity input with the extrafusal muscle at rest (see Granit, 1966; Matsunami and Kubota, 1972). Spontaneous activity from the hypoglossal nerve has been reported by Kawamura and Morimoto (1973). However, increasing stretch of the tongue produced sustained increases in firing rate, which would not be expected of a muscle spindle. Thus, the "stretch receptors" of the tongue do not appear to be muscle spindles. Moreover, even if muscle spindles are present, the function that they are typically adduced for (i.e., tongue position) is not appropriate for them: muscle-spindle systems function so that the length of the specialized muscle fibers (intrafusal muscles) on which the receptor endings are located is continuously adjusted (by a feedback "control" loop through gamma efferent axons) so that a sensitive detection of change in muscle length is always available. Consequently, as Henneman (1974) pointed out, muscle-spindle input would not be a useful indicator of tongue position.

If muscle spindles are often absent, is any sensory input on tongue-muscle length provided beyond that resulting from epithelial receptors? As noted above, neural responses to tongue stretch—initiated by some receptors in the muscle, the fascia, or the submucosa of the tongue—have been recorded from the lateral branch of the hypoglossal nerve by Kawamura and Morimoto (1973). It may be that some of the responses to tapping of the tongue that were recorded by Segundo, Takenaka, and Encabo (1967) from "reticular-formation" nuclei originate in the deep receptors of the tongue.

2.2.2. Jaws

2.2.2.1. Muscle Spindles. The temporal, masseter, medial pterygoid, and digastric jaw muscles, at least, do have muscle spindles typical of skeletal muscle, as noted by Matthews (1972) and Goodwin and Luschei (1974). The cell bodies of these muscle-spindle neurons are located in the nucleus of the trigeminal mesencephalic tract of the hindbrain (see Crosby *et al.,* 1962; Matthews, 1972). Electrophysiological recordings from these units in the domestic cat and in two species of the macaque monkey (*M. mulata* and *M. irus*) by Matsunami and Kubota (1972) and Cody, Lee, and Taylor (1972) showed typical muscle-spindle response patterns to jaw movement. However, no responses suggesting input from joint receptors or from tendon organs of jaw muscles were observed. This absence of joint or tendon responses in the nucleus of the trigeminal mesencephalic tract is not surprising, since only spindle inputs would be expected in this locus.

The significance of the input from the above muscle-spindle afferents of the jaw is questioned by a study by Goodwin and Luschei (1974) of mastication in macaque monkeys before and after destruction of the afferent-neuron cell bodies of the muscle spindles. Chewing rates, chewing patterns, and oral manipulation of food showed no permanent changes after large (80–100%) unilateral or bilateral lesions of the nucleus of the trigeminal mesencephalic tract. This finding suggests that other afferent input is of far greater importance.

2.2.2.2. Nonspindle Inputs. Recordings from various trigeminal nuclei of the brain stem other than the nucleus of the mesencephalic tract indicate that responses do occur to jaw position (e.g., nucleus supratrigeminalis of domestic cat as studied by Jerge, 1963) and to direction of movement of the lower jaw (e.g., main sensory trigeminal nucleus of domestic cat investigated by Eisenman, Landgren, and Novin, 1963). Jerge (1963) has noted that tendon organs, joint receptors, or other structures may be the receptor elements. Whatever the source, an appreciable sensory input is available from deep receptors related to the jaws.

2.3. Areas of the Central Nervous System Receiving Oral Afferent Input

2.3.1. Taste

2.3.1.1. Brain Stem. Hindbrain and midbrain nuclei involved in taste responses have been recognized for many years. The central terminals of relevant cranial nerves (special visceral afferent components of the chorda tympani branch of the facial nerve, the glossopharyngeal nerve, and the vagus nerves) were located from studies of normal and pathological neuroanatomical material (e.g., Kerr, 1962; Crosby *et al.*, 1962), as well as from lesion-degeneration experiments such as those by Rhoton (1968). Several different nuclei were suggested by these studies, including the nucleus of the fasciculus solitarius, the nucleus intertrigeminalis, and the nucleus intercalatus. Functional-anatomical analyses, involving electrophysiological recording in various nuclei during gustatory stimulation and subsequent histological location of the recording site, confirmed the relevance to taste input of several hindbrain nuclei. The gustatory functions of the nucleus of the fasciculus solitarius have been well documented, especially in the laboratory albino rat (see Halpern, 1957, 1959; Pfaffmann, Erickson, Frommer, and Halpern, 1961; Oakley 1962; Halpern and Nelson, 1965). Gustatory functions of the nucleus of the fasciculus solitarius in domestic cat and in primates have been summarized by Halpern (1973, pp. 268–271). For the rabbit, Bava, Innocenti, and Raffaele (1972a) have demonstrated gustatory responses in this nucleus.

Other hindbrain nuclei in which taste responses have been confirmed include the nucleus intertrigeminalis (domestic cat), the parabrachian nucleus (laboratory albino rat), and the nucleus intercalatus (rabbit). See Bava *et al.* (1972b) and Halpern (1973, pp. 271–273) for details. Single-unit studies of the metencephalic taste area of the laboratory albino rat (the parabrachian nucleus or "pontine taste area") are reported by Norgren and Pfaffmann (1975).

2.3.1.2. Forebrain

(A) Anatomical Loci of Responses. Forebrain (i.e., prosencephalon) structures receiving taste input have gradually been elucidated. They include diencephalic elements (Crosby *et al.*, 1962) such as the hypothalamus and thalamus, as Ganchrow and Erickson (1972) observed. For a general summary, see Halpern (1973, pp. 273–276). For the thalamus, taste responses were localized in the medial portion of the ventral nucleus (ventromedial nucleus) in laboratory albino rat and in the cottontop marmoset, a small anthropoid (i.e., simian) primate of the genus *Callithrix* (i.e., *Hapale*) in family Callithricidae. Ruderman, Morrison, and Hand (1972) localized taste responses in the parvocellular division of the ventral posterior medial nucleus of the domestic cat.

Several telencephalic taste regions have been described. A pathway to the substantia innominata, a subcortical nucleus, has been reported by Norgren (1974). In addition, several areas of the cerebral cortex have been found to respond to gustatory stimulation of the tongue. The cortical regions include a portion of the postcentral gyrus homologue in laboratory albino rat and domestic cat (see review by Halpern, 1973, pp. 276–279), the coronal and ectosylvian gyri and the coronal sulcus in the domestic dog (described by Funakoshi, Kasahara, Yamamoto, and Kawamura, 1972), and the presylvian sulcus in domestic cat (reported by Ruderman *et al.,* 1972).

(B) Functional Properties. Individual neurons responsive to a combination of gustatory and thermal and/or mechanical stimulation of the tongue have been regularly observed in hindbrain taste nuclei (e.g., Oakley, 1962; Bernard and Nord, 1971), as they have been in the chorda tympani nerve. However, both multiunit (e.g., see Ruderman *et al.,* 1972; Halpern, 1973, pp. 278–279) and single-unit (Cohen, Landgren, Strom, and Zotterman, 1957; Ganchrow and Erickson, 1972; Funakoshi *et al.,* 1972) responses from rat and cat thalamus and cerebral cortex indicate a separation of taste and somesthetic responses. This scarcity of dual mechanogustatory responses in prosencephalic loci indicates that a filtering or inhibitory mechanism is operating. On the other hand, dual responses to tongue *temperature change* and to gustatory stimuli, a less frequently tested combination, were observed by Cohen *et al.* (1957, see p. 36, Table IV, and Figure 12), in the cerebral cortex of the domestic cat.

2.3.2. Nongustatory Mouth and Jaw Inputs

The regions of the central nervous system in which somesthetic responses from the tongue, from other intraoral tissues, and from the jaws can be recorded have been extensively described and reviewed by Jerge (1967). One major input is through the branches of the trigeminal nerve. The sensory trigeminal-nucleus complex, studied by Kruger and Michel (1962), extends through the hindbrain (rhombencephalon) and into the midbrain (mesencephalon). The trigeminal complex shows a spatial (somatotopic) segregation of "mouth" responses from other head areas (see Nord, 1967; Kerr, Kruger, Schwassmann, and Stern, 1968). In addition, there may be a functional clustering of neurons responsive to particular classes of events, for example, jaw movements and tongue-surface depression, as Eisenman *et al.* (1963) have noted.

Within the trigeminal complex, no responses to gustatory stimuli were observed by Eisenman *et al.* (1963), although responses to very concentrated solutions, above the "gustatory" range, can be expected from the observations of Kawamura *et al.* (1968). The converse (i.e., all nontaste responses' necessarily being of trigeminal origin) is not correct. In a previous section (p. 46), somesthetic responses in the chorda tympani and glossopharyngeal nerves, sometimes in gusto-thermo-mechanoreceptive primary neurons, were described. Nonetheless, some investigators have categorized nonchemosensory responses recorded in nuclei that also show gustatory responses as invariably of trigeminal origin (e.g., Kruger and Michel, 1962; Bava *et al.,* 1972b). This conclusion is moot without functional evidence.

Nontrigeminal brain-stem nuclei also receive sensory input from oral structures (Storey, 1967). These brain-stem regions include the bulbar reticular formation (studied by Segundo *et al.,* 1967), the nucleus supratrigeminalis input (described by Jerge, 1963), and the nucleus intercalatus (Bava *et al.,* 1972b). In addition, the nucleus of the fasciculus solitarius has been shown by Torvik (1956) to receive substantial connections from the trigeminal nerve. This same nucleus has a "somesthetic" input through gustatory axons (see pp. 50–51).

Motor nuclei of the brain stem exhibit clear activation or depression upon stimulation of oral structures. This relationship has been observed in the motor nucleus of the trigeminal nerve and in the hypoglossal nucleus by Kawamura (1964) and by Kawamura and Morimoto (1973). This effect is to be expected on logical grounds, of course, since environmental events in the vicinity of the target muscles, or at the muscles themselves, should affect the motoneurons which output to the muscles. Jaw movement (i.e., jaw-muscle stretch) and mechanical tongue-stimulation produce responses in both the hypoglossal nucleus and the motor nucleus of the trigeminal nerve. The physiological significance of these observations of responses in brain-stem motor nuclei to oral somesthetic stimulation is well demonstrated, especially for the domestic cat. Changes in the contraction of tongue or jaw muscles, alterations in their electromyograms (EMGs) and responses in the motor nerves to the jaw and tongue occur upon touch, deformation, or other events that produce input in the afferent nerves from the tissues of the mouth region of the domestic cat.

Many instances of oral sensory input leading to a modification of tongue- or jaw-muscle activity can be cited. For example, Sauerland and Mizuno (1970) found that reflex-induced activity in the efferent hypoglossal-nerve branch to the tongue protruder-muscle (the genioglossus) is suppressed by prior (5–400 msec) activation of trigeminal afferent-nerve branches from jaw muscles, specifically the digastric or masseteric nerves. Thexton (1973) noted that mechanical deformation (15 g/cm² or more force) of the hard (anterior) palate produced sustained jaw opening with gradual onset of palatal deformation or transient jaw closing with rapid onset of deformation. The jaw opening was accompanied by EMG activity in the temporalis and digastricus muscles of the jaw and rhythmic efferent bursts in the hypoglossal nerve. Kawamura (1964) observed that jaw opening ($\geq$ 4 mm) and closing initiated and ended bursts of activity in the genioglossus muscle if the temporomandibular joint (between the lower jaw and the temporal bone of the skull) was normal, but Lowe and Sessle (1973) found that this was not the case if the joint had been anesthetized. Reflex-induced protrusive movement of the tongue, EMG activity in the genioglossus muscle and anterior intrinsic tongue muscles, and efferent impulses in the "protrusive" branch of the hypoglossal nerve were suppressed by prior (20 msec to 300–500 msec) stimulation of the lingual branch of the trigeminal nerve, the glossopharyngeal nerve, and the superior laryngeal nerve and by afferent activity in nerves from the masseter, digastric, and temporalis muscles of the jaw, the infraorbital nerve, and the teeth (see Sessle and Kenny, 1973; Schmitt, Yu, and Sessle, 1973). Kawamura and Morimoto (1973) observed that forward stretch of the tongue decreases (at 20-g to 30-g stretch load) or eliminates (200-g load) spontaneous efferent impulses in single units from the medial branch of the hypoglossal nerve. Yokota, Suzuki, and Nakano (1974) reported that EMG responses from motor units of extrinsic tongue muscles (styloglossus, genioglossus, hyoglossus) are discretely produced by "gentle stroking" of appropriate areas of the tongue surface, with largely separate areas on both the dorsum and the ventrum related to increased, or decreased, activity in a particular muscle.

3. Motor Input

3.1. Muscular Apparatus

3.1.1. Tongue

The tongue and its muscles have been succinctly described by Kent (1965): "The fleshy part of the tongue . . . is essentially a mucosal sac attached to the basihyoid, stuffed with muscles. . . . The tongue muscle . . . is striated, skeletal, voluntary, somatic, and myotomal by phylogenetic origin" (p. 229). A similar description of the tongue has been provided by Arey (1956). The tongue develops from the first three branchial arches of the mammalian embryo, as Sognnaes (1954) has noted. Development of the anterior division is from the first branchial arch, while the second and third arches yield the posterior division of the tongue.

Anatomically, the internal structure of the mammalian tongue that makes the lingual movements and shape changes possible is a connective-tissue framework into which are insert the muscles that both comprise the main bulk of the tongue and move it (Abd-el-malek, 1939a; Sognnaes, 1954). The framework consists of:

1. The *submucosa*, a connective-tissue layer beneath the outer mucosal membrane of the tongue surface. The submucosa is often particularly firm and thick on the dorsum of the tongue and at the tip of the tongue. Abd-el-malek (1939a) called this anterior thickening the anterior arch (Figure 6).

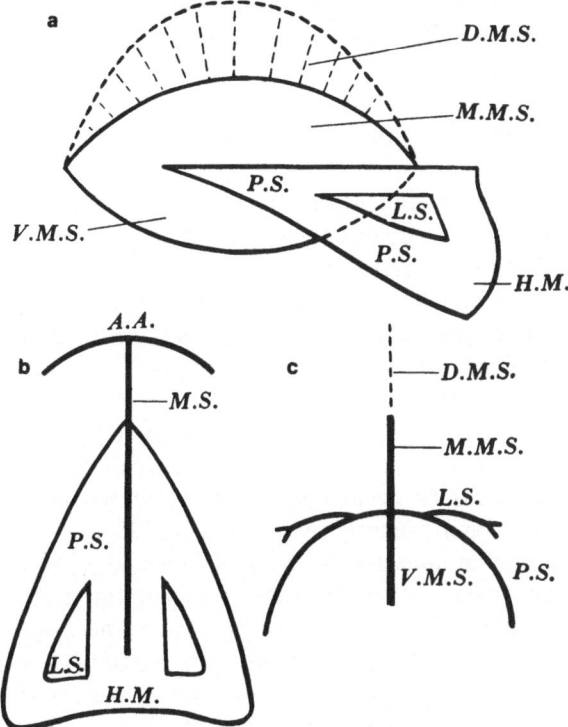

Figure 6. Drawings illustrating the connective-tissue partitions (septa) of the tongue of the cat (*Felis domesticus*): a, left-side view of the tongue; b, frontal section; c, transverse section at the posterior third of the tongue (from Abd-el-malek, 1939a). *A.A.* = anterior arch, a thickening of the lamina propria at the tip of the tongue that receives insertions from the superior longitudinal and inferior longitudinal intrinsic muscles and the genioglossus and styloglossus extrinsic muscles. *D.M.S.* = dorsal part of the median septum, the thinnest part of this septum. *L.S.* = lateral septum, a thick membrane that occurs only in the posterior third of the tongue and joins directly into the hyoglossal membrane. *H.M.* = hyoglossal membrane, which originates from the paramedian and lateral septa of the tongue and extends from the tongue to the body of the hyoid bone. *M.M.S.* = middle part of the median septum. *P.S.* = paramedian septum, the thickest connective-tissue partition in the tongue. It extends from the median septum to the submucosa of the lateral margins of the tongue and extends into the hyoglossal membrane. *V.M.S.* = ventral part of the median septum, a loose connective-tissue structure.

2. *Connective-tissue partitions or septa* (Figure 6). A median septum is generally recognized, while lateral and other septa have also been described. Septa also extend into the several tongue muscles (Tables 2 and 3). An extension of the median septum, the hyoglossal membrane, extends from the base of the tongue to the hyoid bone.

The muscles of the tongue are generally divided into two groups. One group, the *intrinsic* muscles, not only originate wholly within the tongue but also insert primarily within the tongue (Table 3). Activity of the intrinsic muscles changes the shape of the tongue. Furthermore, in concert with the extrinsic muscles (Table 2), the intrinsic muscles also modify tongue motion (Abd-el-malek, 1939a).

The second group of tongue muscles, the *extrinsic* muscles, take origin outside the tongue but have most of their insertions inside the tongue (Table 2; Figure 2). The loci of origin of the extrinsic muscles of the tongue are bones of the skull and hyoid apparatus. These bony origins include the anterior portion of the mandible (i.e., the lower jaw), the posterior skull (i.e., the occipital bone), and various hyoid bones. The lingual insertions of the extrinsic muscles are into the connective-tissue framework of the tongue and into the intrinsic muscles. The primary function of the extrinsic muscles is to move the tongue. The individual muscles can each produce crude motions, but full motions occur only when several extrinsic and intrinsic muscles are active simultaneously (Abd-el-malek, 1939a).

A third group of muscles is of significance with respect to tongue motion, although they neither originate from nor insert into the tongue. Although these muscles do not reach the tongue, the *geniohyoid* muscle (origin: medial surface of the mandible), the *mylohyoid* muscle (origin: medial surface of the mandible), the *sternohyoid* muscle (origin: cartilage of the first rib and, sometimes, the manubrium of the sternum), and the *omo-hyoid* muscle (origin: anterior border of the scapula) insert into the body or the anterior horn of the hyoid bone, with origins on relatively fixed structures (Figures 1 and 2) (see Sisson, 1930; Craigie, 1960; Wischnitzer, 1972). The geniohyoid and mylohyoid muscles thus draw the hyoid bone anteriorly, because the origin is toward the front of the mouth. The sternohyoid and omohyoid, with an origin more aboral than the hyoid bone, draw the hyoid posteriorly. The anatomy and the major actions of these muscles are described by Reighard and Jennings (1963), Wischnitzer (1972), Green (1968), Chiasson (1969), and Romer (1970). Since many extrinsic tongue muscles take origin from the hyoid bone, as Sprague (1943) described (Table 2; Figure 1), movement of the hyoid modifies tongue motion and occurs in conjunction with tongue motions (Ardran and Kemp, 1955).

The geniohyoid, sternohyoid, omohyoid, and mylohyoid muscles are often referred to as part of a set of muscles, the hypobranchial muscles (e.g., see Kent, 1965; Romer, 1970). The hypobranchial muscles, which lie beneath the gills in aquatic vertebrates, give rise to the tongue muscles in the course of the evolution of terrestrial vertebrates. With reference to the location of these hyoid muscles in mammals, Greene (1968) and Friel (1974) classify the geniohyoid and mylohyoid muscles as suprahyoid muscles, while the sternohyoid and omohyoid are infrahyoid muscles. In general, the hypobranchial muscles initially originate, either ontogenetically or phylogenetically, from the anterior trunk muscles.

An additional extrinsic tongue muscle, the palatoglossus, is recognized in humans (*H. sapiens*) (Elliott, 1963; Friel, 1974). Under the name glossopalatus, this muscle is also described in other species (e.g., domestic rabbit) (Craigie, 1960). The palatoglossus originates under the surface of the soft palate and inserts into the sides or base of the tongue. As might be predicted, this muscle elevates the tongue. The palatoglossus muscle, in contrast to other tongue muscles, is innervated through the pharyngeal plexus of the vagus nerve (Cranial X) by the IX, X, and/or XI cranial nerves.

The position of the tongue at any moment is the vector sum of the tension of the several active intrinsic, extrinsic, and hypobranchial muscles. This vector concept has been explicitly presented by Doran and Baggett (1972) in an analysis of the attachments and functions of the genioglossus muscle in several mammals. Vector analysis can be applied whenever the anatomy is adequately described and profits from functional studies. The first step is to analyze the expected tongue motion resulting from the activity of any single extrinsic muscle (Table 2) by considering the origin(s) and insertion(s) of that muscle. For example, the styloglossus muscle of the domestic cat, with origins on the hyoid bone and the rear of the skull (i.e., the occipital bone) and insertion into the apex of the tongue (Figure 6), can be expected to both *retract and elevate* the tongue (Table 2). The retraction would be related to both loci of origin, but the elevation of the tongue would be a function of the occipital origin, since this location is above the tongue. In contrast, the same tongue muscle in the little brown bat originates only from the hyoid bone and therefore *retracts but can be expected not to elevate* the tongue (Table 2). Thus, specific differential predictions of motion can be made. Similarly, the genioglossus muscle, which originates from the center of the lower jaw (i.e., the mandibular symphysis) and inserts into the full length of the tongue, acts to protrude the tongue. This occurs because the origin of the genioglossus is anterior to the tongue, and consequently a contraction pulls the tongue forward. Such analyses must be made for the tongue muscles of a specific mammalian species. After the separate analyses for the muscles of a species are made, the individual, directed motions are combined to yield a final vector sum.

The shape of the tongue also results from activity across an array of muscles, in this case primarily the intrinsic muscles. Thus, contraction of the longitudinal muscles alone, which have attachments at the anterior and posterior limits of the tongue, shortens and thickens the tongue (Table 3). Combined contraction of the longitudinal and transverse muscles shortens and flattens the tongue. As more intrinsic muscles become active, complex shapes can be produced. In general, the connective-tissue framework and the muscles themselves serve to stiffen the tongue (Abd-el-malek, 1939a). The hyoid bone functions as a support for the base of the tongue (Hyman, 1922; Romer, 1970).

3.1.2. Jaws

Specific motions can be attributed to each jaw muscle. In general, *the masseters (Table 1), together with the temporal, the zygomaticomandibularis, and the medial ptergoid muscles, close (adduct) the jaws. The digastric and lateral pterygoid muscles are active during jaw opening (abduction). The masseter and pterygoids are also related to protrusion of the lower jaw. The digastric and zygomaticomandibularis are also related to retraction. These movements combine to produce chewing, biting, grinding, etc.* Møller (1966) reported that during chewing motions in humans, the above muscles, as well as the mylohyoid muscle and the lip muscles (orbicularis oris), are active in a regular pattern. Such patterned sequences of jaw motion characterize, often uniquely, the chewing of many animals (e.g., the American opossum, Crompton and Hiiemae, 1970; Hiiemae and Crompton, 1971; the laboratory albino rat, Weijs, 1973a; the miniature pig, Herring and Scapino, 1973; and the rhesus monkey, Luschei and Goodwin, 1974). However, certain sets of mammalian jaw muscles can not be categorized on a muscle/motion basis. Thus, for the little brown bat (*Myotis lucifugans*), Kallen and Gans (1972) stated that "rather than thinking of pure openings and closings, pure left or right transverse movement, pure protrusion or retursion, one must recognize that all of these processes occur simultaneously. . . . It may be preferable to view the component elements as part of a muscular sling . . . that suspends the moving bony element" (p. 404). The little brown bat may represent the limiting case of a continuum, with various degrees of multimuscle-dependent jaw motion occurring in different mammals.

3.2. Neural Control

3.2.1. Tongue

Most mammalian tongue muscles, both intrinsic and extrinsic, are innervated by branches of the hypoglossal nerve (Cranial XII) (Tables 2 and 3). In the domestic cat (*F. domesticus*), electrical stimulation of hypoglossal branches (not single fibers) to each muscle produces separate tongue movements and/or specific shape changes, while combined stimulation leads to complex events (Abd-el-malek, 1939a). The hypoglossal nerve originates from the hypoglossal nucleus of the hindbrain, as Barnard (1940) and Crosby *et al.* (1962) have noted. In general, the hypoglossal nucleus may be subdivided into regions corresponding to particular tongue muscles. Morimoto, Kato, and Kawamura (1966) observed that electrical stimulation of the hypoglossal nucleus in the domestic cat produces large tongue motions, similar to those described for stimulation of hypoglossal nerve branches. This observation suggests that the electrical stimulation may have directly affected the efferent hypoglossal axons, since stimulation of second-order efferent neurons would be expected to produce tongue-motion patterns different from those of the primary neuron. Each general type of tongue motion—that is, protrusion, retraction, lateral deviation, twisting, and elevation of the tip—could be related to a definite region of the hypoglossal nucleus, although extensive overlap occurred. The region in which stimulation produced tongue retraction was quite large, extending 5 mm rostral to the obex, 2.5 mm lateral to the midline, and occupying more than half the dorsoventral extent of the nucleus. The protrusion region, which was much smaller than the retraction region, fully overlapped with the medial edge of the retraction region but also had a region caudal to the obex and 0.5–1.0 mm lateral to the midline, which was unique to protrusion. Lateral deviation was confined to a small region centered within the tongue-retraction region, while a small tongue-twist region was located primarily caudal to the obex, between 1.0 mm and 2.0 mm lateral to the midline. Stimulation at the caudal limit, 2 mm caudal to the obex, of the tongue-twist region produced elevation of the tip of the tongue. Thus, tongue retraction was most heavily represented. This assessment seems reasonable, since both the styloglossus and the hyoglossus extrinsic tongue muscles produce tongue retraction (Table 2). Retraction, which returns foods and liquids to the mouth, may require finer control than protrusion.

In contrast to the specific and highly localized effect of stimulation in the hypoglossal nucleus of the cat, Bernston and Hughes (1974) reported that complex organized behavior involving the mouth, such as grooming and the ingestion of solid food, could be elicited by electrical stimulation of the medullary reticular formation. Surprisingly, however, no licking of liquid food was ever produced in the cats studied.

A study of hypoglossal-nucleus function in the laboratory albino rat (Zsuzsanna Wiesenfeld, unpublished observations), using highly localized electrical stimulation, found that small, discrete changes in tongue shape and limited motion were produced by low-voltage stimulation, but no large movements were produced, perhaps indicating that a neuronal population presynaptic to the hypoglossal neurons themselves was stimulated. However, somewhat larger motions could be produced by higher stimulus voltages. Therefore, the limited motions may have been due to activation of a limited set of hypoglossal motoneurons.

Wiesenfeld, Halpern, and Tapper (1977) also found that multiunit and single-unit neural activity could be recorded in the hypoglossal nucleus of the laboratory albino rat during voluntary licking of water. The neural activity was recorded with $60-\mu$m platinum–iridium electrodes, which had been previously implanted while the animal was anesthetized. The occurrence of each tongue contact with the metal drinking tube was detected by a solid-state contact-sensitive instrument (see Chapter 2), that is, a drinkometer. In this case, the drinkometer pulse occurred at the end of tongue contact, giving a stop–stop interval as the interlick interval. A high-restriction licking environment was used (Figure 8). Each lick was accompanied by a burst of hypoglossal activity. In addition, the hypoglossal records indicated that aborted licks (see p. 28), in which the direction of tongue motion switched from protrusion to retraction *before* contact with the liquid occurred, were detectable in the hypoglossal nucleus. Single units could sometimes be isolated if wave-form and amplitude criteria were used to identify them. The unit bursts had modal periods of 140–160 msec, which is similar to the interlick interval of rat licking (see Sections 2 and 4 of this chapter and Chapter 2). These single-unit bursts tended to occur during retraction of the tongue from the drinking tube. Specifically, the hypoglossal-unit bursts preceded the end of tongue contact by 20–50 msec. Approximately 30 msec was utilized for tongue retraction (B. P. Halpern and T. L. Nichols, unpublished observations). Consequently, the recorded unit activity began just prior to observed retraction. It may be that the full retraction phase of licking was being monitored.

Tongue movement is affected by regions of the central nervous system beyond the hypoglossal nucleus, of course. For example, large bilateral lesions of frontal cerebral cortex ("motor cortex") in domestic rat produce a near-total inability to secure food pellets from a narrow dispenser with the tongue (Castro, 1972). However, licking water from a drinking tube appeared to be at least qualitatively unchanged. This distinction between two complex tongue-movement patterns may be related to the extensive prelesion practice of licking and the much less practiced (and perhaps less "natural") food-securing task. On the other hand, these data may indicate that there is less cortical involvement in liquid ingestion. In the cat, rhythmic protrusions and retractions of the tongue, at a rate of about 1/sec, are produced by electrical stimulation of the orbital gyrus of the cerebral cortex (Morimoto and Kawamura, 1973).

3.2.2. Jaws

With one exception, the mammalian jaw muscles are innervated by motor (efferent) branches of the mandibular division of the trigeminal nerve (Cranial V). The exception is the posterior portion (belly) of the digastric muscle, which is generally innervated by an efferent branch of the facial nerve (Cranial VII). This posterior portion of the digastric muscle phylogenetically originates from the hyoid (i.e., second visceral) arch. All muscles derived from this arch have Cranial VII innervation. The other jaw muscles originate from the mandibular (first visceral) arch, which has a Cranial V innervation (Kent, 1965; Romer, 1970).

The mylohyoid muscle, a jaw-related muscle, is also important. It raises the floor of the mouth and moves the hyoid bone anteriorly (Hyman 1922); Wischnitzer, 1972; Chiasson, 1969; Friel, 1974). The mylohyoid muscle is a member of the suprahyoid muscles. The mylohyoid takes origin from the inner surface of the mandible and inserts on the body of the hyoid bone and on a median raphe between the two symmetrical mylohyoid muscles. The major phylogenetic origination of the mylohyoid is the mandibular arch. Consequently, it receives trigeminal-nerve innervation. However, the mylohyoid is sometimes described as consisting of anterior and posterior components, with the anterior component, derived from the interhyoideus muscle of the hyoid arch, receiving facial nerve innervation (see Kent, 1965, 1973). Typically, however, the mylohyoid, whether unitary or divisible into anterior and posterior components, receives only trigeminal innervation. This is reported for bat, cat, human, domestic rabbit, and rat by Sprague (1943), Craigie (1960), Reighard and Jennings (1963), Greene (1968), and Friel (1974).

The mandibular branch of the trigeminal nerve, which innervates the jaw and the mylohyoid muscles, originates in the motor nucleus of the trigeminal nerve. The motor nucleus of the trigeminal nerve is located in the pons, the highest division of the hindbrain (Crosby *et al.*, 1962). The individual jaw muscles are represented in distinguishable regions of this nucleus (Elliott, 1963). This nucleus receives sensory (afferent) input from receptors in the jaw muscles and through afferent branches of the trigeminal nerve that reach the motor nucleus as the mesencephalic root of the trigeminal. Muscle-spindle input is common, as Kawamura (1964) and Matsunami and Kubota (1972) have noted.

The facial-nerve innervation of the posterior belly of the digastric probably originates from the dorsal (accessory) facial nucleus of the pons (Crosby *et al.*, 1962).

3.2.3. Hypobranchial Muscles

The hypobranchial muscles originate from a region that in aquatic cartilaginous vertebrates (i.e., elasmobranchs, class Chondrichthyes) is the hypaxial region immediately behind the pharynx and beneath the gills. The intrinsic and extrinsic tongue muscles, which are of hypobranchial classification, have already been discussed.

The remaining hypobranchial muscles of interest to ingestion are the geniohyoid, the sternohyoid, the omohyoid, and the mylohyoid muscles. As noted above, the major or sole innervation of the mylohyoid is through the mandibular branch of the trigeminal nerve. The geniohyoid muscle receives a direct innervation from the hypoglossal nerve. The sternohyoid and omohyoid muscles receive their innervation through a complex path that reflects their origination in the ventral trunk muscles of fish. A descending nerve branch (ramus descendens), apparently carrying innervation both from the hypoglossal nerve and from spinal nerves, reaches these muscles. This nerve branch, the ansa hypoglossi or ansa cervicalis, is a combination of a descending hypoglossal-nerve branch and branches of the first or the first and second cervical (upper spinal) nerves. General treatments of the hypobranchial muscles, as well as discussions of specific groups, are provided by Sprague (1943), Reighard and Jennings (1963), Kent (1965), Meader and Muyskens (1962), Greene (1968), Romer (1970), and Friel (1974).

4. Drinking Behavior

4.1. Licking

4.1.1. Detailed Properties of Licking

A liquid-intake mechanism that requires high-speed repetition of the ingestion act initially seems poorly adapted to mammalian needs because each separate occurrence secures only a few microliters of fluid. Licking/lapping is such a mechanism. The unlikely aspect of this liquid-intake mechanism is especially clear because an apparently superior mechanism—sucking—allows ingestion of many milliliters in a second. Nonetheless, licking/lapping is a common intake pattern (Table 4). Licking/lapping animals include adult opossum (both American and Virginia), domestic cat and dog, laboratory albino rat, hamster, and some bats. A consideration of some of the characteristics of licking/lapping can suggest possible adaptive bases for it. Most currently available data are on liquid licking *per se*. Liquid licking is ingesting liquids by first pressing the tongue against a liquid-supporting surface (a surface with drops or a shallow pool on it) or a liquid dispenser (drop dispenser or drinking tube; see Figure 7) and then delivering a portion of the liquid removed to the mouth. Average licking rates occupy a narrow range across a number of mammals, ranging from 3.5 to 6.5 licks/sec (Table 4). When all aspects of a licking situation are held constant, licking is a relatively stereotyped, low-variability behavior, as Marowitz and Halpern

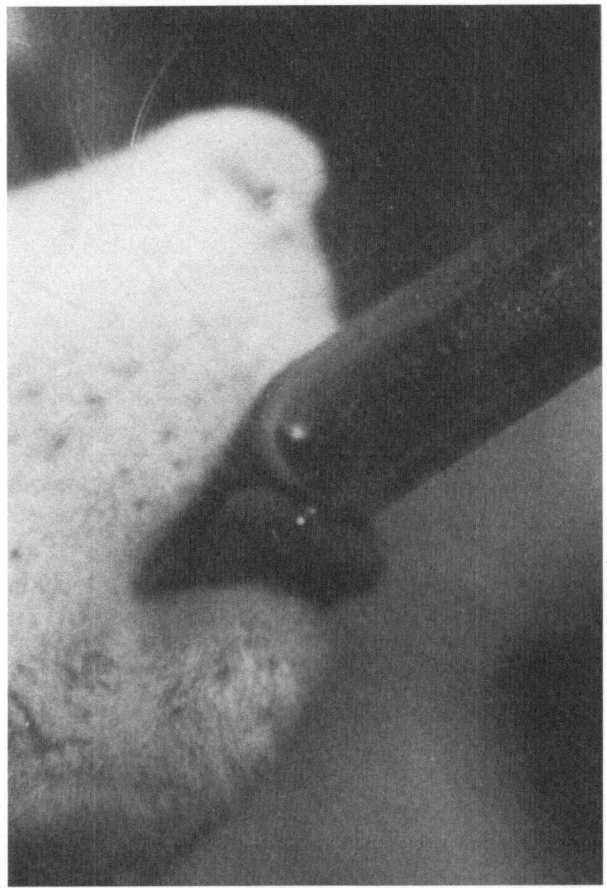

Figure 7.   Photograph of a rat (laboratory albino, *Rattus norvegicus*) drinking water from a stainless-steel drinking tube. The rat was drinking in a low-restriction drinking environment (Marowitz and Halpern, 1973), in which the drinking tube was centered 1 cm outside an elliptical opening 8 × 2.8 cm, centered 10 cm above the floor of the cage in which the rat was located.

(1973) and Halpern (1975) have noted. However, licking is modified by many environmental changes, such as the geometry of a licking environment. Thus, the pattern of licking water from a drinking tube versus an open liquid surface is different in contact duration (time from onset to offset of contact with liquid, or on–off time), interval (time from end of contact with a liquid to start of the next contact, or start–stop interval), and the tongue-movement pattern (Figures 8B, 8C). An analysis of some temporal parameters of licking in several animals is presented in Chapter 3.

Electronic artifacts and distortions are often encountered in attempts

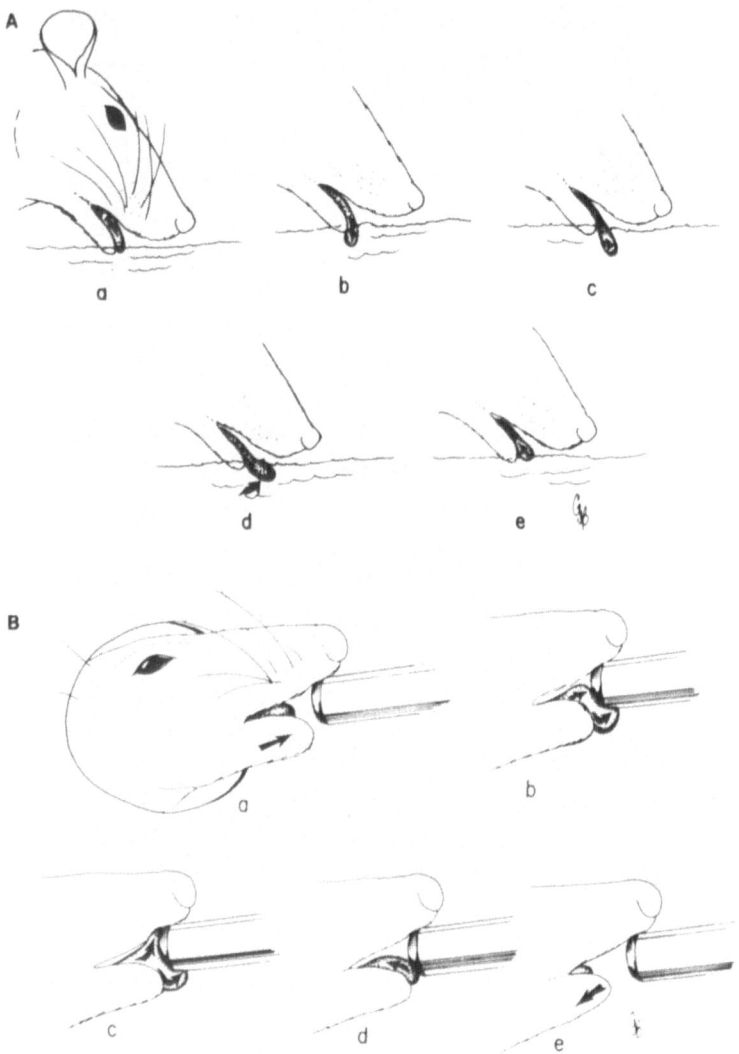

Figure 8. The effect of environmental constraint upon tongue-motion patterns during licking or lapping of water by the rat. The drawings were prepared from enlarged prints obtained from motion pictures taken at 64 frames/sec. The stainless-steel drinking tube (B and C) had a 4-mm tip i.d. A, no restriction. Open container (13 × 9 × 2.3 cm). The intervals are all 15.6 msec. B, low restriction. The opening is a flattened ellipse (8-cm major axis, 2.8-cm minor axis) as in Figure 7. The time intervals separating the drawings are all 15.6 msec. C, high restriction. The opening is a 1.3-cm diameter circle. The intervals are 15.6, 15.6, 15.6, and 31.2 msec between a–b, b–c, c–d, and d–e, respectively. From Marowitz and Halpern (1973).

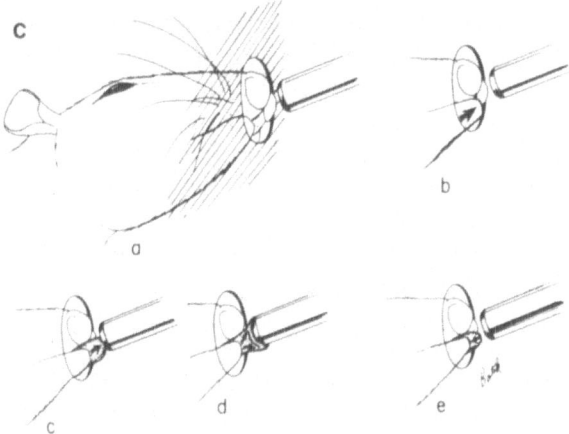

to study licking/lapping. As J. A. W. M. Weijnen points out in Chapter 2, all electronic devices that measure licking/lapping parameters by having the ingesting animal and the liquid complete the circuit have nonzero latencies. That is, the electronic device indicates that lick contact has occurred later than the physical event. Similarly, contact is reported to continue after the tongue is removed from the drinking surface or drinking tube. Commercial drinkometers, as analyzed in Weijnen's Table 2 (Chapter 2), differ appreciably in these start-and-stop latencies. In the present chapter, all drinkometer measurements from experiments done by my colleagues and me are corrected for start-and-stop latency. The correction factors were measured with an optical isolator based upon a light-emitting diode and a phototransistor, with a circuit resistance of less than 100 ohms. Specifically, for Grason–Stadler E4690A drinkometers, used by Tapper and Halpern (1968) and Halpern and Tapper (1971), 20 msec were subtracted from the apparent contact duration and 20 msec added to the apparent stop-start interval. For Grason–Stadler 1241 drinkometers, used by Marowitz and Halpern (1973), Halpern (1975), Halpern and Nichols (1975), and B. P. Halpern and T. L. Nichols (unpublished observations), 25 msec were added or subtracted, as above. Note that interlick interval (i.e., start–start time) and licking rate are not affected by these corrections. An additional source of artifact in electronic measurement exists. This is a liquid bridge between the drinking tube and the tongue. Motion pictures at 200 frames/sec made by B. P. Halpern and T. L. Nichols (unpublished observations) indicate that such a liquid bridge can exist for 10–15 msec in the laboratory albino rat after the tongue has retracted from direct contact with a high-restriction access drinking tube (Figure 8).

Although licking has been examined most thoroughly for the laboratory albino rat, other species have received careful study. The process in the American opossum differs somewhat from that illustrated for the laboratory albino rat in Figure 8B and 8C. A description of licking from shallow pools by the American opossum, based upon cinefluorographic studies, is available from Hiiemae and Crompton (1971). Starting from the waking "rest" position of the jaw (i.e., a slightly open position), which is similar to that of many other mammals, a downward movement of the lower jaw occurs (a downstroke), continuing until a mean gape of 6° is present. The tongue is then rapidly protruded, with the tip pointed down, and contact with the liquid occurs. The tongue now steadily changes shape, becoming flatter. No contact occurs between the jaws and the liquid. The jaw now moves sharply downward, in effect continuing the downstroke that was interrupted while the tongue reached and secured some liquid. The tongue, with liquid on it, next retracts into the mouth. The downstroke continues to completion. Then a rapid upstroke carries the jaw back to the rest position. The secured liquid is moved smoothly backward along the tongue by a series of contraction waves along the length of the tongue and is collected near the epiglottis. This description of liquid licking also fits licking of soft, smooth, nonliquid foods.

Lapping, as defined by Hiiemae and Crompton (1971), differs from licking in that the tongue enters a liquid pool, contacts only the liquid, and then retracts into the mouth. Both lapping and licking probably occur under natural conditions, with licking useful for drops on leaves, for shallow sources, and for soft, nonliquid foods. Lapping involves intake from more extensive liquid sources. Lapping has been studied with the cinefluorographic method in several mammals, including the American opossum and the laboratory albino rat. In the laboratory albino rat, lapping of "barium sulphate impregnated milk," as observed by Hiiemae and Ardran (1968), follows a consistent pattern. The mandible (lower jaw) is first lowered until the incisors are slightly apart. The tongue is then rapidly extended into the liquid. A small volume of liquid moves up the tongue, but most of the liquid remains at the tip. A rapid retraction of the tongue now occurs, with a small portion of liquid ingested. The lapping mechanism appears to depend on tongue movement, not on mandibular movement.

The lapping behavior for milk containing $BaSO_4$, described above, is similar to that observed for water lapping in the laboratory albino rat by Marowitz and Halpern (1973). This latter lapping is shown in Figure 8A. These authors noted that if a relatively large pool of water (13 cm long, 9 cm wide, 2.3 cm deep) was provided, the mandible (i.e., the lower jaw) remained in contact with the liquid across successive lapping cycles. Under these conditions, the tongue extends into the liquid very quickly. Motion pictures permitted measurement of the lapping rates, which ranged from 4.3 to 5.8 laps/sec.

A maintained contact of the lower jaw with the liquid pool is an inconvenience for laboratory measurements of lapping, although such contact may be a common natural event. B. P. Halpern and T. L. Nichols (unpublished observations) found that if a sufficiently small, open liquid surface is provided, each lap is a separate event. In these studies, surfaces of 1 × 1 cm or 1.4 × 2.3 cm were used, with liquid depth of 3.3 cm or 8.3 cm. Electronic (i.e., drinkometer) measurements were taken in addition to motion pictures at 200 frames/sec. The tongue-protrusion portion of a lap, from emergence of the tongue from the mouth to contact with the liquid, required about 20 msec. This initial appearance of the tongue followed the beginning of the lowering of the mandible by approximately 20 msec. Thus, an overall time interval of 40 msec occurred from the initiation of the downstroke of the lower jaw to contact of the tongue with the liquid in the rectangular container. Under these conditions, lapping rates during the first 10 laps ranged from 6.9 to 8.2 laps/sec. This rate is comparable to initial licking rates, as described below on pp. 69–70.

For the American opossum, the initial stages of lapping are similar to those of licking, through the pause in the downstroke (downward movement of lower jaw). The subsequent tongue motions, however, differ from those of licking (Hiiemae and Crompton, 1971). During lapping, after the tongue protrudes into the liquid, it quickly begins to move backward, often accompanied by additional opening of the lower jaw. The tongue then reverses direction, moves rapidly forward through the liquid, and then is quickly withdrawn with an upward movement. The final forward and upward movement of the tongue ripples the liquid and carries some of the liquid away on the tongue. The jaw begins an upstroke as the tongue starts to retract. With the tongue retracted into the mouth and the upstroke completed, the jaw is held in a occlusional position (power stroke) while the liquid moves proximally on the tongue, either to the molar region or to the area of the epiglottis. As in licking, these latter, intraoral phases of liquid movement are a series of wavelike crests and troughs along the length of the tongue. Overall, the tongue-motion sequence during lapping in the American opossum appears to be similar to that in the laboratory albino rat.

Licking/lapping has a number of advantages as a liquid-intake mechanism. One such advantage is the ability to utilize small, discrete drops of liquid (Hulse and Sutter, 1970). This ability is directly related to the small volume (5–10 μl) secured by each lick. While this small amount necessitates many intake cycles for an appreciable volume, it also permits cessation of ingestion after small intakes. With sufficient motivation, licking contact can end after a single lick (as Halpern and Tapper, 1971, and

Scott, 1974, have observed), providing a significant advantage in the avoidance of toxic liquids. Indeed, it may be that even the small volume that flows on the tongue during a single lick (Hiiemae and Ardran, 1968; B. P. Halpern and T. L. Nichols, unpublished observations) need not be ingested. This could be the case if the liquid can be recognized, a rejection decision made (see pp. 27–29) and the tongue position during retraction modified so that the liquid on its surface is not delivered into the mouth.

Rejection of the volume of a single lick or lap, rather than swallowing, conveys an advantage only if little adsorption of toxins occurs during the period of contact of the liquid with the tongue. The preswallowing contact period is briefer than 91 msec, which was determined by subtraction of the time of the extension phase of licking from the interlick interval. The interlick interval at the beginning of licking is about 131 msec, of which approximately 40 msec would be devoted to a downstroke of the mandible and protrusion of the tongue (B. P. Halpern and T. L. Nichols, unpublished observations).

How much absorption can occur in ≤91 msec? This rate is dependent upon the nature of the lingual epithelium. The epithelial layer of the dorsum of the anterior division of the rat's tongue is about 78 μm thick in general and about 31 μm thick on the fungiform papillae. These dimensions are based upon Mistretta (1971, Figure 1). A similar epithelial thickness (41 μm) is shown for fungiform papillae in a laboratory albino rat in Bernard and Halpern (1968, Figure 12A). One especially important factor would be the thickness of the outermost layer of the lingual epithelium, the stratum corneum (i.e., the keratin layer). Mistretta (1971) reported a thickness of keratin of 30–50 μm on the dorsum of the tongue in general, but a 5-μm thickness on fungiform papillae. Bernard and Halpern (1968) showed a 8-μm keratin layer on a fungiform papilla. The fungiform papillae, which represent elevated, well-vascularized regions of very low keratinization, are quite common on the anterolateral portions of the anterior division of the tongue. Therefore, some potential for liquid penetration exists while a liquid contacts the tongue. Hellekant (1965a) has demonstrated that ethyl alcohol can penetrate the tongue of the domestic cat sufficiently to produced neural activity in the lingual branch of the trigeminal nerve. This response is expected, since (as noted on p. 47) many trigeminal axons reach the fungiform papillae. The time required to penetrate to the axons is significant. The latencies ranged from 4 to 12 sec. This rate contrasts with latencies of much less than 1 sec for chorda tympani nerve responses in domestic cat to alcohol and to other aqueous solutions (Hellekant, 1965b). Thus, for this licking/lapping species, little penetration would be expected in less than 100 msec. This result is reasonable, since, as Ham (1974) points out, all stratified epithelium absorbs relatively poorly. Consequently, rejection of the first drop licked or lapped, after tongue contact but before swallowing, would be a useful strategy.

Not all licking/lapping animals actually drink water. Thus, Hudson (1964) notes that some rodents that live in arid habitats eat a dry diet and drink no water under natural conditions. However, such rodents may drink in the laboratory (Boice, 1967). Other desert rodents eat primarily succulents (i.e., fleshy plants), which are indigenous to arid regions and which retain and store water (Bridgwater and Kurtz, 1963), and do not normally drink (Hudson, 1964; Magalhael, 1967). A parallel situation could exist in some carnivores, if the liquid content of their diet is adequate.

The general temporal patterns of licking are probably invariant across most licking mammals. However, the following statements are based upon the laboratory albino rat (*R. norvegicus*) unless otherwise noted.

4.1.2. General Characteristics of Licking

a. *The licking rate is highest at the beginning of licking and decreases after the first few licks.* This characteristic was observed in inexperienced weanlings by Schaeffer and Premack (1961) and experienced adults by Hulse and Suter (1970). A similar decrease occurs in hamsters (Schaeffer and Huff, 1965) and several other mammals (Cone, 1974).

b. *The licking rate decreases rapidly at the beginning of a period of licking, then more slowly as licking continues* (Snyder and Hulse, 1961; Davis, 1973; Schaeffer and Premack, 1961; Halpern, 1975). This nonlinear decrease was also seen in domestic cat (*F. domesticus*) by Schaeffer and Huff (1965).

c. *The inverse relationship between the licking/lapping rate and the number of licks/laps emitted is independent of the liquid being drunk and of the degree of access restriction.* This inverse variation occurs with water (Keehn and Arnold, 1960; Snyder and Hulse, 1961; Schaeffer and Premack, 1961), viscous and nonviscous saccharin (Wilcove and Allison, 1972), NaCl (Halpern, 1975), cellulose suspended in water (Halpern, 1975), and sucrose (Davis, 1973; B. P. Halpern and T. L. Nichols, unpublished observations). With continued licking, a decline in licking rate results both from highly restricted access, in which only the tip of the tongue contacts the drinking tube during a lick (Snyder and Hulse, 1961; Halpern, 1975), and from low-restriction, open-surface lapping (open container 1 cm in diameter—Schaeffer and Premack, 1961; open container 1 × 1 cm or 1.4 × 2.3 cm—B. P. Halpern and T. L. Nichols, unpublished observations). The relationship is also seen in domestic cat, *F. domesticus* (Schaeffer and Huff, 1965).

d. *The maximum instantaneous licking rate is about 9-10 licks/sec.*

This instantaneous rate was measured in inexperienced weanlings drinking water by Schaeffer and Premack (1961) and in adults drinking sucrose solutions by Davis (1973). In cat, *F. domesticus,* Schaeffer and Huff (1965) observed that maximum rate (water or milk) is 5 licks/sec. For young (< 35 days old) gerbils, *M. unguiculatus,* maximum rate for water is 11 licks/sec (Pierson and Schaeffer, 1975).

e. *The licking rate (central tendency) decreases about 20% during the first 15–20 seconds of licking.* A 20% decrease by the 70th lick in rats was reported by Halpern (1975). This corresponds to 14 sec of licking (B. P. Halpern and T. L. Nichols, unpublished observations).

f. *The pattern of tongue movement changes with the degree of restriction of access to the liquid* (Marowitz and Halpern, 1973; B. P. Halpern and T. L. Nichols, unpublished observations) (see Figure 8). In a high-restriction environment (a drinking tube 2 mm outside a 1-cm opening), the tongue moves in a straight line to the drinking-tube opening, flattens against the opening, and then returns to the mouth (30 msec from maximum flattening to return to the mouth). With a low-restriction environment (drinking tube 1 cm outside an "elliptical" opening 8 × 2.8 cm), the tongue first reaches the *bottom* of the drinking tube (Figure 7), moves up to and across the opening while pressed against the drinking tube, and is then retracted into the mouth. For a "no-restriction" lapping environment (an open container 13 cm long, 9 cm wide, 2.3 cm deep, placed on a large surface on which the rat also stands), the lower jaw is placed in continuous contact with the liquid, while the tongue moves down and out over the lower teeth, then swings forward, and finally returns to the mouth.

g. *The mean licking rate differs widely among different species of rodents (from 6.5/sec in the murid R. norvegicus, Norway rat, to 3.5/sec in cricetid Dipodymus ordii, kangaroo rat).* No sex differences have been reported, nor has domestication in *R. norvegicus* changed the licking rate (Boice, 1967). Different species of the same genus had similar rates (e.g., *Noetoma micropus, N. albigula, N. mexicana:* 5.3, 5.3, 4.5, respectively). Free-living liquid-intake patterns ranging from the nondrinking mode of a desert rodent (e.g., *D. ordii,* kangaroo rat) to obligatory fluid intake do not predict licking rate in the laboratory. In domestic cat, the mean rate is 3.7/sec (Schaeffer and Huff, 1965), while Cone (1974) described a 5.4/sec mean rate in the New Zealand white rabbit.

h. *Some of the liquid removed from a drinking tube or a drop dispenser by the tongue contact of a lick cycle is lost during each retraction of the tongue.* This finding has been reported by Halpern (1975) and B. P. Halpern and T. L. Nichols (unpublished observations) for drinking-tube studies with distilled water, NaCl, and sucrose, all containing a dye (0.05% red #40) that permits visualization of the liquid in high-speed movies. It has also been reported with distilled water containing cellulose, as well as with water in a drop dispenser (S. Hulse, personal communication). Direct measurements by D. Baron and B. P. Halpern (unpublished observations) indicate that in a high-restriction licking environment, 0.1 μl/lick (range: 0.04–0.2 μl/lick) are lost during the first 10 sec of licking.

i. *No liquid is lost during lapping.* This efficient ingestion has been observed with distilled water containing dye (B. P. Halpern and T. L. Nichols, unpublished observations).

j. *Licking-contact duration (on–off time) with a drinking tube may change during licking.* An increased duration was observed with low-restriction (Hulse and Suter, 1970) and high-restriction (Marowitz and Halpern, 1973; Halpern, 1975) drinking of water. A decrease in duration was found with a viscous nutritive liquid, Nutrament, while with saccharin solutions the duration of contact may remain constant (Allison, 1971) or decrease (Wilcove and Allison, 1972).

k. *The volume removed from a drinking tube by each lick (μl/lick) may remain constant while contact duration and interval (off–on time) change.* This volume constancy was observed with saccharin solutions by Allison (1971) and Wilcove and Allison (1972).

l. *Licking rate (licks/sec or on–on time) may vary with the concentration or the chemical characteristics of the liquid.* Davis (1973) reported a direct variation with concentration for sucrose, glucose, and fructose from a drinking tube. Effects of sucrose concentration and of the presence of saccharin have been seen by Cone, Wells, Goodson, and Cone (1975). Differences between licking rates for solutions of quinine and sucrose have been noted by Volo (1975), with lower rates for quinine (open liquid surface, i.e., Richter tube).

m. *The volume removed by each lick may change with the chemical composition or the concentration of the liquid.* Volume decreased with increased Na–saccharin concentration and with Nutrament versus aqueous Na–saccharin (Allison, 1971). For quinine, sucrose, or water licked from an open surface (Richter tube), the volumes for quinine were smaller than for the other two liquids (Volo, 1975).

n. *Consistent licking-rate differences occur between animals* (Keehn and Arnold, 1960; Allison, 1968). These differences are discussed for a number of mammals by Cone (1974).

o. *The volume removed by each lick may change as a function of the presence or absence of dry food.* Ad libitum, animals showed a mean volume of 3.4 μl/lick, and the same animals gave a mean volume of 3.6 μl/lick (not significantly different) during the 2-hr feeding period of a 22-hr food-deprivation schedule. However, for the remaining 22-hr, the volume was 5.2 μl/lick (significant difference) (Wilson and Barboriak, 1970). The metal drinking tube with a 3-mm opening was used.

p. *Electronic drinkometers have finite "on" and "off" latencies.* This subject is discussed in Chapter 2. That chapter, and Chapter 3 suggest mechanoelectronic and optical alternatives.

4.2. Sucking

What are the structural characteristics of a mammal's head and mouth that facilitate liquid intake from an extended surface through sucking? First of all, the mammal should have an elongated head, or a snout, with the mouth located on the anteroventral portion. The nose should be dorsal or in a dorsal portion of the snout. These locations are desirable because either the entire mouth or at least the lips must be immersed in, or brought into contact with, the liquid for direct sucking from a liquid surface to occur. Respiration continues during sucking, with brief interruptions at each swallow (Balint 1948; B. P. Halpern and T. L. Nichols, unpublished observations). However, for respiration to continue during a typical sequence in suction drinking and swallowing, there must be adequate separation between the mouth and the nose. Another common feature of suction drinkers is a movable lip. These anatomical considerations, and in particular the need for movable lips, apply to free-living mammals drinking from streams, pools, etc., either directly or with the aid of accessory devices.

However, in contrast to those of adult suction-drinking animals, quite different needs exist for newborn and infant mammals that suckle-feed from a nipple. In the newborn human, the tip of the tongue and the lower lip remain in contact, moving dorsally and ventrally together in a rocking motion during each suckle cycle (Bosma, 1967). The nipple penetrates deeply into the oral cavity, with the lower lip and the tongue tip pressing from below and the upper lip and the gum from above. This pressure expresses liquid from the nipple or teat, while intraoral suction aids with the suckling process. Thus, the lips and the gums function together in the suckle, with the tip of the tongue operating as part of the lower-lip–lower-gum unit. Therefore, typical suckling can be expected to occur in mammals that are unable to carry out suction drinking as adults. For example, suction drinking is precluded in primates of the suborder Lemuroidea, since the upper lip is attached to the gum of the upper jaw, as Sonntag (1925, p. 705) and Montagu (1945, p. 15) have noted.

For weaned (postsuckling) mammals, an "ideal" head and lips configuration for suction-drinking is shown clearly in the pig (Figures 9 and 10). With the type of head arrangement described above, liquid can be sucked in without an appreciable loss of forward vision and with the nostrils and external ears maintaining an orientation similar to that found under non-drinking conditions. A comparable snout, with the nostrils set far back from the mouth, occurs in the "gelada baboon," genus *Theropithecus*. This genus, which contains baboonlike primates, is part of the family Cercopithecidae of the superfamily Cercopithecoidea. The "true" baboons, genus *Papio* of the same superfamily (Cercopithecoidea), all have doglike

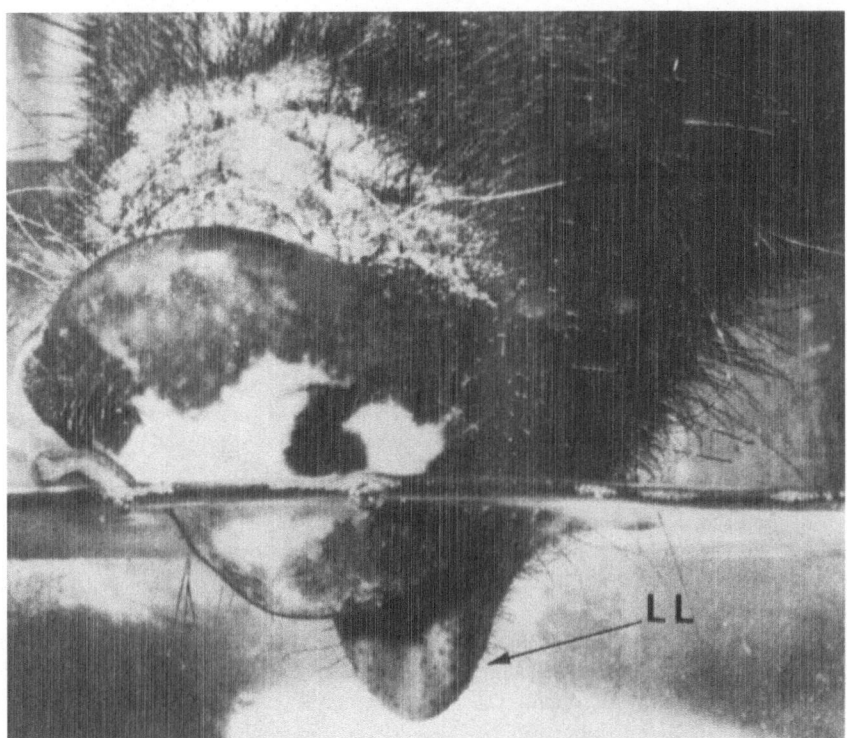

Figure 9. Photograph of miniature pig (*Sus scrofa*) sucking water. The snout is immersed in the water nearly up to the external nares. The lower lip (LL) protrudes by about 1 cm and is formed into a trough shape. Except anteriorly, the lips are closed. The tongue does not leave the oral cavity. Periods of suction drinking are typically 3-6 sec long. From Herring and Scapino (1973).

snouts, as do the baboonlike Celebese apes (genus *Cynopithecus*), but neither of these latter genera have their nostrils set back appreciably from the mouth (Montagu, 1945).

However, in fact, the possession of an elongated head or snout *per se* does not distinguish between suction drinkers and the lickers/lappers. Thus, for example, rat, hamster, opossum, and rabbit have elongated heads but are all licking/lapping drinkers (Table 4). In the laboratory albino rat at least, the lower jaw is placed below the liquid surface when it is lapping (Figure 8A) from a large, extended liquid surface, as Marowitz and Halpern (1973) have noted. However, the location of the external nares in this rat is such that if the entire mouth were placed in the liquid, aspiration of liquid into the external nares would be relatively likely (Figure 7; Figure 8A). In the domestic dog, only the tongue contacts the liquid during lapping (B. P. Halpern, unpublished observation). As in the rat, the location of the

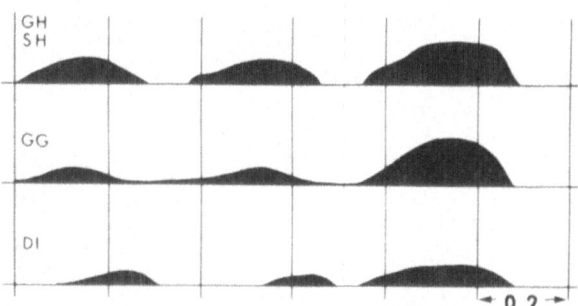

Figure 10. A diagram representing the temporal pattern and relative amplitude of the electromyographic (EMG) activity recorded from the geniohyoid (GH), sternohyoid (SH), genioglossus (GG), and digastric (DI) muscles of a miniature pig (*Sus scrofa*) during the drinking of water (suction drinking). Insulated 63.5-μm stainless-steel wire was inserted into these muscles, either acutely through a hypodermic needle, which was then withdrawn, or during general anesthesia. To the left of the vertical broken line are two suction cycles. These rhythmic bursts, each 100–300 msec long and separated by 50–150 msec, occur in groups of two to six. Each such group was terminated by a swallow, shown to the right of the broken vertical line. The solid, vertical time lines are separated from each other by 0.2 sec. Modified from Herring and Scapino (1973).

external nares precludes sucking, since contact of the lips with an extended liquid surface would also bring the nares into contact with the liquid.

Some difficult-to-classify drinking situations also exist. These include the in-flight "ram-scoop" drinking of some bats (Table 4), as well as the "tongue pumping" of the common vampire bat, *D. rotundus* (see Wimsatt, 1959; Gimbrone, 1965).

In humans and other primates of the suborder Anthropoidea (i.e., Simiae), a clear violation of the proposed correlation between an elongated head and/or snout and suction drinking is seen (Table 4). Humans are almost exclusively suction drinkers, as Ardran and Kemp (1955) and B. P. Halpern and T. L. Nichols (unpublished observations) have observed, yet they have no snout whatsoever. Some suction-drinking anthropoid primates of the superfamily Hominoidea do have elongated heads (e.g., the chimpanzee, genus *Pan* of the family Pongidae) but many hominoid primates have very little snout structure present (e.g., the gibbon, *Hylobates lar* of the family Pongidae; see Schultz, 1969). For a human to drink from an extended liquid surface such as a pool of water or a stream, the location of the mouth on the head requires that the surface of the face be brought parallel to the liquid surface, with the lips contacting the liquid. The eyes are thus placed in such a position that forward vision is greatly restricted, to about 1 m from the mouth (B. P. Halpern, unpublished observation). The external ears face the water, with the external auditory meatus no more than about 16 cm above the liquid surface. The external nares are very close to the liquid surface. Thus, vision and smell are temporarily almost lost, while hearing is reduced in effectiveness. If this sensory loss were not bad enough, simultaneously the back of the head and the neck is exposed. The undesirable aspects of humans' drinking directly from a body of water were recognized in the Old Testament (I Judges 7:5-7; see Catholic Biblical Association of America, 1970).

Primates have overcome the problem of suction drinking in several ways. The simplest involves the use of natural cuplike depressions in trees, which often fill with water (Jolly, 1972; Reynolds and Reynolds, 1965). This permits drinking by nonhuman primates in a relatively safe environment. Often, a hand is dipped into the "bowl" and pulled out, and water is sucked (and licked) from the fingers. However, adult chimpanzees do also drink directly from streams (Reynolds and Reynolds, 1965; Goodall, 1965; Van Lawick-Goodall, 1968, 1971), as do other primates (Schultz, 1969). A second solution, useful with extended or flowing sources of water, explicitly requires the hands. The cupped hands of a human can produce a quite-effective drinking bowl, which can be filled with water (about 40–50 ml) and then brought up to the lips for suction drinking (B. P. Halpern, unpublished observations). This method permits liquid intake with a relatively normal orientation of the head. Similar manual scooping is said to occur in nonhuman primates (Schultz, 1969). Other methods of suction drinking in humans involve tools. One tool is the drinking vessel; another, the drinking straw. Drinking vessels are a cultural artifact found in almost every culture. Thus, a seminomadic "Old Stone Age" people studied in the forests of Bolivia (the Siriono, studied in 1940–1941) had no metal or stone implements, no drawing, painting, or written language, and no domestic animals and could not "make" fire, but they regularly used cut sections of bamboo stem or dried, hollowed calabash (gourd) fruit as drinking containers "in camp" (Holmberg, 1969). During hunting trips, they used patuju leaves to remove water from a "water hole," thus avoiding the hazards of direct lip-contact drinking. Strawlike devices are also relatively widely dispersed in human societies and are, in effect, an artifical snout. Thus, a cultural solution to the drinking problem produced by flat faces, as well as perhaps pre-cultural solutions involving the anatomical and manipulative characteristics of hands and limbs, exists in humans. What of tool use in other primates, however? Drinking-tool use, specifically a mass of rolled or chewed leaves that are repeatedly placed in water and then squeezed (and, perhaps, sucked) in the mouth, has been observed in the chimpanzee (Van Lawick-Goodall, 1968, 1971; Jolly, 1972). After removing the leaves, chimpanzees will also insert a twig into water and then lick it.

The problem of drinking can be solved in quite a different way: by the elimination of drinking. This may seem to be an unlikely mechanism for those mammals that are only moderately successful at producing a concentrated urine. However, animals dwelling in a rain forest may have a diet that, when primarily vegetable in nature, has a high water content. Thus fruits, shoots, buds, etc. (Jolly, 1972), may well provide enough liquid intake so that separate drinking of water rarely if ever occurs. This has been observed with several anthropoid primates, including the mountain gorilla, *Gorilla gorilla beringei* (Schaller, 1965) howler monkeys, *Alouatta palliata* (Carpenter, 1965); and the langur monkey, *Presbytis entellus* (Jay, 1965). The giraffes (genus *Giraffa*, family Gifaffidae, suborder Ruminantia, order Artiodactylia) rarely drink, obtaining water from their diet of leaves (M. Dickie, personal communication). About two gallons per *week* (14 ml/kg/ wk) is drunk by free-living giraffes (Walker, 1964). This is a small intake indeed. Dietary water is clearly important, since when a dry diet is fed to giraffes, as in a zoo, water intakes of four gallons per *day* are reported by Owen (1841). Small mammals may also require little or no drinking when eating fresh greens. Thus, the guinea pig, *Cavia porcellus*, can be maintained on appropriate fresh greens without any separate water (Alvord, 1968).

Similarly, the diet of carnivores that make direct kills (as contrasted to scavenger carnivores) is also rich in liquid. Thus, it could be that drinking of separate water is an uncommon event for some carnivores under natural circumstances. Nonetheless, such carnivores would frequently encounter liquids and pastes in the course of their normal diet. Consequently, licking would not be unknown.

In contrast, some carnivores, swine, chiropterans, and nondesert rodents, as well as many primates, do seek out and drink water *per se* under natural conditions. Active search for, and ingestion of, water also exists in African antelope. In this context, *Antelope* means Bovidae, excluding the Caprinae, or goats (Jarman, 1974). In general, the appearance of Bovidae (a family in the suborder Ruminantia; includes buffalo, cattle, sheep, gazelle, etc.) at sources of water is a common observation. This also seems to be the case with other grass eaters, such as the horse (*Equus caballus*, family Equidae, order Perissodactyla). With reference to primates, chimpanzees have been seen drinking from streams by Goodall (1965), while hamadryad baboons (*Papio hamadryas*) dig holes for water (Jolly, 1972). Many primates and artiodactyls are suction drinkers (Table 4). This method of liquid ingestion permits intake of a large volume of liquid in a relatively short period of time, particularly when the normal liquid source is a volume of liquid in which the lips or the snout can be immersed and suction can be effectively used. The intake rates can be very high. Rates of 40 ml/sec have been observed in humans, although 10–13 ml/sec is more common (B. P. Halpern and T. L. Nichols, unpublished observations). Suction-drinking rates above 100 ml/sec have been reported for the donkey and the camel (suborder Tylopoda of the order Artiodactyla) (Schmidt-Nielsen, 1964).

Drinking in humans has received some detailed study. In particular, the characteristics of suction drinking from a glass have been described. An individual human sip has a contact duration of approximately 1.0 sec and an overall cycle length (i.e., time from lifting a drinking vessel off a surface to return of the drinking vessel to the surface) of approximately 4.5 sec (Halpern, 1975; Halpern and Nichols, 1975; B. P. Halpern and T. L. Nichols, unpublished observations). When multiple sips are taken from a drinking vessel, approximately 1.5 sec elapses between sips, although oral contact with the liquid is often maintained throughout. Using 4.5 sec as a very conservative estimate of the total duration of a single sip cycle and 12 ml as the typical volume of a single sip, an intake rate of approximately 2.7 ml/sec is obtained. However, an estimated human intake rate of 5 ml/sec (with 1-sec contact, 1.5-sec interval, and 12-ml volume, 12 ml/2.5 sec = 4.8 ml/sec) is probably more realistic. This rate is 120 times greater than the typical laboratory rat's intake rate of 40 μl/sec (Table 4). From the point of view of total water need, such differences in intake rate are not unreasonable, considering the approximately 140-fold difference in mass between a 0.5-kg rat and a 70-kg adult human. However, as a stimulus for gustatory neural responses, the human flow rate might appear to be much higher. This would be so because of the sensitivity of the phasic component of gustatory neural responses to flow rate (Switzky, 1965; Bealer and Smith, 1975).

The above calculations would suggest an approximately 120-fold higher liquid-ingestion flow rate in the human compared to the laboratory albino rat. However, since human liquid ingestion actually occurs in sips approximately 1 sec long, the human gustatory-stimulus actual flow rate is approximately 12 ml/sec. In contrast, the rat, at laboratory limiting conditions (i.e., 6 μl/lick volume and a liquid contact duration of 55 msec/lick; see Halpern 1975), produces a gustatory-stimulus flow rate of 0.109 ml/sec (109 μl/sec). Therefore, a gustatory-stimulus flow-rate difference of $\gg$ 20:1 is seen between human and rat (but < 120:1). Another variable that should be considered is the tongue-surface area wet by the liquid. For a laboratory rat, a 5-μl, 55-msec artificial lick wets 5 mm^2 of the tongue (Halpern and Marowitz, 1973), while a natural lick under restricted-access conditions (Marowitz and Halpern, 1973) wets about 17 mm^2 of the rat's tongue before the tongue retracts into the mouth (B. P. Halpern and T. L. Nichols, unpublished observations). This rate results in a flow for the rat of 6.4 μl/mm^2/sec. In humans, the anterior 5 cm of the tongue has a "rectangularized" tip width of approximately 5 cm, yielding an area of approximately 2500 mm^2. The human 12 ml/sec flow rate over this area produces a flow of 4.8 μl/mm^2/sec, similar to the flow in the rat. Detailed consideration of the taste-bud density at the tip of the anterior portion of the tongue of rat and human may change these figures somewhat, but the overall picture seems to be one in which comparable taste responses can be expected in rats and humans during ingestion, since the flows are estimated to differ by a factor of 1.25. A large flow difference would have been expected to cause quite different taste responses, since rat gustatory neural responses increase with stimulus-liquid flow rate (Switzky, 1965; Bealer and Smith, 1975), as does human-judged taste intensity (Meiselman, Bose, and Nykvist, 1972).

The position and the movement sequence of the tongue in relationship to the palate and the teeth during the ingestion of liquid and of a paste by humans has been described in detail by Ardran and Kemp (1955). The liquid (an aqueous barium-sulfate emulsion) was drunk from a drinking vessel (glass) or through a straw; the paste (a viscous barium-sulfate-water mixture) was sucked from a spoon. The following is the basic ingestion sequence for drinking from a glass (quoted from Ardran and Kemp, 1955, pp. 256–257):

1. "The glass is raised to the mouth, is grasped by the lips, and tilted until the fluid level reaches the lips, the upper lip dipping beneath the fluid."
2. "The tongue fills most of the mouth cavity proper, being opposed to the hard [and soft (author's note)] palate, the gums, and the teeth. . . . the hyoid bone is raised from its position of rest."
3. ". . . the tongue is withdrawn from the hard palate from before backward and fluid is drawn . . . into the space created, apposition between the tongue and soft palate being maintained."
4. "When a large quantity of barium emulsion is drawn into the mouth [a large sip (author's note)] the mandible is lowered, thus creating more space between the superior surface of the tongue and the hard palate, and the forepart of the soft palate is drawn forward to preserve palatoglossal closure . . . the hyoid bone falls slightly."
5. "The barium contained in the mouth is swallowed in the usual manner, the hyoid being elevated and drawn forwards."
6. If sipping continues, "the tilt of the glass must be increased to keep the fluid level at the lips; one bolus succeeds another without reconstitution of the airway until . . . the subject pauses."

In drinking from a straw, steps 2 thru 5 above are followed. Steps 1 and 6 differ, however:

1. "The straw is adjusted so that its proximal end is just inside the mouth cavity proper. . . . the lips are closed around the straw and the head may be slightly bowed forwards."
6. "If suction continues, the sequence of movements is repeated again and again. The second bolus begins to enter the mouth as the last of the first bolus passes thru the fauces. . . . The mode of passage

of each successive bolus through the pharynx is similar to . . .
drinking."

4.3. Swallowing

The final stage in the liquid-intake sequence is swallowing, the movement of ingesta from the oral cavity through the esophagus and into the
stomach. In animals with complex stomachs (i.e., ruminants), liquids follow
the course of solids in adults, initially entering a portion of the rumen
(Dukes, 1943). However, in young ruminants, liquids may pass directly to
the omasum, especially during suckling.

Swallowing is an active process, although gravity may also be involved
(Davenport, 1961). Typically, a coordinated sequence of tongue, pharynx,
hypopharynx, and esophageal activity produces the swallow. This sequence
produces a pressure wave, with no suction components (Kawasaki, Ogura,
and Takenouchi, 1964), which moves the bolus. Swallowing has been extensively studied by many workers. Valuable reference sources include books
by Davenport (1961) and Graber (1963); the review by Møller (1966); and
specific articles by Mosher (1927), Ardran and Kemp (1955), Abd-el-malek
(1955), Ardran *et al.* (1958), Kawasaki *et al.* (1964), and Hendrix (1974).

Before the swallowing sequence begins, considerable accumulation of
ingesta may occur between the dorsum of the tongue and the soft palate, as
Ardran *et al.* (1958) and Hiiemae and Ardran (1968) reported. This accumulation is not obligatory, however. In human, swallowing may occur with
no accumulation on the posterior division of the tongue.

Tongue position in humans during the preparation of a liquid bolus has
been described by Ardran and Kemp (1955) and Abd-el-malek (1955). If a
small volume of liquid is involved, it is usually contained in a depression
formed by an elevation of the anterolateral margins of the dorsum of the
tongue (Figure 11). During swallowing, the forepart of the tongue presses
against the hard palate, while the posterior portion of the tongue moves
away from the soft palate. A small-volume liquid bolus primarily contracts
the midline dorsum during the swallow. However, if an appreciable volume
is rapidly poured into the mouth, as is done with a wineskin or a Lebanese
water-cooling bottle (an *abrik*; J. Bassili, personal communication), the
sides of the tongue are involved. Thus, as was previously suggested in Section 4.2, pressure ingestion of liquid in the adult human, in common with
suckling in the infant, may particularly involve the lateral taste organs. This
"special case" in the human may be contrasted with swallowing in the laboratory rat. Hiiemae and Ardran (1968) reported that preswallowing accumulation of liquid occurs at the lateral margins of the tongue in normal
drinking by the laboratory rat. Therefore, the lateral taste organs, the foliate

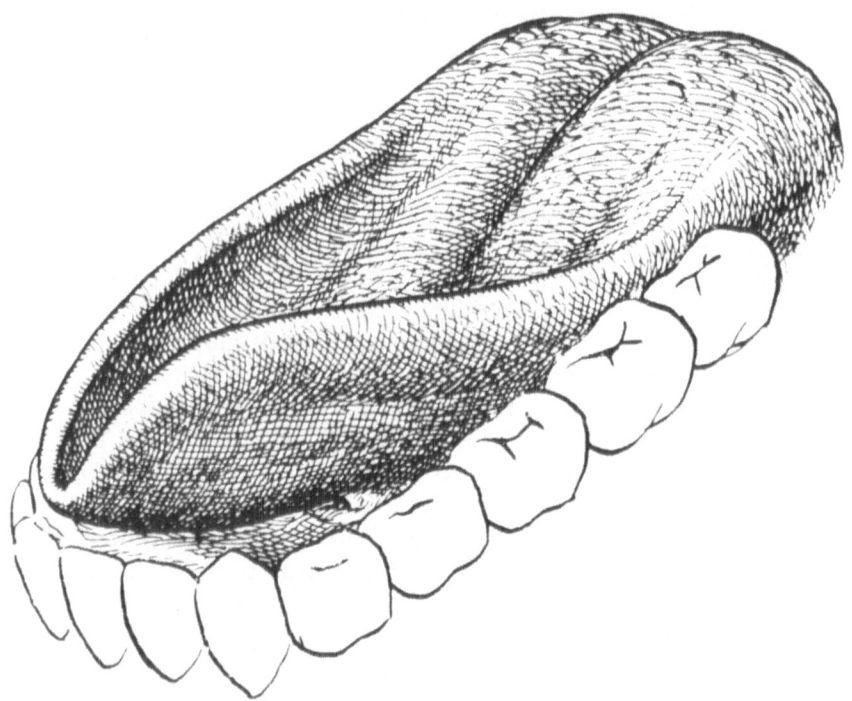

Figure 11.   Diagram of the adult human tongue during the first stage of swallowing a liquid bolus, prepared from direct observation and photographs through spaces provided by missing teeth, the lips being parted by small retractors. This tongue shape is produced primarily by contraction of the genioglossus and the hyoglossus extrinsic tongue muscles (Table 2). Reproduced from Abd-el-malek (1955).

papillae, would routinely be involved. A significant difference would be that the taste buds of the fungiform papillae of the anterior division of the rat's tongue would be stimulated during each lick/lap ingestion, but in the human, the anterior-division taste buds may be bypassed during suckling and pressure drinking. In such cases, the extralingual taste buds of the rear of the mouth may provide gustatory input that supplements that initiated by the circumvallate- and foliate-papillae taste buds. Such extralingual stimulation may be particularly prominent during swallowing.

ACKNOWLEDGMENTS

Many colleagues have helped me with the preparation of this chapter. The initial outline was read and commented on by Drs. Abd-el-malek, Carl Gans, Albert I. Farbman, Howard Moskowitz, Herbert L. Meiselman, Ingliss Miller, Bruce Oakley, and the editors of this volume. Successive drafts of the chapter were reviewed and discussed by Drs. Vernon Brightman, Avis Cohen, Howard Evans, Marion Frank, R. Jeffrey Dean, Kathrine Houpt, Robert Johnston, Don Justesen, Lewis Lipsitt, Joseph Mendelson, Richard Moon, Carl Pfaffmann, Jan Weijnen, Zsuzsanna Wiesenfeld, and Donna Zahorik. Important references were brought to my attention by the above colleagues, as well as by Drs. John Lee Smith, William L. Brown, Jr., and Mary Dickie. Ms. Donna Baron helped in many ways, Mr. William Hemsath made the electronic measurements of the drinkometers, and Ms. Linda Chandler typed seemingly unending drafts. Ms. Pauline Halpern advised on the mathematical content. Original data in this chapter were collected with support from NSF-BNS74-00878.

References

Abd-el-malek, S., 1939a, A contribution to the study of the movements of the tongue in animals, with special reference to the cat, *J. Anat.* **73**:15-31.

Abd-el-malek, S., 1939b, Observations of the morphology of the human tongue, *J. Anat.* **73**:201-210.

Abd-el-malek, S., 1955, The part played by the tongue in mastication and deglutition, *J. Anat.* **89**:250-255.

Allison, J., 1968, Individual differences in eating and drinking in the rat, *Psychon. Sci.* **13**: 31-32.

Allison, J., 1971, Microbehavioral features of nutritive and nonnutritive drinking in rats, *J. Comp. Physiol. Psychol.* **76**:408-417.

Allison, J., and Castellan, N. J., 1970, Temporal characteristics of nutritive drinking in rats and humans, *J. Comp. Physiol. Psychol.* **70**:116-125.

Alvord, J. R., 1968, Development and control of licking behavior in the guinea pig. M.Sc. thesis, Utah State University.

Ardran, G. M., and Kemp, F. H., 1955, A radiographic study of movements of the tongue in swallowing, *Dent. Practitioner* **5**:252-263.

Ardran, G. M., Kemp, F. H., and Ride, W. D. L., 1958, A radiographic analysis of mastication and swallowing in the domestic rabbit: *Oryctolagus cuniculus, Proc. Zool. Soc. London* **130**:257-274.

Arey, L. B., 1956, "Developmental Anatomy," Saunders, Philadelphia, 680 pp.

Balint, M., 1948, Individual differences of behavior in early infancy and an objective method for recording them: I. Approach and the method of recording, *J. Genetic Psych.* **73**:57-79.

Barbour, R. W., and Davis, W. H., 1969, "Bats of America," University of Kentucky Press, Lexington, 286 pp.

Barnard, J. W., 1940, The hypoglossal complex of vertebrates, *J. Comp. Anat.* **72**:489-524.

Bartoshuk, L. M. B., 1965, Effects of adaptation on responses to water in cat and rat, Ph.D. thesis, Brown University.

Basmajian, J. V., 1974, "Muscles Alive" (3rd Ed.), Williams and Wilkins Co., Baltimore, 525 pp.

Bava, A., Innocenti, G. M., and Raffaele, R., 1972a, Effects exerted by stimulation of glossopharyngeal taste buds on the nucleus intercalatus and adjoining medullary structures, *Arch. Fisiol.* **69**, Fasc. II-III (IV):131-159.

Bava, A., Innocenti, G. M., and Raffaele, R., 1972b, Effects exerted by trigeminal and fastigial stimulation on Staderini's nucleus intercalatus. *Arch. Sci. Biol.* **56** (Fasc. 1):13-34.

Bealer, S. L., and Smith, D. V., 1975, Multiple sensitivity to chemical stimuli in single human taste papillae. *Physiol. Behav.* **14**:795-799.

Beaudreau, D. E., and Jerge, C. R., 1968, Somatotopic representation in the Gasserian ganglion of tactile peripheral fields in the cat, *Arch. Oral Biol.* **13**:247-256.

Beidler, L. M., 1965, Comparison of gustatory receptors, olfactory receptors, and free nerve endings, *Cold Springs Harbor Symp. Quant. Biol.* **30**:191-200.

Beidler, L. M., 1969, Innervation of rat fungiform papilla, *In* "Olfaction and Taste-III." (C. Pfaffmann, ed.), Rockefeller University Press, New York, pp. 352-369.

Beidler, L. M., and Smallman, R. L., 1965, Renewal of cells within taste buds, *J. Cell. Biol.* **27**:263-272.

Benjamin, R. M., Halpern, B. P., Moulton, D. G., and Mozell, M. M., 1965, The chemical senses, *Ann. Rev. Psych.* **16**:381-416.

Bernard, R. A., 1962, An electrophysiological study of taste reception in the calf, Ph.D. thesis, Cornell University.

Bernard, R. A., 1964, An electrophysiological study of taste reception in peripheral nerves of the calf, *Am. J. Physiol.* **206**:827-835.

Bernard, R. A., and Halpern, B. P., 1968, Taste changes in Vitamin A deficiency, *J. Gen. Physiol.* **52**:444-464.

Bernard, R. A., and Nord, S. G., 1971, A first-order synaptic relay for taste fibers in the pontine brain stem of the cat, *Brain Res.* **30**:349-356.

Bernston, G. G., and Hughes, H. C., 1974, Medullary mechanisms for eating and grooming behaviors in cats, *Exp. Neurol.* **44**:255-265.

Boice, R., 1967, Lick rates and topographies as taxonomic criteria in southwestern rodents, *Psychonomic Sci.* **9**:431-432.

Bosma, J. F., 1967, Human infant oral function, *In* "Symposium on Oral Sensation and Perception" (J. F. Bosma, ed.), Charles C Thomas, Springfield, Ill., pp. 98-110.

Boudreau, J. C., Anderson, W., and Oravec, J., 1975, Chemical stimulus determinants of cat geniculate ganglion chemoresponsive group II unit discharge, *Chem. Senses Flav.* **1**:495-517.

Boudreau, J. C., Bradley, B. E., Bierer, P. R., Kruger, S., and Tsuchitani, C., 1971, Single unit recording from the geniculate ganglion of the facial nerve of the cat, *Exp. Brain Res.* **13**:461-488.

Boudreau, J. C., and Tsuchitani, C., 1973, "Sensory Neurophysiology," Van Nostrand Reinhold, New York, 470 pp.

Bradley, R. M., 1971, Tongue topography, *in* "Handbook of Sensory Physiology" (Vol. 4, Chemical Senses, Part 2) (L. M. Beidler, ed.), Springer-Verlag, New York, pp. 1-30.

Bradley, R. M., 1972, Development of the taste bud and gustatory papillae in human fetuses, *in* "Third Symposium on Oral Sensation and Perception" (J. F. Bosma, ed.), Charles C Thomas, Springfield, Ill., pp. 137-162.

Bridgwater, W., and Kurtz, S., 1963, "The Columbia Encylopedia" (3rd ed.), Columbia University Press, New York, 2388 pp.

Brightman, V. J., 1976, The vallate foliate complex and suckling behavior, *Anat. Rec.* **184**:363.

Brown, P. B., Maxfield, B. W., and Moraff, H., 1973, "Electronics for Neurobiologists," MIT Press, Cambridge, Mass., 543 pp.

Butterfield, E. C., and Siperstein, G. N., 1972, Influence of contingent auditory stimulation upon non-nutritive suckle, *in* "Third Symposium on Oral Sensation and Perception" (J. F. Bosma, ed.), Charles C Thomas, Springfield, Ill., pp. 313-334.

Carleton, A., 1938, Observations on the problem of the proprioceptive innervation of the tongue, *J. Anat.* **72:**502-507.

Carpenter, C. R., 1965, The howlers of Barro Colorado Island, *in* "Primate Behavior" (I. DeVore, ed.), Holt, Rinehart, and Winston, New York, pp. 250-291.

Castro, A. J., 1972, The effects of cortical ablations on tongue usage in the rat, *Brain Res.* **45:**251-253.

Catholic Biblical Association of America, 1970, "The New American Bible," P. J. Kennedy and Sons, New York, p. 297.

Chiasson, R. B., 1969, "Laboratory Anatomy of the White Rat" (2nd ed.), Wm. C. Brown, Dubuque, Iowa, 81 pp.

Cody, F. W. J., Lee, R. W. H., and Taylor, A., 1972, A functional analysis of the components of the mescencephalic nucleus of the fifth nerve in the cat, *J. Physiol.* **226:**249-261.

Cohen, M. J., Landgren, S., Strom, L., and Zotterman, Y., 1957, Cortical reception of touch and taste in the cat, *Acta Physiol. Scand.* **40:**Supp. 135.

Cone, A. L., Wells, R., Goodson, L., and Cone, D. M., 1975, Changing lick rate of rats by manipulating deprivation and type of solution, *Psych. Rec.* **25:**491-495.

Cone, D. M., 1974, Do mammals lick at a constant rate? *Psych. Rec.* **24:**353-364.

Cone, D. M., Cone, A. L., Golden, A. J., and Sanders, S. L., 1973, Differential lick rates in opossum: A challenge to the invariance hypothesis, *Psych. Rec.* **23:**343-347.

Cone, D. M., Golden, A. J., and Sanders, S. L., 1971, "Psychonomic Society Program Abstract," p. 16.

Conger, A. D., and Wells, M. A., 1969, Radiation and aging effects on taste structure and function, *Radiat. Res.* **37:**31-49.

Cooper, S., 1953, Muscle spindles in the intrinsic muscles of the human tongue, *J. Physiol.* **122:**193-202.

Craigie, E. H., 1960, "Bensley's Practical Anatomy of the Rabbit," University of Toronto Press, Toronto, 391 pp.

Crompton, A. W., and Hiiemae, K., 1970, Molar occlusion and mandibular movements during occlusion in the American opossum, *Didelphis marsupialis* L., *Zool. J. Lin. Soc.* **49:** 21-47.

Crook, C. K., and Lipsitt, L. P., 1976, Neonatal nutritive sucking: Effects of taste stimulation upon sucking rhythm and heart rate, *Child Devel.* **47:**518-522.

Crosby, E. C., Humphrey, T., and Lauer, E. W., 1962, "Correlative Anatomy of the Nervous System," Macmillan, New York, 731 pp.

Darian-Smith, I., Isbister, J., Mok, H., and Yokota, T., 1966, Somatic sensory cortical projection areas excited by tactile stimulation of the cat: A triple representation, *J. Physiol.* **182:**671-689.

Davenport, H. W., 1961, "Physiology of the Digestive Tract," Year Book Medical Publishers, Chicago, 221 pp.

Davis, J. D., 1973, The effectiveness of some sugars in stimulating licking behavior in the rat, *Physiol. Behav.* **11:**39-45.

Dean, H. W., 1954, Epithelium, *in* "Histology" (R. O. Greep, ed.), Blakiston, New York, pp. 63-82.

De Lorenzo, A. J. D., 1963, Studies on the ultra-structure and histophysiology of cell membranes, nerve fibers and synaptic junctions in chemoreceptors, *in* "Olfaction and Taste" (Y. Zotterman, ed.), Pergamon Press, Oxford, pp. 5-17.

Doran, G. A., and Baggett, H., 1972, The genioglossus muscle: A reassessment of its anatomy in some mammals, including man, *Acta Anat.* **83:**403-410.

Dukes, H. H., 1943, "The Physiology of Domestic Animals," Comstock, Ithaca, N.Y., pp. 282-288.

Eaton, T. H., 1951, "Comparative Anatomy of the Vertebrates," Harper & Brothers, New York, 340 pp.

Eisenman, J., Landgren, S., and Novin, D., 1963, Functional organization in the main sensory trigeminal nucleus and in the rostral subdivision of the nucleus of the spinal trigeminal tract in the cat, *Acta Physiol. Scand.* **59**:Supp. 214.

Ellenberger, W., and Baum, H., 1932, "Handbuch der Vergleichenden Anatomie der Haustiere," Julius Springer, Berlin, 1101 pp.

Elliott, H. C., 1963, "Textbook of Neuroanatomy," Lippincott, Philadelphia, 542 pp.

Erickson, R. P., 1968, Stimulus coding in topographic and non-topographic afferent modalities: On the significance of the activity of individual sensory neurons, *Psychol. Rev.* **75**:447–465.

Erickson, R. P., Doetsch, G. S., and Marshall, D. A., 1965, The gustatory neural response function, *J. Gen. Physiol.* **49**:247–263.

Farbman, A. I., 1965, Electron microscope study of the developing taste bud in rat fungiform papilla, *Dev. Biol.* **11**:110–135.

Farbman, A. I., 1971, Development of the taste bud, *in* "Handbook of Sensory Physiology" (Vol. IV, Chemical Senses, Part 2, Taste) (L. M. Beidler, ed.), Springer-Verlag, New York, pp. 51–62.

Farbman, A. I., 1972, The taste bud: A model system for developmental studies, *in* "Developmental Aspects of Oral Biology" (H. C. Slavkin and L. A. Bavetta, eds.), Academic Press, New York, pp. 109–123.

Farbman, A. I., and Allgood, J. P., 1971, Innervation, sensory receptors and sensitivity of the oral mucosa, *in* "Current concepts of the histology of the oral mucosa" (C. A. Squier and J. Meyer, eds.), Charles C Thomas, Springfield, Ill., pp. 250–273.

Faull, J. R., and Halpern, B. P., 1972, Taste stimuli: Time course of peripheral nerve response and theoretical models, *Science* **178**:73–75.

Frank, M., 1968, Single fiber responses in the glossopharyngeal nerve of the rat to chemical, thermal, and mechanical stimulation of the posterior tongue, Ph.D thesis, Brown University.

Frank, M., 1972, Taste responses of single hamster chorda tympani nerve fibers, *in* "Olfaction and Taste," IV (D. Schneider, ed.), Wissenschaftliche Verlagsgesellschaft MBH, Stuttgart, Germany, pp. 287–293.

Frank, M., 1974, The classification of mammalian afferent taste nerve fibers, *Chem. Senses Flavor* **1**:53–60.

Frank, M., 1975, Response patterns of rat glossopharyngeal taste neurons, *in* "Olfaction and Taste," V (D. A. Denton and J. P. Coghlan, eds.), Academic Press, New York, pp. 59–64.

Friel, J. P., 1974, "Dorland's Illustrated Medical Dictionary" (25th ed.), Saunders, Philadelphia, 1748 pp.

Funakoshi, M., Kasahara, Y., Yamamoto, T., and Kawamura, Y., 1972, Taste coding and central perception, *in* "Olfaction and Taste," IV (D. Schneider, ed.), Wissenschaftliche Verlagsgesellschaft MBH, Stuttgart, Germany, pp. 336–342.

Ganchrow, D., and Erickson, R. P., 1972, Thalamocortical relations in gustation, *Brain Res.* **36**:289–305.

Geddes, L. A., 1972, "Electrodes and the Measurement of Bioelectric Events," Wiley-Interscience, New York, 364 pp.

Gibson, J. J., 1967, The mouth as an organ for laying hold on the environment, *in* "Symposium on Oral Sensation and Perception" (J. F. Bosma, ed.), Charles C Thomas, Springfield, Ill., pp. 111–136.

Gimbrone, M. A., 1965, Feeding mechanisms of the vampire bat, *Desmodus Rotundus Murinus*, B.A. honors thesis, Cornell University.

Goodall, J., 1965, Chimpanzees of the Gombe Stream Reserve, *in* "Primate Behavior" (I. DeVore, ed.), Holt, Rinehart, Winston, New York, pp. 425-473.

Goodwin, G. M., and Luschei, E. S., 1974, Effects of destroying spindle afferents from jaw muscles on mastication in monkeys, *J. Neurophysiol.* 37:967-981.

Graber, T. M., 1963, "Orthodontics," W. B. Saunders, Philadelphia, pp. 120-121 (swallowing), 238-240 (suckling), p. 259 (swallowing).

Granit, R. (ed.), 1966, "Muscle Afferents and Motor Control," Wiley, New York, 466 pp.

Grant, P. G., 1973, Lateral pterygoid: Two muscles? *Am. J. Anat.* 138:1-10.

Graziadei, P. P. C., 1969, The ultrastructure of vertebrate taste buds, *in* "Olfaction and Taste," III (C. Pfaffmann, ed.), Rockefeller University Press, New York, pp. 315-330.

Greene, E. C., 1968, "Anatomy of the Rat," Hafner, New York, 370 pp.

Grossman, R. C., and Hattis, B. F., 1967, Oral mucosal sensory innervation and sensory experience, *in* "Symposium on Oral Sensation and Perception" (J. F. Bosma, ed.), Charles C Thomas, Springfield, Ill., pp. 5-62.

Guth, L., 1957, The effects of glossopharyngeal nerve transection on the circumvallate papilla of the rat, *Anat. Rec.* 128:715-726.

Hafez, E. S. E., 1962, "The Behavior of Domestic Animals," Williams and Wilkins, Baltimore, 619 pp.

Hafez, E. S. E., and Schein, M. W., 1962, The behavior of cattle, *in* The Behavior of Domestic Cattle" (E. S. E. Hafez, ed.), Williams and Wilkins, Baltimore, pp. 247-296.

Halpern, B. P., 1957, Electrical activity in the medulla oblongata following chemical stimulation of the rat's tongue, M.Sc. thesis, Brown University.

Halpern, B. P., 1959, Gustatory responses in the medulla oblongata of the rat, Ph.D. thesis, Brown University.

Halpern, B. P., 1973, The use of vertebrate laboratory animals in research on taste, *in* "Methods of Animal Experimentation," IV (W. I. Gay, ed.), Academic Press, New York, pp. 225-362.

Halpern, B. P., 1975, Temporal patterns of liquid intake and gustatory neural responses, *in* "Olfaction and Taste," V (D. A. Denton and J. P. Coghlan, eds.), Academic Press, New York, pp. 47-52.

Halpern, B. P., Bernard, R. A., and Kare, M. R., 1962, Amino acids as gustatory stimuli in the rat, *J. Gen. Physiol.* 45:681-701.

Halpern, B. P., and Marowitz, L. A., 1973, Taste responses to lick duration stimuli, *Brain Res.* 57:473-478.

Halpern, B. P., and Nelson, L. M., 1965, Bulbar gustatory responses to anterior and to posterior tongue stimulation in the rat, *Am. J. Physiol.* 209:105-110.

Halpern, B. P., and Nichols, T. L., 1975, Human drinking: Orbicularis oris EMG patterns and liquid contact durations, *Physiologist* 18:238.

Halpern, B. P., and Tapper, D. N., 1971, Taste stimuli: Quality coding time, *Science* 171:1256-1258.

Ham, A. W., 1974, "Histology" (7th ed.), Lippincott, Philadelphia, 1006 pp.

Hamosh, M., and Scow, R. O., 1973, Lingual lipase, *in* "Fourth Symposium on Oral Sensation and Perception (Development in the Fetus and Infant)" (J. F. Bosma, ed.), U.S. Department of Health, Education, and Welfare, National Institutes of Health, Bethesda, Md., pp. 311-322.

Harper, H. W., Hay, J. R., and Erickson, R. P., 1966, Chemically evoked sensations from single human taste papillae, *Physiol. Behav.* 1:319-325.

Hellekant, G., 1965a, The effect of ethyl alcohol on non-gustatory receptors of the tongue of the cat, *Acta Physiol. Scand.* 65:243-250.

Hellekant, G., 1965b, Electrophysiological investigations of the gustatory effect of ethyl alcohol: II. A single fiber analysis in the cat, *Acta Physiol. Scand.* 64:398-406.

Hendrix, T. R., 1974, The motility of the alimentary canal, *in* "Medical Physiology" II, (13th ed.) (V. B. Mountcastle, ed.), Mosby, St. Louis, pp. 1210-1221.

Henkin, R. I., 1967, Sensory mechanisms in familial dysautonomia, *in* "Symposium on Oral Sensation and Perception" (J. F. Bosma, ed.), Charles C Thomas, Springfield, Ill., pp. 341-349.

Henneman, E., 1974, Peripheral mechanisms involved in the control of muscle, *in* "Medical Physiology," I (13th ed.) (V. B. Mountcastle, ed.), Mosby, St. Louis, pp. 617-635.

Herring, S. W., and Scapino, R. P., 1973, The physiology of feeding in miniature pigs, *J. Morph.* **141** (4):427-460.

Hiiemae, K. M., and Ardran, G. M., 1968, A cinefluorographic study of mandibular movement during feeding in the rat (*Rattus norvegicus*), *J. Zool. Lond.* **154**:139-154.

Hiiemae, K., and Crompton, A. W., 1971, A cinefluorographic study of feeding in the American Opossum, *Didelphis marsupialis, in* "Dental Morphology and Evolution" (A. A. Dahlburg, ed.), University of Chicago Press, Chicago, pp. 299-334.

Hiiemae, K., and Jenkins, F. A., 1969, The anatomy and internal architecture of the muscles of mastication in *Didelphis marsupialis, Postilla* **140**:1-49.

Hildebrand, M., 1974, "Analysis of Vertebrate Structure," Wiley, New York, 710 pp.

Holmberg, A. R., 1969, "Nomads of the Long Bow," Natural History Press, Garden City, N.Y., 294 pp.

Hosokawa, H., 1961, Proprioceptive innervation of striated muscles in the territory of cranial nerves, *Tex. Rep. Biol. Med.* **19**:405-464.

Hudson, J. W., 1964, Water metabolism in desert mammals, *in* "Thirst" (M. J. Wayner, ed.), Macmillan, New York, pp. 211-235.

Hulse, S. H., and Suter, S., 1970, Emitted and elicited behavior: An analysis of some learning mechanisms associated with fluid intake of rats, *Learn. Motiv.* **1**:304-315.

Hyman, L. H., 1922, "A Laboratory Manual for Comparative Vertebrate Anatomy," University of Chicago Press, Chicago.

Iggo, A., and Leek, B. F., 1967, The afferent innervation of the tongue of sheep, *in* "Olfaction and Taste," II (T. Hayashi, ed.), Pergamon, Oxford, pp. 493-507.

Ishiko, N., 1970, Local taste specificity within lingual nerve fields in the cat, *Brain Res.* **24**:343-346.

Ishiko, N., 1974, Local gustatory functions associated with segmental organization of the anterior portion of cat's tongue, *Exp. Neurol.* **45**:341-354.

Jakinovich, W., Jr., and Oakley, B., 1975, Comparative gustatory responses in four species of gerbilline rodents, *J. Comp. Physiol.* **99**:89-101.

Jarman, P. J., 1974, The social organization of antelope in relation to their ecology, *Behavior* **48**:215-267.

Jay, P., 1965, The common langur of north India, *in* "Primate Behavior" (I. DeVore, ed.), Holt, Rinehart, and Winston, New York, pp. 197-249.

Jerge, C. R., 1963, The function of the nucleus supratrigeminalis, *J. Neurophysiol.* **26**:393-402.

Jerge, C. R., 1967, The neural substratum of oral sensation, *in* "Symposium on Oral Sensation and Perception" (J. F. Bosma, ed.), Charles C Thomas, Springfield, Ill., pp. 63-83.

Jolly, A., 1972, "The Evolution of Primate Behavior," Macmillan, New York, pp. 130-131, 292.

Kallen, F. C., and Gans, C., 1972, Mastication in the little brown bat, *Myotis lucifugus, J. Morph.* **136**:385-420.

Kawamura, Y., 1964, Recent concepts of the physiology of mastication, *in* "Advances in Oral Biology," I (P. H. Staples, ed.), Academic Press, London, pp. 77-109.

Kawamura, Y., and Morimoto, T., 1973, Neurophysiological mechanisms related to reflex control of tongue movements, *in* "Fourth Symposium on Oral Sensation and Perception"

(J. F. Bosma, ed.), U.S. Department of Health, Education, and Welfare, Washington, D.C., pp. 206–221.

Kawamura, Y., Okamoto, J., and Funakoshi, M., 1968, A role of oral afferents in aversion to taste solutions, *Physiol. Behav.* 3:537–542.

Kawasaki, M., Ogura, J. H., and Takenouchi, S., 1964, Neurophysiologic observations on normal deglutition: I. Its relationship to the respiratory cycle, *Laryngoscope* 74:1747–1765.

Kay, R. H., 1964, "Experimental Biology," Reinhold, New York, pp. 21–35.

Kaye, H., 1972, Effect of variations of oral experience upon suckle, *in* "Third Symposium on Oral Sensation and Perception" (J. F. Bosma, ed.), Charles C Thomas, Springfield, Ill., pp. 261–292.

Keehn, J. D., and Arnold, E. M. M., 1960, Licking rates of albino rats, *Science* 132:739–741.

Kemper, B. I., 1972, Discussion, *in* "Third Symposium on Oral Sensation and Perception" (J. F. Bosma, ed.), Charles C Thomas, Springfield, Ill., pp. 372–374.

Kent, G. C., 1965, "Comparative Anatomy of the Vertebrates," Mosby, St. Louis, 457 pp.

Kent, G. C., 1973, "Comparative Anatomy of the Vertebrates," Mosby, St. Louis, 414 pp.

Kerr, F. W. L., 1962, Facial, vagal, and glossopharyngeal nerves in the cat, *Arch. Neurol.* 6:264–281.

Kerr, F. W. L., Kruger, L., Schwassmann, H. O., and Stern, R., 1968, Somatotopic organization of mechanoreceptor units in the trigeminal nuclear complex of the macaque, *J. Comp. Neurol.* 134:127–144.

Kitchell, R. L., 1961, Neural response patterns in taste, *in* "Physiological and Behavioral Aspects of Taste" (M. R. Kare and B. P. Halpern, eds.), University of Chicago Press, Chicago, pp. 39–49.

Kron, R. E., Stein, M., Goddard, K. E., and Phoenix, M. D., 1967, Effect of nutrient upon the sucking behavior of newborn infants, *Psychosomat. Med.* 29:24–32.

Kruger, L., and Michel, F., 1962, A single neuron analysis of buccal cavity representation in the sensory trigeminal complex of the cat, *Arch. Oral Biol.* 7:491–503.

Kubota, K., 1966, Comparative anatomical and neurohistological observations on the tongue of the Japanese Pika (*Ochotona hyperborea yezoensis*, Kishida), *Anat. Rec.* 154:1–12.

Kubota, K., and Hayama, S., 1964, Comparative anatomical and neurohistological observations on the tongues of pigmy and common marmosets, *Anat. Rec.* 150:473–486.

Kubota, K., and Iwamoto, M., 1967, Comparative anatomical and neurohistological observations on the tongue of the slow loris (*Nycticebus coucang*), *Anat. Rec.* 158:163–176.

Kubota, K., and Togawa, S., 1966, Comparative anatomical and neurohistological observations on the tongue of the Japanese dormouse (*Glirus japonicus*), *Anat. Rec.* 154:545–552.

Lipsitt, L. P., 1977, Taste in human neonates: Its effect on sucking and heart rate, *in* "Taste and Development: The Genesis of Sweet Preference" (J. M. Weiffenbach, ed.), U.S. Government Printing Office, Washington, D.C., pp. 125–142.

Lowe, A. A., and Sessle, B. J., 1973, Tongue activity during respiration, jaw opening, and swallowing in cat, *Can. J. Physiol. Pharm.* 51:1009–1011.

Luschei, E. S., and Goodwin, G. M., 1974, Patterns of mandibular movement and jaw muscle activity during mastication in the monkey, *J. Neurophysiol.* 37:954–966.

MacConaill, M. A., and Basmajian, J. V., 1969, "Muscles and Movement," Williams and Wilkins, Baltimore, 325 pp.

Magalhael, J., 1967, The golden hamster, *in* "The UFAW Handbook of the Care and Management of Laboratory Animals" (W. Lane-Peter, A. N. Worden, B. F. Hill, J. S. Paterson, and H. G. Vevers, eds.), Williams and Wilkins, Baltimore, pp. 327–339.

Marlow, C. D., Winkelmann, R. K., and Gibilisco, J. A., 1965, General sensory innervation of the human tongue, *Anat. Rec.* 152:503–512.

Marowitz, L. A., and Halpern, B. P., 1973, The effects of environmental constraints upon licking pattern, *Physiol. Behav.* **11**:259–263.

Matsunami, K., and Kubota, K., 1972, Muscle afferents of trigeminal mesencephalic tract nucleus and mastication in chronic monkeys, *Jap. J. Physiol.* **22**:545–555.

Matthews, P. B. C., 1972, "Mammalian Muscle Receptors and Their Central Actions," Edward Arnold, London, 630 pp.

McCutcheon, N. B., and Saunders, J., 1972, Human taste papilla stimulation: Stability of quality judgments over time, *Science* **175**:214–216.

Meader, C. L., and Muyskens, J. H., 1962, "Handbook of Biolinguistics." Part One. Revised Edition. H. C. Weller, Toledo Speech Clinic, Toledo, Ohio. pp. 98–100, 150–163.

Meiselman, H. L., Bose, H. E., and Nykvist, W. E., 1972, Effect of flow rate on taste intensity responses in humans, *Physiol. and Behav.* **9**:35–38.

Meiselman, H. L., and Halpern, B. P., 1970, Effects of *Gymnema sylvestre* on complex tastes elicited by amino acids and sucrose, *Physiol. Behav.* **5**:1379–1384.

Mendelson, J., Zielke, S., Werner, J. S., and Freed, L. M., 1973, Effects of airstream acessibility on airlicking in the rat, *Physiol. Behav.* **11**:125–130.

Miller, I. J., Jr., 1971, Peripheral interactions among single papilla inputs to gustatory nerve fibers, *J. Gen. Physiol.* **57**:1–25.

Miller, I. J., Jr., 1974, Branched chorda tympani neurons and interactions among taste receptors, *J. Comp. Neurol.* **158**:155–166.

Miller, I. J., Jr., 1975, Mechanisms of lateral interactions in rat fungiform taste receptors, *in* "Olfaction and Taste," V (D. A. Denton and J. P. Coghlan, eds.), Academic Press, New York, pp. 217–221.

Miller, I. J., Jr., 1977, Gustatory receptors of the palate, *in* "Food Intake and Chemical Senses" (Y. Katsuki, M. Sato, S. Takagi and Y. Oomura, eds.) Tokyo University Press, Tokyo, in press.

Miller, I. J., Jr., and Preslar, A. J., 1975, Spatial distribution of rat fungiform papillae, *Anat. Rec.* **181**:679–684.

Mistretta, C., 1970, A study of rat chorda tympani discharge patterns in response to lingual stimulation with a variety of chemicals, Ph.D. thesis, Florida State University, 222 pp.

Mistretta, C. M., 1971, Permeability of tongue epithelium and its relation to taste, *Am. J. Physiol.* **220**:1162–1167.

Mistretta, C. M., 1972, Topographical and histological study of the developing rat tongue, palate, and taste buds, *in* "Third Symposium on Oral Sensation and Perception" (J. F. Bosma, ed.), Charles C Thomas, Springfield, Ill., pp. 163–187.

Møller, E., 1966, The chewing apparatus, *Acta Physiol. Scand.* **69**:Supp. 280.

Moncrieff, R. W., 1967, "The Chemical Senses," Chemical Rubber, Cleveland, 760 pp.

Montagna, W., 1959, "Comparative Anatomy," Wiley, New York, 397 pp.

Montagu, M. F. A., 1945, "An Introduction to Physical Anthropology," Charles C Thomas, Springfield, Ill., pp. 325.

Morimoto, T., Kato, I., and Kawamura, Y., 1966, Studies on functional organization of the hypoglossal nucleus, *J. Osaka Dent. Univ.* **6**:75–87.

Morimoto, T., and Kawamura, Y., 1973, Properties of tongue and jaw movements elicited by stimulation of the orbital gyrus in the cat, *Arch. Oral Biol.* **18**:361–372.

Mosher, H. P., 1927, X-ray study of movements of the tongue, epiglottis and hyoid bone in swallowing, followed by a discussion of difficulty in swallowing caused by retropharyngeal diverticulum, postcricoid webs and exostoses of cervical vertebrae, *Laryngoscope* **37**:235–262.

Mountcastle, V. B., 1968, Physiology of sensory receptors: Introduction to sensory processes, *in* "Medical Physiology," II (12th ed.) (V. B. Mountcastle, ed.), Mosby, St. Louis, pp. 1345–1371.

Nejad, M. S., 1962, Factors involved in the mechanism of stimulation of gustatory receptors and bare nerve endings of the tongue of the rat, Ph.D. thesis, Florida State University.

Nord, S. G., 1967, Somatotopic organization in the spinal trigeminal nucleus, the dorsal column nuclei and related structures, *J. Comp. Neurol.* **130**:343-356.

Norgren, R., 1974, Gustatory afferents to ventral forebrain, *Brain Res.* **81**:285-295.

Norgren, R., and Pfaffmann, C., 1975, The pontine taste area in the rat, *Brain Res.* **91**:99-117.

Oakley, B., 1962, Microelectrode analysis of second order gustatory neurons in the albino rat, Ph.D. thesis, Brown University.

Oakley, B., 1967a, Altered taste responses from cross-regenerated taste nerves in the rat, *in* "Olfaction and Taste," II (T. Hayashi, ed.), Pergamon, Oxford, pp. 535-547.

Oakley, B, 1967b, Altered temperature and taste responses from cross-regenerated sensory nerves in the rat's tongue, *J. Physiol.* **188**:353-371.

Oakley, B., 1969, Taste preference changes following cross-innervation of rat fungiform taste buds, *Physiol. Behav.* **4**:929-933.

Oakley, B., 1972, The role of taste neurons in the control of the structure and chemical specificity of mammalian taste receptors, *in* "Olfaction and Taste," IV (D. Schneider, ed.), Wissenschaftliche Verlagsgesellschaft MBH, Stuttgart, pp. 63-69.

Oakley, B., 1974, On the specification of taste neurons in the rat tongue, *Brain Res.* **75**:85-96.

Oakley, B., 1975, Receptive fields of cat taste fibers, *Chem. Senses Flavor* **1**:431-442.

Ochs, S., 1965, "Elements of Neurophysiology," Wiley, New York, pp. 138-144.

Ogawa, H., Sato, M., and Yamashita, S., 1973, Variability in impulse discharges in rat chorda tympani fibers in response to repeated gustatory stimulations, *Physiol. Behav.* **11**:469-479.

Owen, R., 1841, Notes on the anatomy of the nubian giraffe, *Trans. Zool. Soc. London* **2**:217-248.

Pfaffmann, C., 1941, Gustatory afferent impulses, *J. Cell. Comp. Physiol.* **17**:243-258.

Pfaffmann, C. 1952, Taste preference and aversion following lingual denervation, *J. Comp. Physiol. Psychol.* **45**:393-400.

Pfaffmann, C., 1959, The sense of taste, *in* "Handbook of Physiology (Section 1: Neurophysiology. Vol. I)" (H. W. Magoun, ed.), American Physiological Society, Washington, D.C., pp. 507-533.

Pfaffmann, C., 1975, Phylogenetic origins of sweet taste, *in* "Olfaction and Taste," V (D. A. Denton and J. P. Coghlan, eds.), Academic Press, New York, pp. 3-10.

Pfaffmann, C., Erickson, R. P., Frommer, G. P., and Halpern, B. P., 1961, Gustatory discharges in rat medulla and thalamus, *in* "Sensory Communication" (W. A. Rosenblith, ed.), Wiley, New York, pp. 455-473.

Pierson, S. C., and Schaeffer, R. W., 1975, Lick rate development in infant mongolian gerbils, *Bull. Psychon. Soc.* **5**:47-48.

Poulos, D. A., and Benjamin, R. M., 1968, Response of thalamic neurons to thermal stimulation of the tongue, *J. Neurophysiol.* **31**:28-43.

Poulos, D. A., and Lende, R. A., 1970, Response of trigeminal ganglion neurons to thermal stimulation of oral-facial regions: I. Steady-state response, *J. Neurophysiol.* **33**:508-517.

Reighard, J., and Jennings, H. S., 1963, "Anatomy of the Cat," Holt, Rinehart, Winston, New York, 486 pp.

Reynolds, V., and Reynolds, F., 1965, Chimpanzees of the budongo forest, *in* "Primate Behavior" (I. DeVore, ed.), Holt, Rinehart, Winston, New York, pp. 368-424.

Rhoton, A. L., Jr., 1968, Afferent connections of the facial nerve, *J. Comp. Neurol.* **133**:89-100.

Richter, C. P., 1956, Salt appetite of mammals: Its dependence upon instinct and metabolism, *in* "L'Instinct dans le Comportement des Animaux et de l'Homme" (J. Autuori, *et al.*, eds.), Masson et Cie, Paris.

Rollin, H., 1973, Geschmacksausfälle nach Operation von Kleinhirnbrückenwinkeltumoren, *Hals-Nasen-Ohrenklinik* **21**:237-240.

Romer, A. S., 1970, "The Vertebrate Body," Saunders, Philadelphia, 601 pp.

Ruderman, M. I., Morrison, A. R., and Hand, P. J., 1972, A solution to the problem of cerebral cortical localization of taste in the cat, *Exp. Neurol.* **37**:522-537.

Sameroff, A. J., 1973, Reflexive and operant aspects of sucking behavior in early infancy, *in* "Fourth Symposium on Oral Sensation and Perception" (J. F. Bosma, ed.), U.S. Dept. Health, Education, and Welfare, Bethesda, Md., pp. 135-151.

Sato, M., 1973, Gustatory receptor mechanism in mammals, *Adv. Biophys.* **4**:103-152.

Sato, M., 1975, Response characteristics of taste nerve fibers in macaque monkeys: Comparison with those in rat and hamster, *in* "Olfaction and Taste," V (D. A. Denton and J. P. Coghlan, eds.), Academic Press, New York, pp. 23-26.

Sato, M., Yamashita, S., and Ogawa, H., 1969, Afferent specificity in taste, *in* "Olfaction and Taste," III (C. Pfaffmann, ed.), Rockefeller University Press, New York, pp. 470-487.

Sauerland, E. K., and Mizuno, N., 1970, Protective mechanism for the tongue: Suppression of genioglossal activity induced by stimulation of trigeminal proprioceptive afferents, *Experientia* **26**:1226.

Scalzi, H. A., 1966, The cytoarchitecture of gustatory receptors from the rabbit under normal and experimental conditions, Ph.D. thesis, University of Massachusetts.

Schaeffer, R. W., and Huff, R., 1965, Lick rates in cats, *Psychon. Sci.* **3**:377-378.

Schaeffer, R. W., and Premack, D., 1961, Licking rates in infant albino rats, *Science* **134**:1980-1981.

Schaller, G. B., 1965, The behavior of the mountain gorilla, *in* "Primate Behavior" (I. DeVore, ed.), Holt, Rinehart, Winston, New York, pp. 324-367.

Schmidt-Nielsen, K., 1964, "Desert Animals," Oxford University Press, London, pp. 67-69, 90-93, 106-107.

Schmitt, A., Yu, S.-K. J., and Sessle, B. J., 1973, Excitatory and inhibitory influences from laryngeal and orofacial areas on tongue position in the cat, *Arch. Oral Biol.* **18**:1121-1130.

Schultz, A. H., 1969, "The Life of Primates," Weidenfeld and Nicolson, London, 281 pp.

Scott, T. R., 1974, Behavioral support for a neural taste theory, *Physiol. Behav.* **12**:413-417.

Segundo, J. P., Takenaka, T., and Encabo, H., 1967, Somatic sensory properties of bulbar reticular neurons, *J. Neurophysiol.* **30**:1221-1238.

Sessle, B. J., and Kenny, D. J., 1973, Control of tongue and facial motility: Neural mechanisms that may contribute to movements such as swallowing and sucking, *in* "Fourth Symposium on Oral Sensation and Perception" (J. F. Bosma, ed.), U.S. Department of Health, Education, and Welfare, Bethesda, Md., pp. 222-231.

Seto, H., 1972, The sensory innervation of the oral cavity in the human fetus and juvenile mammals, *in* "Third Symposium on Oral Sensation and Perception" (J. F. Bosma, ed.), Charles C Thomas, Springfield, Ill., pp. 35-75.

Sisson, S., 1930, "The Anatomy of Domestic Animals," W. B. Saunders, Philadelphia, 930 pp.

Snyder, H. L., and Hulse, S. H., 1961, Effect of volume of reinforcement and number of consummatory responses on licking and running behavior, *J. Exp. Psych.* **61**, 474-479.

Sognnaes, R. F., 1954, The oral cavity, *in* "Histology" (R. O. Greep, ed.), Blakiston, New York, pp. 458-511.

Sonntag, C. F., 1920, The comparative anatomy of the tongues of the mammalia: I. General description of the tongue, *Proc. Zool. Soc. London*, 115-129.

Sonntag, C. F., 1921, The comparative anatomy of the tongues of the mammalia: III. Family 2. Cercopithecidae: With notes on the comparative physiology of the tongues and stomachs of the langurs, *Proc. Zool. Soc. London*, 277-322.

Sonntag, C. F., 1922, The comparative anatomy of the tongues of the mammalia: VII. Cetacea, sirenia, and ungulata, *Proc. Zool. Soc. London,* 639-657.

Sonntag, C. F., 1923, The comparative anatomy of the tongues of the mammalia: VIII. Carnivora, *Proc. Zool. Soc. London,* 129-153.

Sonntag, C. F., 1924, The comparative anatomy of the tongues of the mammalia: X. Rodentia, *Proc. Zool. Soc. London,* 725-741.

Sonntag, C. F., 1925, The comparative anatomy of the tongues of the mammalia: XII. Summary, classification and phylogeny, *Proc. Zool. Soc. London,* 701-762.

Sprague, J. M., 1943, The hyoid region of placental mammals with especial reference to the bats, *Am. J. Anat.* **72**:385-472.

Storey, A., 1967, Extra-trigeminal sensory systems related to oral function, *in* "Symposium on Oral Sensation and Perception" (J. F. Bosma, ed.), Charles C Thomas, Springfield, Ill., pp. 84-97.

Switzky, H. N., 1965, "Physical Variables in Taste Stimulation" M.Sc. thesis, Brown University. 73 pp.

Tapper, D. N., and Halpern, B. P., 1968, Taste stimuli: A behavioral categorization, *Science* **161**:708-710.

Taylor, A., 1969, A technique for recording normal jaw movements in conscious cats, *Med. Biol. Eng.* **7**:89-90.

Thexton, A. J., 1969, Characteristics of reflex jaw opening in the cat, *J. Physiol. London* **201**:678-688.

Thexton, A. J., 1973, Oral reflexes elicited by mechanical stimulation of palatal mucosa in the cat, *Arch. Oral Biol.* **18**:971-980.

Torvik, A., 1956, Afferent connections to the sensory trigeminal nuclei, the nucleus of the solitary tract and adjacent structures, *J. Comp. Neurol.* **106**:51-141.

Van Lawick-Goodall, J., 1968, The behaviour of free-living chimpanzees in the Gombe Stream Reserve, *Anim. Behav. Monog.* 1 (Part 3):161-311.

Van Lawick-Goodall, J., 1971, "In the Shadow of Man," Dell, New York, 304 pp.

Vance, W. B., 1966, Water intake of partially ageusic rats, *Life Sci.* **5**:2017-2021.

Vance, W. B., 1970, Oral infusion and taste preference behavior in the white rat, *Psychon. Sci.* **18**:133-134.

Volo, A. M., 1975, Gustatory control of licking behavior in rats, Program, Eastern Psychological Association, 46th Annual Meeting, p. 66 (Abstract).

Walker, E. P., 1964, "Mammals of the World," II, Johns Hopkins University Press, Baltimore, Md., pp. 647-1500.

Walker, W. F., Jr., 1967, "A Study of the Cat," Saunders, Philadelphia, 181 pp.

Wang, M. B., and Frank, M., 1970, unpublished observations, reported in: Pfaffmann, C., Physiological and behavioral processes of the sense of taste, *in* "Taste and Smell in Vertebrates" (G. E. W. Wolstenholme and J. Knight, eds.), J. and A. Churchill, London, pp. 35-37.

Weijs, W. A., 1973a, Functional morphology of the masticatory apparatus of the albino rat, *Acta Morphol. Neerl.-Scand.* **11**:162.

Weijs, W. A., 1973b, Morphology of the muscles of mastication in the albino rat, *Rattus norvegicus* (Berkenhout, 1769), *Acta Morphol. Neerl.-Scand.* **11**:321-340.

Wiesenfeld, Z., Halpern, B. P., and Tapper, D. N., 1977, Licking behavior: Evidence for a hypoglossal oscillator. *Science,* in press.

Wilcove, W. G., and Allison, J., 1972, Macro- and microbehavioral response to viscosity among rats licking saccharin, *Psychon. Sci.* **26**:161-163.

Williams, P. L., and Warwick, R., 1975, "Functional Neuroanatomy of Man," Saunders, Philadelphia, 1194 pp.

Wilson, A. S., and Barboriak, J. J., 1970, Lick volume determined by food schedules in rats, *Psychon. Sci.* **20**:271-272.

Wimsatt, W. A., 1959, Portrait of a vampire, *Ward's Bull.* **32** (Spring):35-39, 62-63.

Winkelmann, R. K., 1962, The mammalian end-organ in oral tissue of the cat, *J. Dent. Res.* **41**:207-212.

Wischnitzer, S., 1967, "Atlas and Dissection Guide for Comparative Anatomy," Freeman, San Francisco, 178 pp.

Wischnitzer, S., 1972, "Atlas and Dissection Guide for Comparative Anatomy, 2nd Edition," W. H. Freeman and Co., San Francisco, pp. 132-133, 142, 148-153.

Wolff, P. H., 1972, The interaction of state and non-nutritive sucking, *in* "Third Symposium on Oral Sensation and Perception" (J. F. Bosma, ed.), Charles C Thomas, Springfield, Ill., pp. 293-312.

Yamada, K., 1966, Gustatory and thermal responses in the glossopharyngeal nerve of the rat, *Jap. J. Physiol.* **16**:599-611.

Yamada, K., 1967, Gustatory and thermal responses in the glossopharyngeal nerve of the rabbit and cat, *Jap. J. Physiol.* **17**:94-110.

Yamamoto, T., and Kawamura, Y., 1975, Dual innervation of the foliate papillae of the rat: An electrophysiological study, *Chem. Senses Flavor* **1**:241-244.

Yokota, T., Suzuki, K., and Nakano, K., 1974, Reflex control of extrinsic tongue muscle activities by lingual mechanoreceptors, *Jap. J. Physiol.* **24**:73-91.

Zotterman, Y., 1935, Action potentials in the glossopharyngeal nerve and in the chorda tympani, *Skand. Arch. Physiol.* **72**:73-77.

2

The Recording of Licking Behavior

Jan A. W. M. Weijnen

1. Introduction

A contact-sensitive electronic instrument was described by Pilgrim (1948), but it was only after the publication of "An electronic drinkometer" (Hill and Stellar, 1951) that contact sensors became routine tools in studies requiring the recording of water intake. These devices are still commonly referred to as *drinkometers*; however, preference here will be given to the terms *lick* or *contact sensors*. The principle of the device is simple. One lead from the instrument makes contact with the drinking tube; the other one goes to the metal floor of the test chamber. Each time the animal completes the circuit, switching is accomplished in the output circuit of the sensor, and the switch can be used for counting or for some other means of registration.

2. What Current Level is Acceptable in the Sensing of Licking Behavior?

The current that is needed to operate a lick sensor should not have any artifactual behavioral consequences for the subject. Segal and Oden (1969), using a Grason–Stadler lick sensor, reported that the current did not influence water intake in schedule-induced polydipsia. But Martonyi and Valenstein (1971) observed an effect of the current in taste-preference

Jan A. W. M. Weijnen • Department of Psychology, Physiological Psychology Section, Tilburg University, Tilburg, The Netherlands.

studies. More of a very mildly bitter quinine solution was consumed when the Lehigh Valley sensor (LV 221-05) was switched on. The instruments that were used in both studies operate at current intensities $\leq 1\ \mu$A (short-circuit values, indicated by the manufacturers). Weijnen and Spijkers (unpublished results) found that rats could discriminate between water and water with concomitant electrical tongue-stimulation at a current intensity of 5 μA. Negative results were obtained with a 1-μA current. The electrical stimulation was used as a conditioned stimulus in a passive-avoidance situation (see Chapter 5, Section 3.6.3.1).

A reinforcing effect of lick-contingent electrical stimulation of the tongue without concomitant water intake could be observed at currents as low as 1 μA. At 0.5 μA, a significant effect was no longer observed in experiments with unselected rats (see Chapter 5). Recent studies with animals that were selected for their superior current-licking performance suggest even lower reinforcement thresholds in some animals.

Aversive properties of the lick-sensor current during water intake can be demonstrated at intensities that have positively reinforcing properties if the electric current is supplied without water (current licking). Water intake, measured in 10-min two-choice tests, was suppressed by the addition of a 100-μA current during drinking from one of the bottles; the nonelectrified bottle was preferred. However, no difference was found with 50 μA (see Chapter 5, Section 2.3.2.2).

A definitive statement as to the current level that is acceptable in the sensing of licking behavior cannot be given at present. More research is required. The answer might be dependent on the application to which the sensor is put. It is not impossible that the current-detection threshold differs when the animal is licking water, a flavored solution, or an empty tube (airlicking). The threshold for anodal tongue stimulation is probably not the same as for cathodal stimulation.

Passing a small electric current through an animal is not the only way to operate a lick sensor. Until now, however, all commercially available instruments have depended on this principle. Later in this chapter, attention will be drawn to alternative solutions that are based on the interruption of a light beam falling on a photocell and on the detection of licking behavior with a phonograph cartridge or a pressure transducer.

For a good knowledge of the intensity of the current that is passing through the subject, information is required on the:

1. Resistance of the animal.
2. Capacitance of the input circuit of the sensor.
3. Design of the lick sensor.

2.1. The Rat Considered as a Resistor in the Input Circuit of a Contact Sensor

The electrical resistance of the rat can be quite variable. There are two factors that play an important role in this respect:

1. The quality and the area of the rat's contact with the floor of the test chamber. The resistance is low in rats standing with wet feet on a metal plate, whereas high values can be reached in animals standing on a dirty and dry grid floor.
2. The resistance also varies with the intensity of the current. Campbell and Teghtsoonian (1958) have shown that the resistance of skin and tissue decreases with increasing current intensities.

An illustration of the influence on the animal's resistance of the size of the contact with the test-chamber floor and of the intensity of the measuring current is given in Figure 1. This figure is based on measurements in a test chamber with a dry and clean floor, performed with rats that were kept on a bedding of wood shavings.

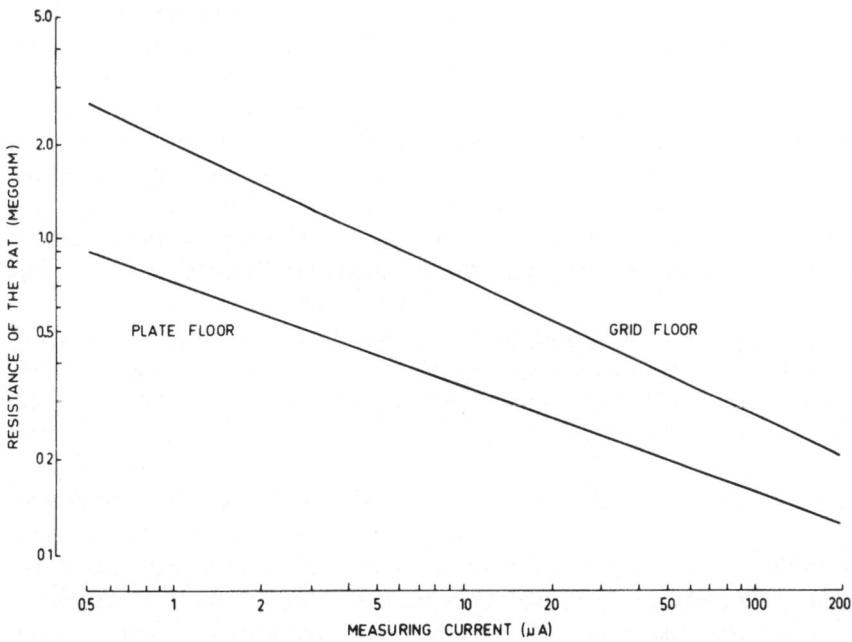

Figure 1. Smoothed representation of the resistance of rats as a function of the intensity of the current that was used to measure the resistance, for two types of floors of the test chamber. The rats contacted the floor with their hindfeet only.

Regarding the housing conditions, Campbell and Masterson (1969) reported: "Animals housed directly on dry sawdust or shavings have a higher skin resistance than animals housed in wire-mesh cages where dirt and moisture accumulate on the mesh" (p. 8).

A convenient way to determine the resistance of the rat as a function of the intensity of the measuring current is to measure the voltage generated across a resistor in series with the animal. The voltage measurements were performed with an oscilloscope (Tektronix Type 502, 1 MΩ input impedance). A 10-kΩ series resistor was used. Conversion of the voltage into the resistance of the rat was established by a comparison of the measured value to a calibration curve that was constructed when precision resistors of known value were substituted for the rat. Ohm's law was used to calculate the current passing through the rat (measured voltage divided by the value of the measuring resistance). To obtain current intensities above 40 μA, a 150-V dc supply was used with a variable-series resistor to set the nominal current level. A 30-V dc source was used to determine resistance values at medium current intensities. Resistance values for the lowest intensities of the measuring current were obtained with a low-voltage (1–10 V dc); for these measurements the oscilloscope was connected directly in series with the rat. Most of the data were obtained during water-licking sessions. Current-licking results were also included, but only for the higher intensities. The animals were licking at a tube passing through the ceiling of the test chamber (see Figure 4 later in this chapter). Therefore, only the hindfeet of the rats could be in contact with the floor of the test chamber. It is likely that lower resistance values will be found with rats that contact the floor with all four feet. When the low voltage was used, the resistance of the animal increased within periods of continuous licking, presumably because of polarization. As a rule, the median oscilloscope reading over approximately 25 licks was estimated. Large inter- and intrasubject variations in the calculated resistance occurred; no attempt has been made to measure these variations in a systematic way, as the method was unsuitable for such a study. Figure 1 should be regarded only as a graph that is useful in estimating the subject's resistance over a wide range of current intensities. Lines were fitted by eye to the means of medians of observations at various current levels. The graph is based on results of experiments scattered over a period of 15 months under variable conditions. In accordance with the results of Campbell and Teghtsoonian (1958), the resistance of the animals decreased with increasing intensities of the current that was used to measure the resistance. At current levels that are employed by commercially available sensors, the estimated animal resistance is in the megohm range.

A substantial part of the total resistance of the rat is made up of the resistance of the feet that are in contact with the floor of the test chamber.

Therefore, a higher resistance value can be expected if the contact area with the floor is reduced, as happens with rats standing on a mesh or grid floor instead of on a metal plate. In this study 2.4-mm grids were employed, spaced 12.7 mm apart (center to center). The resistance of rats that completed the input circuit of the sensor by licking the drinking tube or licking rod was lower with a test chamber equipped with a plate floor, compared with values obtained with a grid floor, for all current intensities (see Figure 1). A further decrease in subject's resistance was observed after the feet of the animals were wetted. When the hind feet of some rats were wetted with urine, a drop in the resistance to 75 kΩ was observed (plate floor, 55 μA, mean of four animals).

2.2. Capacitance of the Input Circuit of the Lick Sensor

Information about the intensity of electrical stimulation is frequently given in terms of either short-circuit values or values that are measured with a dummy load. With both ways of expressing the intensity, knowledge of the actual resistance of the subject and of the stability of the current supply is essential. Even if this knowledge is available, the electrical stimulation still might not be effectively controlled. Current-intensity readings made with an ammeter are not necessarily adequate for the determination of the actual current passing through the rat. The capacitance of the input circuit of the contact sensor may have a large influence on current intensity that will not be revealed by the relatively slow-acting ammeter.

Long connections between lick sensors and test chambers with shielded leads or leads with two or more conductors can result in a high capacitance of the input circuit (Table 1). The contribution of the test chamber itself can easily be overlooked. A large (700 ml) flat-sided water bottle mounted directly against the aluminum test compartment can be equivalent to over

Table 1. *A Comparison of the Capacitance of Some Arbitrarily Chosen Samples of Electric Leads*

Sample	Capacitance/meter
2 patch cords, hanging free from other wires or metal objects	3 pF
2 patch cords, loosely twisted	30 pF
2 conductors in multicore cable	50, 60, 80 pF
Power cord (2 conductors)	60, 100, 120 pF
Shielded lead	30[a], 67, 100, 130 pF

[a] Special low-capacitance shielded lead.

100 pF when filled up; in empty state only a few pF will be measured. Capacitive discharge of voltage through a rat upon the closing of the sensing circuit may result in a large transient of current through the animal (cf. Campbell and Teghtsoonian, 1958). Results of current-licking experiments have shown that a short duration of electrical stimulation of the tongue, at an adequate intensity, can in itself have reinforcing properties (see Chapter 5, Section 2.4). This initial elevation in current intensity is illustrated in Figure 2. A subject with a resistance of 1 MΩ (replaced by the oscilloscope in the figure) receives electrical stimulation at an intensity of 0.97 μA, as measurement with an ammeter would show. However, this only holds for the asymptotic intensity. Observation of the oscilloscope shows that the current in circuit A of Figure 2 is initially much higher because of capacitative discharge from the shielded lead. Although this capacitance does not completely recharge to 30 V after breaking of the contact during trains of, say, 100-msec pulses at 6 Hz, the current intensity is still elevated during the first part of the pulse. In circuit B, this problem is not encountered. However, the risk of touching electrified objects that are not protected by a current-limiting resistor is greater with the type-B circuit. Many lick

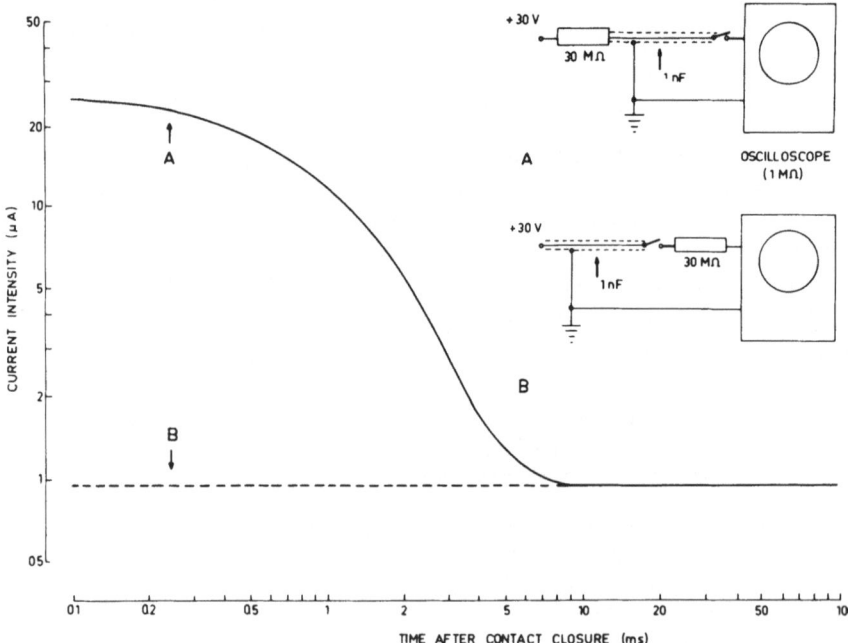

Figure 2.   Influence of the capacitance of the input circuit of the contact sensor on the intensity of the current as a function of time after contact closure for two types of circuits.

sensors are operated by animals standing on a grounded floor of the test chamber (a situation resembling circuit A), and the current intensity is regulated by a large resistor.

The author of this chapter has been confronted with the effect of a high capacitance of the input circuit after having moved a test chamber to a distant room in the laboratory. A long multicore cable was used to connect the licking rod with a circuit equivalent to a 150-V dc supply with a 3-MΩ series resistor. Lick-contingent electrical stimulation of the tongue (with 50 μA) had been reinforcing to the animal when short leads were used. The capacitance of the long cables, however, resulted in strong aversive reactions of the rats upon making contact with the licking rod. There can be other consequences of the capacitance of the input circuit. A lick sensor may stay activated for some time after the animal has broken contact with the instrument. For the "Type II" sensor, which is described later, this delay is 22 msec at 50 pF, increasing to 110 msec at 1000 pF (1 nF). The time between onset of the tongue contact with the drinking tube and the beginning of the output of the sensor was more constant: 7–8 msec. It is clear that the duration of the output signal of a lick sensor (if not fixed to a standard length by the designer) needs correction if exact information on the contact time is required. The correction factor varies with the type of sensor used and with the capacitance of the input circuit. In publications on lick duration, however, the output of the lick sensor has usually been mistakenly presumed to be identical to the duration of the contact of the tongue with the fluid of the drinking tube. Interlick-interval (on–on time) measurements are, of course, not affected by a constant difference in input and output duration of the sensor; but a high capacitance of the input circuit can result in a decrease of the maximum operating frequency of the sensor. Other transients that occur when the tongue makes and breaks contact with the licking rod or the drinking tube can be derived from the capacitative and inductive properties of the rat itself (Ramsay, Knapp, and Zeiss, 1970). In an exploratory experiment, an estimate was made of the capacitative properties of the rat. The current passing through the animal was monitored on an oscilloscope. During the charging of the capacitance of the licking rat, the voltage across the series resistor that was used for monitoring fell to an asymptotic value in a few milliseconds. The asymptotic current intensity was determined by the source voltage and the total resistance of the circuit. A 30-V dc supply was used with a 200-kΩ series resistor to set the nominal current intensity. By successive approximation, a circuit was found that resulted in a similar pattern to the display on the oscilloscope (Figure 3). Only two rats were used, both at the same effective current intensity (75 μA). The estimated value of the capacitance of the rat, 5–10 nF, compares well with the 10 nF reported by Campbell and

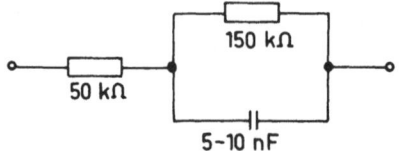

Figure 3. A circuit resulting in a pattern of display on the oscilloscope that was similar to the display induced by a licking rat.

Masterson (1969). They employed phase-shift measurements with 60 Hz ac to calculate the capacitance. Unlike the resistance of the subject, the capacitance is independent of the current intensity.

The effect of the capacitance of the rat on the intensity of the current passing through the animal might be of importance in experiments employing a circuit as shown in insert A of Figure 2, under conditions where the capacitance of the sensor circuit is high. The lower the capacitance of the circuit and the more the current is stabilized, the less the influence of the rat capacitance on the current intensity. Effects attributable to the inductive properties of the rat are difficult to measure and were neglected in this study.

3. The Technique of Recording Licking Behavior

3.1. Recording the Lick Response

To ensure that every single lick is registered, it is essential that after each lick the sensor circuit be broken, otherwise the instrument stays activated and no new licks are recorded. This condition occurs if, for example, a rat puts a front paw on the spout. But even if the animal can reach the drinking spout with only its tongue, there is a possibility that the rat will not break the circuit between licks because of a small water bridge from its tongue to the spout. Another problem in lick recording is the chance that the animal may not be able to reach the water because of a retracted meniscus of the water in the opening of the spout. This condition may be encountered with small apertures, especially if the position of the drinking tube is not vertical. Under these circumstances, licks are recorded without any water intake if the spout is made of conductive material, otherwise nothing is registered or ingested.

Various methods have been reported in the literature to achieve adequate recording of licking behavior. Typically, the tip of the drinking tube is positioned at some distance behind a circular or rectangular opening in a side wall of the test chamber (Allison, 1971; Cone and Cone, 1973; Hulse,

1967). These solutions seem straightforward, but they can modify the behavior that is being measured: lick frequency, contact time, and volume per lick (Hill and Stellar, 1951; Corbit and Luschei, 1969; Marowitz and Halpern, 1973; Hutcheson and Mills, 1974). Marowitz and Halpern investigated licking behavior under three degrees of environmental restraint. Motion-picture analysis revealed the effect of the constraint upon tongue motion (see Chapter 1, Figure 8).

For the study of the reinforcing effects of lick-contingent electrical stimulation of the tongue (see Chapter 5), another method of recording was developed (Figure 4). Ball-point watering tubes (Atco Manufacturing Company), made of stainless steel, were used for water licking. These tubes allow an easy flow of water without drippage. The glass sleeve that is shown in the same figure limited activation of the contact sensor to, virtually only, tongue contacts. For the recording of current licking behavior, a stainless-

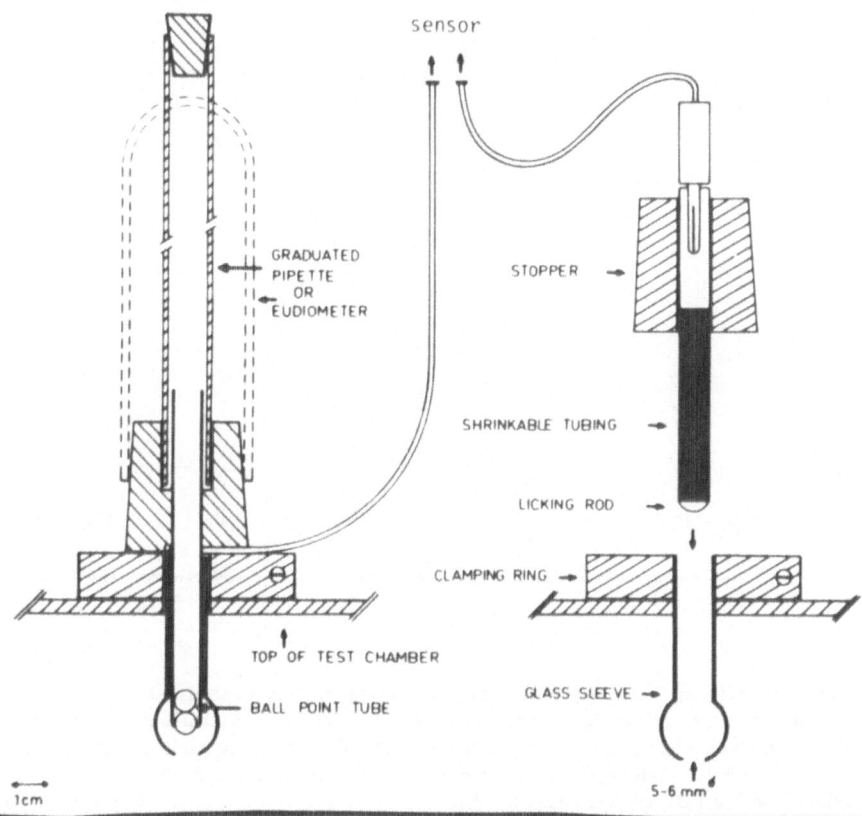

Figure 4.   Diagrammatic representation of the water- and current-licking equipment.

steel rod with the same dimensions as the ball-point tube was used. Shrink-
able tubing was fitted on the drinking tube and the licking rod to ensure good
electrical insulation from the glass sleeve, which can become moist inside.
Both the tube and the rod were put through a rubber stopper and could be
brought into position quickly when lowered into the glass sleeve, which was
fixed to the test chamber with a clamping ring. A critical point in the
arrangement is the depth of the tube or rod in the glass sleeve. If the
distance to the lower end of the glass sleeve is too small, failures to break
the circuit between licks may occur. The consequences of this condition are
clearly illustrated by the distributions of the interlick intervals and of the
lick durations shown in Figure 5 (interlick interval defined as the period of
time between the onset of two subsequent licks, or on–on time. Lick duration is an equivalent of on–off time). Subsidiary peaks occur at multiples of
the principal modal interlick interval. Peaks are also seen in the lick-
duration distribution at the modal lick-duration plus n times the modal
interlick interval. Similar harmonics can be obtained with other techniques

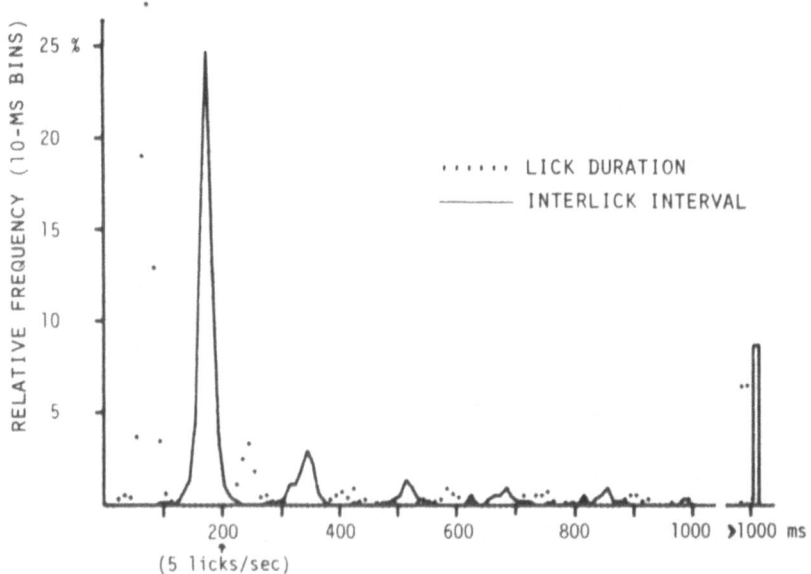

Figure 5.   Selected illustration of the lick-duration and interlick-interval distribution of a rat
licking water from a drinking spout (diameter of the orifice, 2.5 mm) that was recessed 2 mm
in the glass sleeve (see Figure 4). Out of the first 1000 licks that were made by the rat, which
was kept on a 85% water-deprivation schedule, 500 lick durations and interlick intervals were
sampled with a PDP lab 8/e computer (BASIC/RT program). The output of a Type II (0.4
μA) sensor, which is described in this chapter, was employed for the measurements. Lick duration was corrected for the difference between the moment of closing and opening of the input
and output circuit of the sensor (see Section 2.2).

of recording licking behavior (Corbit and Luschei, 1969; Crawford, 1970). It is obvious that periodic failure to break the input circuit between licks results in a reduction of the number of lick responses that are measured. If subsidiary peaks occur in the distribution of interlick intervals, but not in the distribution of lick durations, then most likely not every tongue movement has been sensed. The cause can be sensor failure (e.g., if resistance of the input circuit is critically high) but also abortive attempts to reach the drinking tube. Recording of airlicking can create sensing problems, especially when the availability of the airstream has not been made lick-contingent and when a relatively high air pressure is used (Oatley and Dickinson, 1970). The tongue might only occasionally touch the tube or make contact more than once during a single lick movement: one contact while the tongue is protruded and another while it is retracted (Chapter 4). It might be possible, too, that an interruption of the contact of the tongue with the tube occurs when it passes over the orifice.

The effect of recessing the drinking tube in the glass sleeve (1–4 mm) on the lick-duration and interlick-interval distribution can be seen in Figure 6. Recessing the tube in the sleeve puts a constraint on licking behavior, resulting in a slowing down of its rate. This effect can be reliably reproduced and is observed during water licking both from ball-point tubes and from drinking spouts (diameter of the orifice, 2.5 mm) and is also obtained in current-licking studies.

Experience has taught that recessing the drinking tube 3 mm during water licking and 2 mm during current licking or airlicking ensures that single licks are measured independently in most cases.

It is clear from these results, and from earlier reports on the subject, that the results of licking-behavior studies are highly dependent on the method used to collect the data. An adequate description of the testing situation in the methods section of publications on licking behavior is therefore essential. Extrapolation of results to licking behavior, in general, should be tried only with caution.

3.2. Measuring Time Spent Licking

In his chapter on airlicking and cold licking (Chapter 4), Mendelson has reported the occurrence of double-contact licks. To overcome this problem, he advocates measuring the time spent licking instead of the number of licks. One can achieve this by operating/resetting an electronic timer via the output contacts of the lick sensor. The output contacts of the timer are used to activate a running-time counter. Setting the timer at such a value (200–250 msec) that its output contacts just stay closed during periods of uninterrupted licking results in a fairly accurate way of measur-

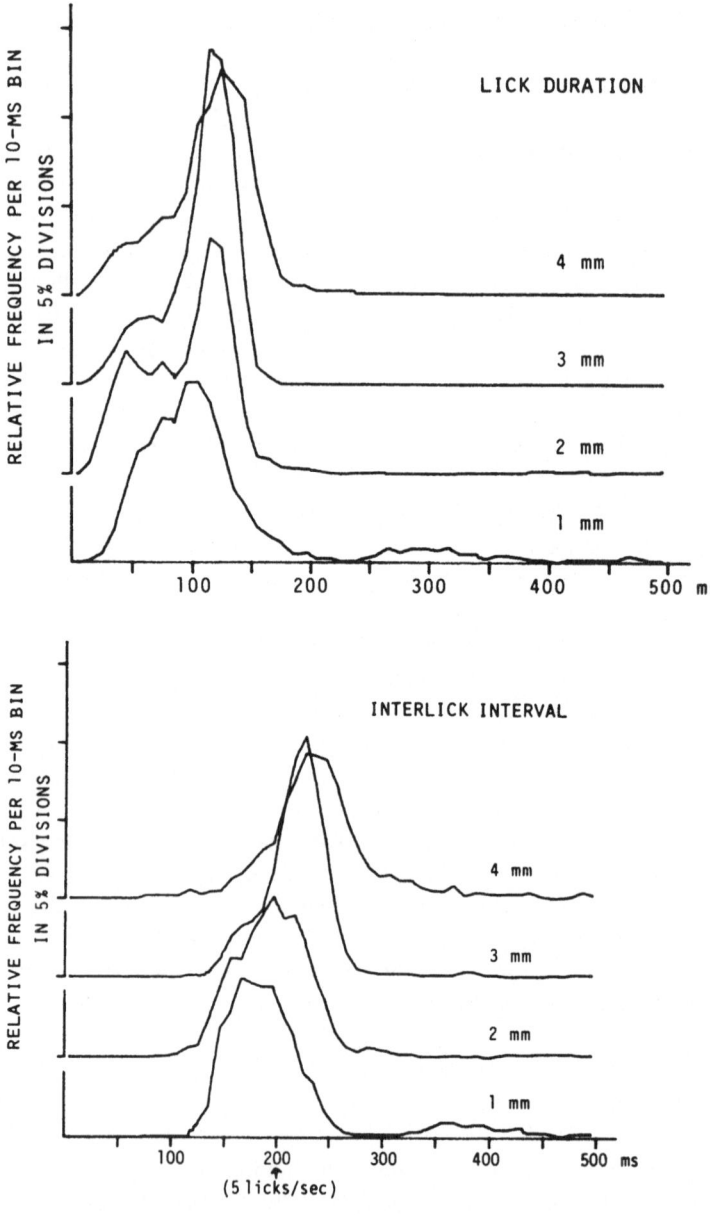

Figure 6.   Average distribution of lick duration and of interlick intervals of four rats drinking water from a ball-point tube that was recessed 1, 2, 3, or 4 mm in the glass sleeve. For details see legend to Figure 5.

ing the time spent licking. The same circuit is useful in activating an air valve in lick-contingent airlicking experiments.

4. Lick Sensors

The following survey of lick sensors begins with a brief description of the sensors that are currently commercially available. Then more detailed information is supplied on lick sensors that were employed in the study of lick-contingent electrical stimulation of the tongue (Chapter 5). Sensors that pass a stable low current through the animal as well as instruments with an adjustable current value (tongue stimulators) are described. Finally, an account is given of alternative solutions to the sensing of licking behavior that do not require the passage of an electric current through the subject.

4.1. Commercially Available Sensors

Sensors that record licking behavior can be bought from several companies that manufacture behavioral-research equipment. A consumer report, based on a comparative investigation, would be interesting but has not been attempted. The data presented in Table 2 were supplied by the manufacturers with the aid of a questionnaire. Choosing among these sensors is difficult, since the relevant criteria are still lacking (see Section 2). The risk of relatively high transient currents increases when a higher voltage is used in the high-impedance input circuit and also when the capacitance of the input circuit increases (see Section 2.2).

4.2. Lick Sensors That Were Used for the Study of Current-Licking Behavior

In the first study of lick-contingent electrical stimulation of the tongue, a BRS-Foringer DO-101 sensor was used (Slangen and Weijnen, 1972). The current in the input circuit of this instrument is highly dependent upon the variable resistance of the animal in the test situation. Under certain conditions it might have been possible for the animal to have been exposed to a few tenths of microamperes (Figure 9). For subsequent studies of current licking, better regulation of the current intensity was needed (Weijnen, 1972). (Also see Chapter 5.)

Table 2. *Commercially Available Sensors*

	Manufacturer						
	Coulbourn Instr. Lehigh Valley, Pa.	Grason–Stadler Concord, Mass.	Grason–Stadler Concord, Mass.	Janssen Scientific Instr. Beerse, Belgium	Lafayette Instr. Lafayette, Ind.	Scientific Prototype New York, N.Y.	Tech. Serv. International Beltsville, Md.
Model	S26-01	E4690A	1241-9450	1.0056[a]	58008	101-K	SI-251 SI-451
Power supply	+/-12 V dc	28 V dc 117 V ac	+/-12 V dc	12 V dc	115 V ac	117 V ac	-12 V dc
Shielded input necessary?	No	b	b	c	No	No	Yes
High-impedance circuit characteristics:							
Voltage	-12 V dc	-22 V dc	5 V dc	12 V dc	10 V dc	9 V dc	-9.2 V dc
Resistance	22 MΩ	22 MΩ	4.7 MΩ	200 MΩ	110 MΩ	44 MΩ	10 MΩ
Intensity of current in input circuit:							
Short-circuit	0.5 μA	1 μA	1 μA	0.06 μA	0.09 μA	0.2 μA	0.9 μA
0.1-MΩ load	0.5 μA	1 μA	1 μA	0.06 μA	0.09 μA	0.2 μA	0.9 μA
1-MΩ load	0.5 μA	1 μA	0.9 μA	0.06 μA	0.09 μA	0.2 μA	0.8 μA
10-MΩ load	d	d	d	0.057 μA	0.083 μA	0.17 μA	0.46 μA
Maximum load for reliable operation	3 MΩ	1 MΩ	1 MΩ	80 MΩ	70 MΩ	22 MΩ	10 MΩ
Output:							
Logic signal	*			*		*	*
Relay contacts		*	*		*		
Delay between beginning of input and output signal	6 msec	5-10 msec	2 μsec	0.4 μsec[a]	7-10 msec	8 msec	<1 msec

[a] The output of the sensor can be processed by the JSI pulse conditioner. This instrument has facilities to reject lick-contact durations and intervals, between two successive licks, that are smaller than a presettable value. This feature implicates an extra input–output delay that is equal to the preset value.

[b] Special input cable provided with the sensor.

[c] Very short and low-capacitance input leads are necessary.

[d] This load exceeds the maximum load for reliable operation.

The first sensor that was developed for tongue stimulation, with con-comitant lick recording, was the Type I sensor (Figure 7; Table 3). This instrument was a modification of a multipurpose pulse-forming circuit designed by Bintz and Zucker (1970). The current intensity was limited to 65 μA (short-circuit value). An animal that completed the circuit received about 40-50 μA (Figure 9). The sensor produced a 35-msec closure of the output circuit upon each lick. The recording of licking behavior with a sta-ble current of considerably lower intensity was achieved with the Type II sensor, an adapted version of an unpublished design by Anthoni (Figure 8; Table 3). The short-circuit current intensity of this sensor is 0.4 μA (the intensity depending somewhat on the characteristics of the transistors used). Variations in the resistance of the animal completing the circuit have little effect on the current intensity (Figure 9). Producing stabilized currents of relatively high intensities for tongue stimulation with concomitant lick recording was more complicated. The Type I performance was too much influenced by variations in the resistance of the rat. The effect of these variations on the intensity of the current going through the subject can be reduced if a large resistor is put in series with the animal; however, a higher-source voltage is needed to produce the same nominal current level. A 150-V dry battery was used. The extent of the resulting current-intensity regulation can be seen in Figure 9. A nominal current intensity of 12.5 μA, obtained when a 12-M Ω resistor is put in series with the rat, is less

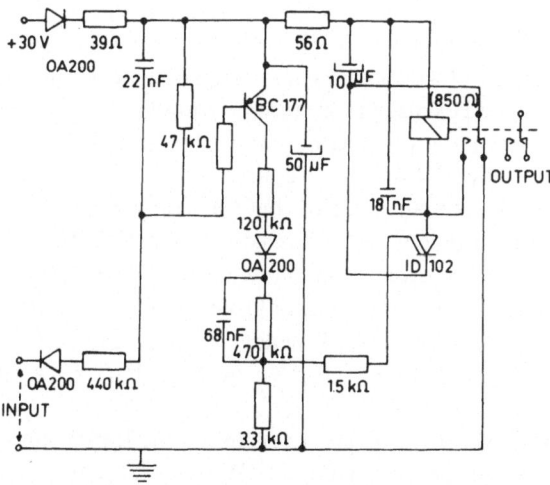

Figure 7. Type I lick sensor, based on a design by Bintz and Zucker (1970). The current intensity in the input circuit is equivalent to the current generated by 30 V with the load in series with 465 kΩ.

Table 3. Characteristics of Type I and Type II Sensors

Model	Type I	Type II
Power supply	30 V dc	30 V dc[a]
Shielded input necessary?	No	No
High-impedance circuit characteristics:		
Voltage	30 V dc	30 V dc
Resistance	465 kΩ	69 MΩ
Intensity of current in input circuit:		
Short-circuit	65 μA	0.43 μA
0.1-MΩ load	53 μA	0.43 μA
1-MΩ load	20 μA	0.43 μA
10-MΩ load	[b]	0.38 μA
Maximum load for reliable operation	1.2 MΩ	60 MΩ
Output:		
Logic signal		
Relay contacts	*	*
Delay between beginning of input and output signal	7 msec	6 msec

[a] Operates satisfactorily at 24 V, too. This results in some reduction of the input current intensity and of the maximum load for reliable operation.
[b] This load exceeds the maximum load for reliable operation.

influenced by changes in the resistance of the subject than is a nominal current of 200 μA.

It was not necessary to design a new type of sensor to record licking behavior with concomitant stimulation of the tongue at adjustable current intensities. Figure 10 illustrates how a voltage-divider circuit made it possible to employ a Type II sensor as a "tongue stimulator."

When a current passes through an animal as it completes the input circuit, the electrical stimulation is—of course—not limited to the tongue. The feet and all tissue in the current path between tongue and feet are affected too. Aspects of this problem are discussed in Chapter 5. In order to prevent inadvertent exposure to direct contact with 150 V or 120 V, the voltage divider circuit and the lick sensor should be built as a self-contained unit. References to the use of the Type II sensor in Chapter 5 include the current intensity, for example, Type II (50 μA), or Type II (0.4 μA) if no voltage divider was used.

The performance of the Type II sensor can be further improved by the use of a reed relay in the output circuit. With selected transistors (BC 157 or 2N5447, h$_{FE}$ at 1 μA > 150), the reed relay, and a 6-V dc power supply, a sensor has been constructed that needed a 0.1-μA current in its input circuit. The input impedance was 60 M Ω.

4.3. Lick Sensors That Do Not Require the Passage of an Electric Current through the Animal

There are two important reasons for avoiding the use of electrically operated lick sensors:

1. The use of such a sensor, in certain situations, may be incompatible with another variable which is being studied.
2. The current flow through the subject might have an undesirable effect on the behavior of the animal.

King (1969) and King, Justesen, and Simpson (1970) needed an instrument that could be used in the presence of microwaves. For this purpose, animal-to-metal contact had to be avoided. Martonyi and Valenstein (1971) observed that rats that could choose between water and a quinine solution drank significantly more of the mildly bitter solution when lick sensors were used than when they were switched off.

The use of a photo lick sensor solved the problem in both studies. This type of sensor (and other instruments that are discussed later) is very useful,

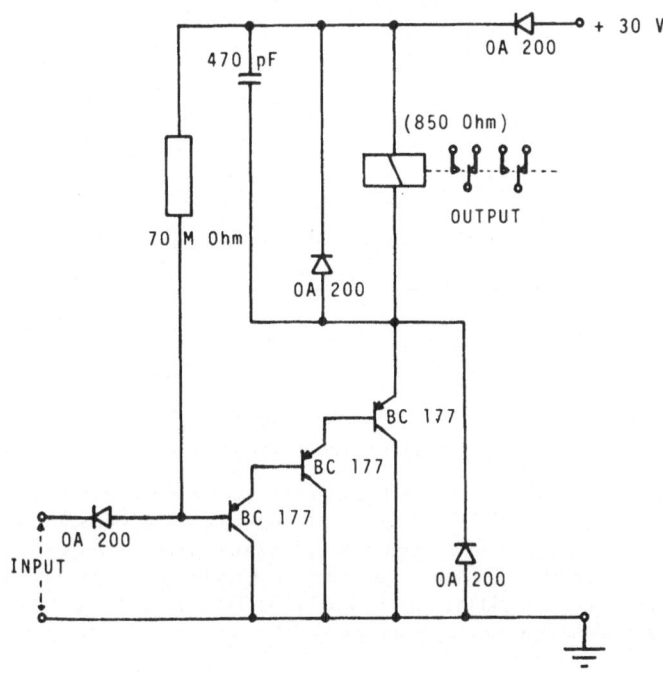

Figure 8. Type II lick sensor, adapted from an unpublished design by Anthoni.

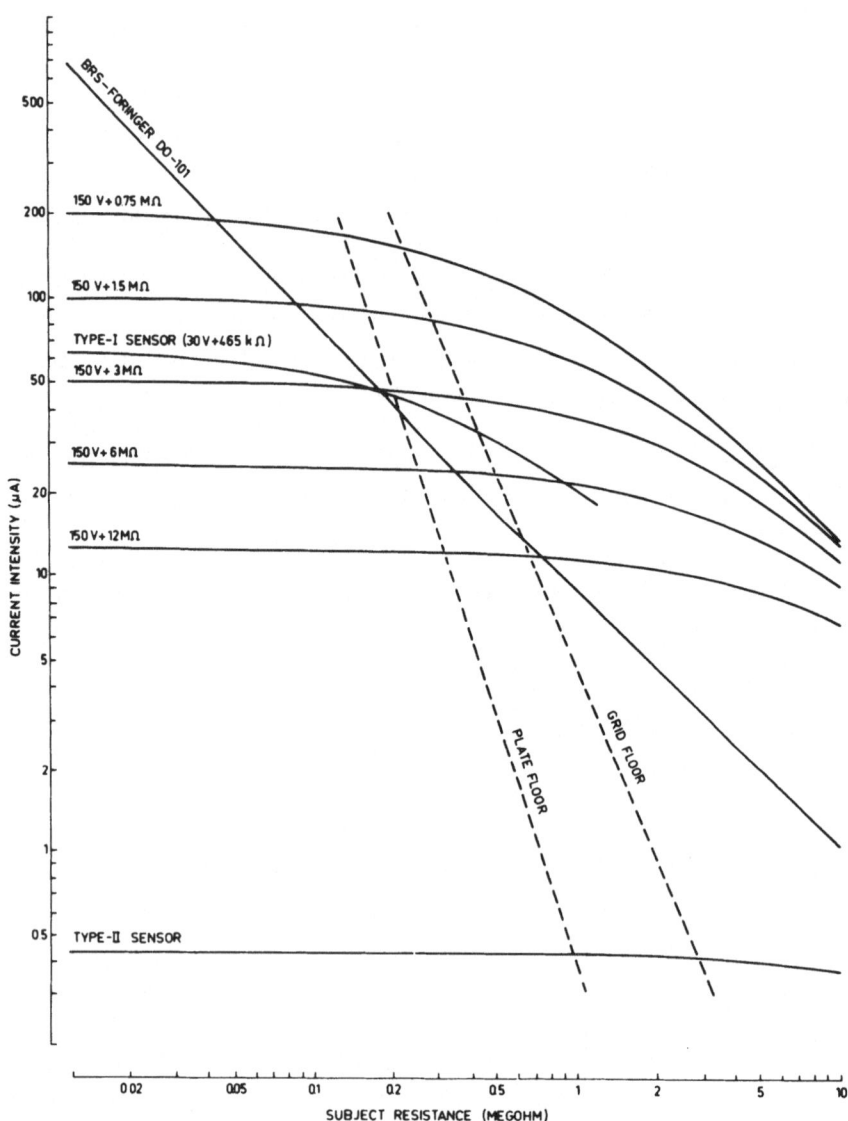

Figure 9. Current-intensity versus subject-resistance relationships for the Type I and Type II sensors, the BRS-Foringer DO-101, and for circuits employing a 150-V dc source with a large resistance in series with the animal. An estimate of the median current passing through the rat for each of the stimulation systems can be made by a reading of the current value at the intersection of the function and the lines marked "plate floor" or "grid floor" (see Figure 1). Large deviations of the estimated value can be expected with the BRS-Foringer instrument.

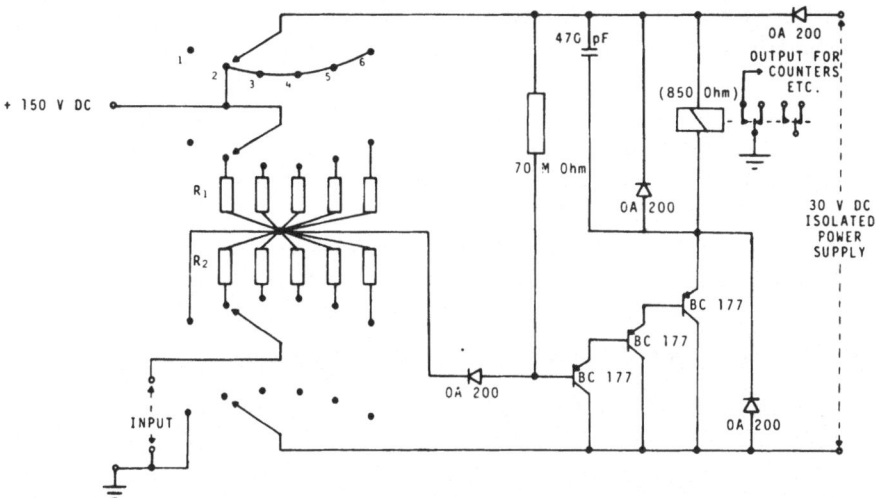

Figure 10. Application of the Type II sensor as a "tongue stimulator." This is achieved by operation of the instrument through a voltage divider. The nominal current in the input circuit is determined by the total resistance of the two components (R1 + R2) of the voltage divider. Connecting 150 V to ground through these resistors results in a voltage drop of 30 V,[1/(1 + 4) × 150 V], at the point constituting the original input of the Type II sensor. This is equivalent to lowering this input to "0" V with reference to its own power supply and therefore operates the sensor. The power supply that is used to operate this sensor must be isolated from the supply that is used to operate other parts of the behavioral-research equipment. In position 1 of the rotary switch, the minus of this isolated supply is grounded, and the sensor operates as a normal Type II (0.4 μA) sensor. In the other positions, the ground connection is broken, and the "plus" of the supply is connected to 150 V dc. For the current-licking study, sets of R_1 and R_2 were selected to generate 12.5, 25, 50, 100, and 200 μA to stimulate the tongue (short-circuit values). In order to prevent inadvertent exposure to direct contact with 150 V or 120 V, the voltage-divider circuit and the sensor should be built as a self-contained unit.

too, in research that requires concurrent electrical recording or stimulation techniques.

4.3.1. Photo Lick Sensors

With these sensors, the licking behavior of an animal is recorded through the interruption of a light beam falling on a photocell. Each time that the tongue is protruded and retracted, the light beam should be broken once. As is true for "conventional" sensors, photo lick sensors also require careful control of the drinking situation to ensure the adequate recording of licking behavior. In the designs that have been developed, care had to be taken that changes in the ambient light conditions would not interfere with the performance of the instrument. Construction details have been published in the papers cited above (see also Chapter 3). A modification of the sensor described by Martonyi and Valenstein has been published by Hutcheson and Mills (1974). They used an infrared light-emitting diode. Their photo lick sensor is characterized by its small size, fast switching, and low power consumption and therefore negligible ambient heating.

An interesting test of the performance of a photo lick sensor would be the analysis of data collected simultaneously with this sensor and a conventional device (number of licks, lick duration, and interlick-interval times). Until now, the performance of these two types of sensors has been compared only indirectly (King, Justesen, and Simpson, 1970).

4.3.2. Lick Sensors Employing a Phonograph Cartridge

Eisman (1969) has described a method of recording licking behavior in which the lick response is picked up by a crystal phonograph cartridge. The cartridge is mounted with its stylus in contact with the drinking tube. A slightly different approach has been used by Davis (1973). A fine wire attached to the phonograph cartridge is displaced by the tongue of the animal when it passes across the tip of the drinking tube. The digital output of the phonograph-cartridge sensor requires amplification and filtering. Eisman reported to have successfully used a commercially available audio detection relay for this purpose.

It is not clear from the above-cited publications whether or not with this recording method each lick response results in one, and only one, lick registration, and whether it successfully rejects movements of the animal or other vibrations in the test environment.

4.3.3. Pressure-Sensitive Lick Sensor

For the measurement of lick-response execution in the rat, Vrtunski and Wolin (1974) designed a system employing a pressure transducer. Since their principal interest was in the force that the rat's tongue exerts upon the watering tube, relatively complex equipment was necessary for their study. If only counting of licking responses is needed, then one might consider a simplified version of their technique. To this end, a watering tube is mounted on a pressure transducer; it is also connected with flexible tubing to a graduated pipette. The (amplified) output of the transducer is fed into an adjustable voltage-level detector (Schmitt trigger). The output of the detector can be interfaced with recording equipment, such as event markers and counters.

5. Conclusion

The behavioral consequences of the current that is needed to operate most of the available lick sensors are still poorly understood.

It is desirable for a lick sensor that does not require the passage of an electric current through the animal to become commercially available. A definite need exists for such an instrument. A photo lick sensor might be a good choice. The sensing head of the device should be easy to install in existing test chambers and might include a (factory-fitted) watering tube. The construction of the instrument should permit adequate recording of licking and drinking behavior with only a minimal constraint on the execution of the response. Developmental research should be guided by a direct comparison of the performance of the new sensor with the output of a conventional one (number of licks, lick duration, and interlick-interval times). Other methods of calibrating lick sensors are discussed by Justesen in Chapter 3 of this book.

References

Allison, J., 1971, Microbehavioral features of nutritive and nonnutritive drinking in rats, *J. Comp. Physiol. Psychol.* **76**:408–417.

Bintz, J., and Zucker, M., 1970, A multifunction pulseforming circuit, *J. Exp. Anal. Behav.* **13**:161–162.

Campbell, B. A., and Masterson, F. A., 1969, Psychophysics of punishment, *in* "Punishment and Aversive Behavior" (B. A. Campbell and R. M. Church, eds.), Appleton-Century-Crofts, Meredith Corporation, New York, pp. 3–42.

Campbell, B. A., and Teghtsoonian, R., 1958, Electrical and behavioral effects of different types of shock stimuli on the rat, *J. Comp. Physiol. Psychol.* **51**:185–192.

Cone, A. L., and Cone, D. M., 1973, Variability in the burst lick rate of albino rats as a function of sex, time of day, and exposure to the test situation, *Bull. Psychon. Soc.* **2**:283–284.

Corbit, J. D., and Luschei, E. S., 1969, Invariance of the rat's rate of drinking, *J. Comp. Physiol. Psychol.* **69**:119–125.

Crawford, M. L. J., 1970, Shock-avoidance and shock-escape drinking in rats: Rate of licking, *Psychon. Sci.* **21**:304–305.

Davis, J. D., 1973, The effectiveness of some sugars in stimulating licking behavior in the rat, *Physiol. Behav.* **11**:39–45.

Eisman, E., 1969, Monitoring drinking: A method compatible with the recording of electrophysiological phenomena, *Behav. Res. Methods Instrum.* **1**:300–301.

Hill, J. H., and Stellar, E., 1951, An electronic drinkometer, *Science* **114**:43–44.

Hulse, S. H., 1967, Licking behavior of rats in relation to saccharin concentration and shifts in fixed-ratio reinforcement, *J. Comp. Physiol. Psychol.* **64**:478–484.

Hutcheson, J. S., and Mills, K. C., 1974, A compact and inexpensive drinkometer for use with small animals, *Physiol. Behav.* **13**:179–181.

King, N. W., 1969, Effects of low-level microwave irradiation upon reflexive, operant, and discrimination behaviors of the rat, doctoral dissertation, University of Kansas.

King, N. W., Justesen, D. R., and Simpson, A. D., 1970, The photo-lickerandum: A device for detecting the licking response, with capability for near-instantaneous programming of variable quantum reinforcement, *Behav. Res. Methods Instrum.* **2**:125–129.

Marowitz, L. A., and Halpern, B. P., 1973, The effects of environmental constraints upon licking pattern, *Physiol. Behav.* **11**:259–263.

Martonyi, B., and Valenstein, E. S., 1971, On drinkometers: Problems and an inexpensive photocell solution, *Physiol. Behav.* **7**:913–914.

Oatley, K., and Dickinson, A., 1970, Air drinking and the measurement of thirst, *Anim. Behav.* **18**:259–265.

Pilgrim, F. J., 1948, A simple electronic relay for counting, timing, or automatic control, *J. Psychol.* **26**:537–540.

Ramsay, D. A., Knapp, J. Z., and Zeiss, J. C., 1970, Transients in "constant-current" generators, *Behav. Res. Methods Instrum.* **2**:122–124.

Segal, E. F., and Oden, D. L., 1969, Effects of drinkometer current and of foot shock on psychogenic polydipsia, *Psychon. Sci.* **14**:13–14.

Slangen, J. L., and Weijnen, J. A. W. M., 1972, The reinforcing effect of electrical stimulation of the tongue in thirsty rats, *Physiol. Behav.* **8**:565–568.

Vrtunski, P., and Wolin, L. R., 1974, Measurement of licking response execution in the rat, *Physiol. Behav.* **12**:881–886.

Weijnen, J. A. W. M., 1972, Lick-contingent electrical stimulation of the tongue; its reinforcing properties in rats under dipsogenic conditions, doctoral dissertation, University of Utrecht (University Microfilms, Ann Arbor, Mich. Order No. 76-6040).

3

Classical and Instrumental Conditioning of Licking: A Review of Methodology and Data

Don R. Justesen

1. Introduction and Overview

One night several years ago I was part of an audience that was being entertained by the comedian Shelley Berman. During the delivery of a monologue by which he was regaling the audience with an impression of a beleaguered child psychologist, Berman suddenly interjected a comment about experimental psychologists: ". . . you know, those people who study rats for reasons of their own!" I confess that the feelings he invoked in me then about the sometimes seemingly abstruse activities of my discipline were later aroused again by Doctors Weijnen and Mendelson when they asked me to write a chapter on the licking behaviors of rodents. Subsequently, after reading more than 300 papers on licking that have appeared in the biopsychological literature during the past 25 years, I came to realize that the seeming abstruseness of the subject matter actually lay in the eye of the beholding psychologist.

The rodent's licking response is a microcosm of adaptive behavior. Perhaps in no other bit of behavior does one see such a close functional and

Don R. Justesen ● Laboratories of Experimental Neuropsychology, U.S. Veterans Administration Hospital, and Department of Psychiatry, Kansas University Medical Center, Kansas City, Kansas.

temporal juxtaposition of the triad of activities without which survival of the complex organism would be impossible: reflexive, instrumental, and consummatory responding. From the perspective of the thirsty rat in its home cage in the laboratory, its approach to the spout of a water bottle is instrumental in the obtaining of water, as is the tongue that extends to the orifice of the spout. The instrumental (operant) lick is quickly followed by an orderly succession of licks at a rate of six to seven per second; the succession of licks is a reflex *elicited* by—is respondent to—water that contacts the tongue. Both the operant lick and the respondent series of licks that follow are consummatory maneuvers.

This "genetic unity" of operant, respondent, and consummatory activity characterizes the rat's licking response (cf. King, Justesen, and Simpson, 1970, with Ferster and Skinner, 1957, p. 7), but what of higher mammals, including man? As Halpern indicates in Chapter 1, *Homo sapiens* and other primates as consumers of fluids are classified as "suckers," not "lickers." Does this mean that consummatory licking is an evolutionary rung no longer clung to by the highest primate? Not at all. According to Eibl-Eibesfeldt (1970), the human neonate does rely on oral suction for the ingestion of milk during its first few weeks of life. But subsequently, suction gives way to a form of pressure licking in which the tongue is alternately elevated to and then withdrawn from the nipple in a sustained, rhythmic sequence. One can readily observe that the older infant in nursing at the bottle or the breast is consuming fluid while the corners of the mouth are open—a feat that would not be possible if oral suction had not been replaced by pressure licking by the tongue. As the human being passes from infancy to childhood, he or she again comes to rely on an oral vacuum—sipping or sucking—in the ingestion of fluids, but the licking response is never entirely abandoned. In particular, repetitive rhythmic movements of the tongue are often seen in young children during the development of speech. It is currently a moot question whether a sustained rhythmic licking of fluids akin to that observed in the rat will occur in the mature human being who is not permitted to sip from a glass or suck from a straw, but the data on guinea pigs (Alvord, 1968; Alvord, Cheney, and Daley, 1971) augur the possibility. Alvord and his colleagues observed that guinea pigs suck from the same kind of waterspout that engenders licking in the rat but that the animals are readily trained to lick by the interposition of a shield with a small aperture between an animal and the source of water. After such training, the cumulative records of the guinea pig's licking to intermittent (fixed-ratio) schedules of reinforcement are almost undistinguishable from those of the rat (cf. Alvord *et al.*, 1971, with Justesen, Levinson, and Daley, 1967).

In the greening of the adult years of the human being, one has good reason to suspect that the tongue lick often comes to play a prominent role in another form of consummation. The data of Masters and Johnson (1966, 1970) suggest that the tongue is more often than not a better instrument than the penis in promoting the orgasmic response in the female, for whom "the most important thing is to maintain a steady uninterrupted rhythm of [lingual] stimulation [of the clitoris]" (Kathy Kelly, unpublished, cited in Barbach, 1975, p. 166). Viewed from an ontological and sociological perspective, the human licking response is one that earlier provides nutritive consummation for the individual and later, as befits a happy ending, often progresses to a role in the sexual consummation of another person.

However much the licking response is retained as an operative or latent component of the human being's behavioral repertoire, its expression in the rat and other mammals has provoked the scientific curiosity of many psychologists. In this chapter, I shall review published fruits of this curiosity as regards, especially, the respondent and operant characteristics of the rat's licking response to water and to other aqueous solutions. Largely excluded in this emphasis is a large number of related reports on airlicking, cold licking, and licking to electrical stimulation of the tongue. These topics receive cursory mention by me but are covered in depth in Chapters 4 and 5.

Because there is considerable variation in the terms that have been used by different investigators to denote different components of the tongue lick in studies of respondent and operant conditioning, I shall adopt for the sake of clarity and consistency the terminology of King *et al.* (1970). A major distinction is made, first of all, between *integrant* and *adjunct* licking. The former is exemplified by those instances in which the tongue is *directly responsible for* in the ingestion of fluid, as is the case, for example, when a feral rat licks water from a pond or a laboratory rat licks from a cup or from a drinking spout that is continuously replenished with fluid from a reservoir such as a water bottle. The adjunct case is that in which licking is detected by one means and reinforced by another; Hulse (1960), for example, electrically detected the contact of the rat's tongue with a flush brass sensor that was situated just below an orifice of a plastic tube from which a droplet of water could be expelled into the subject's mouth by an appropriately programmed electromechanical pump. With continuous reinforcement, where each lick is reinforced by liquid, there may be little difference[1] between the integrant and adjunct cases but only in the latter case is it possible to reinforce licks on an intermittent or delayed basis, i.e., every nth lick or after n (milli-)seconds of time. Also exclusive to the adjunct case is the capability of programming negative reinforcement (cf. Williams and Teitelbaum, 1956, with Crawford, 1970) or positive reinforcement (Mac-

---

[1] One virtually unavoidable difference is the ~15 or more milliseconds in delay of reinforcement in the adjunct case. See Section 3.2.1 for discussion.

Donald, 1972) of the tongue lick without requiring concomitant and continuous ingestion of water. Integrant licking of fluids in conjunction with adjunctive presentations of food pellets is a special compound case that is currently identified as *schedule-induced polydipsia* (SIP). It is because the term *adjunctive* is so strongly identified with SIP methodology that the back-formation *adjunct* is used by King *et al.* (1970) and by me to differentiate simple "dry" licking from integrant "wet" licking.

A distinction is also made between *local* and *global* rates of licking. Local rates are those that characterize a period of uninterrupted, continuous licking. Global rates are those identified with relatively long periods of licking, from minutes to hours, during which an animal might spend considerable time in activities other than licking. Finally, a distinction in the adjunct case is necessary in comparisons of local rates of licking *immediately before and after* reinforcement by a fluid. A drop of water that contacts the tongue *elicits a burst* of licks the rate of which is higher than the local rate before reinforcement (cf., e.g., Hulse and Suter, 1968, 1970, with Hulse, 1966). In this chapter, the term *burst rate* is synonymous with the local rate of the respondent licking that occurs to and is measured immediately following contact of the tongue with fluid.

The term *reinforcement* has connotations from both classical and instrumental conditioning, and both of the partially overlapping sets of meanings are applicable to lingual conditioning. It is difficult to avoid the confusion that may arise from equivocation of nonoverlapping meanings, since there are both pure and mixed cases of each set in the experimental literature on licking. Perhaps by reading examples of each, the reader may make the necessary distinctions. The case of pure respondent conditioning of the tongue lick is found, for example, in the study of DeBold, Miller, and Jensen (1965), who implanted rats with a chronic intraoral cannula that dispensed water directly onto the tongue. The dispensing of the water was always under control of the investigator; that is, it was not made contingent upon licking. The case of pure operant conditioning of the tongue lick is found in MacDonald (1972), who trained rats to lick at a contact sensor for positive reinforcement by water from a remote dispenser; that is, licks were detected by a "dry" waterspout, and receipt of a water reinforcer from another, spatially removed spout was made contingent upon them.

In the great majority of the published reports on licking behavior, the mode of reinforcement is a mix of classical and instrumental cases. For example, Justesen *et al.* (1967) observed rats responding to a fixed-ratio schedule in which each series of 49 "dry licks" at an electrical contact-sensor was succeeded on the 50th lick by a reinforcer of sugar water that was posited directly on the tongue. The reinforcer in this case was both operant and respondent in nature. While ostensibly the instrumental reinforcer of the rat's operant licking, the sugar water, upon contacting the tongue, unfailingly *elicited* a burst of licks. Ordinarily a rat performing under such a contingency would be characterized as responding to a "fixed-ratio-50" (FR-50) schedule of reinforcement; however, if 10 of each series of 50 licks were elicited, the schedule veridically would be that of FR-40. Because there are both operant and respondent components of high-demand, intermittent schedules of aqueous reinforcement of the tongue lick, I shall refer to them as *nominal* schedules. For example, for the case just cited, the rats performed on a *nominal* FR-50 schedule. The case of continuous reinforcement (CRF), in which all licks but the initial lick are both respondently and operantly reinforced by an aqueous solution, is particularly confusing, since a nominal FR-10 schedule may veridically be a CRF schedule. In anthropomorphic terms, for the sake of illustration: the rat approaching a water dispenser that reinforces every tenth lick may "volunteer" the first lick, then find that nine more licks occur automatically in rapid order. While one might make a logical argument that nominal CRF is actually FR-*1/10* or FR-*1/8*, etc., it seems best to retain a conventional nomenclature of scheduling that is prefixed by the term *nominal*. Summary descriptions of integrant and adjunct licking in terms of global and local rates of responding are given in Figure 1.

	Global rate *of responding* (in licks/sec)	*Local rate* *of responding* (in licks/sec)
Integrant licking	Averaged rate of licking at a waterspout for direct ingestion of fluid by each lick over lengthy periods that may include intervals of nonresponding. Operant and respondent licks are inseparable because each lick after the first is elicited by fluid.	Averaged rate of an uninterrupted series of licks. Operant rates and respondent rates are inseparable because each lick is reinforced by a quantum of fluid that would in isolation elicit a burst of 3–10 or more licks.
Adjunct licking	Averaged rate of licking at a contact sensor for continuous or intermittent, near-immediate or delayed reinforcement by fluid, food pellet or brain stimulation, etc., over lengthy periods that may include intervals of nonresponding. In high demand, intermittent schedules of fluid reinforcement there are both purely operant licks (after an elicited burst is run-off) and respondent licks (immediately after aqueous reinforcement of a prior lick).	Averaged rate of an uninterrupted series of licks. Given highly intermittent schedules of aqueous reinforcement, the local respondent-rate (burst-rate) is that which occurs immediately after reinforcement. The local operant rate is that which occurs after the elicited burst has been runoff but before a subsequent reinforcer is presented.

Figure 1.   Examples of global and local rates of responding for the case of integrant (direct or "wet") licking and adjunct (indirect or "dry") licking. Note that experimental isolation of operant and respondent licks can take place only in the adjunct case.

My original aim in setting out to write this chapter was to summarize the data on classical and instrumental conditioning of licking behavior. I expanded that aim to include methods of detecting and reinforcing the lick response when it became apparent from scrutiny of the literature that licking device and licking datum are so closely bound together that a sensible accounting of one necessitates description of the other.

2. Methods of Detecting and Reinforcing the Lick Response

2.1. Techniques of Measurement

2.1.1. Electrical Detection

If licks are to be detected in a reliable and accurate manner, a means is needed to measure contact of the tongue with the fluid to be drunk or with a contact sensor through which programming of liquid or some other reinforcer can be arranged. The first data on electrically detected licks were reported by Stellar and Hill (1952). The investigators placed a small voltage from a dc (direct-current) supply between the wire-mesh floor of a rat's cage and a water reservoir that was intubated with a drinking spout. When a rat's tongue contacted the water in the orifice of the spout, a flow of electrons (or ions) of about 0.3 μA would pass from the mesh floor (and through the rat) to the spout and thence via a conductor to the control grid of a thermionic valve that, in the wake of a reduced anode current, caused a relay in the anode circuit to open. Switch contacts on the relay permitted a kymographic marker to be activated on the occasion of each make-and-break contact by a rat with the waterspout.

While the thermionic valve has given way to the transistor, and the kymograph to more modern devices for recording and displaying licking responses, the basic methodology of electronic lick-detection of Stellar and Hill has persisted and is still the most widely used today. Variations of the theme of electrical detection have been developed by DeBold *et al.* (1965), who implanted chronic lick-sensors in the oral cavities of rats; by Deaux and Patten (1964), who fitted the rat with a removable detector–reinforcer harness that gives the animal the appearance of a diminutive baseball umpire (see Figure 2); and by Hulse (1960, p. 1), who employed an ac (alternating-current) detector to avoid "ionization at the tongue."

Slangen and Weijnen (1972) discovered that rats will lick a spout for weak electrical stimulation of the tongue *as the sole reinforcer*, and Weijnen (1972) found that both ac and dc at 10 μA are positively reinforcing. It is doubtful that dc $\leq$10 μA would irreversibly injure the tongues of rats by

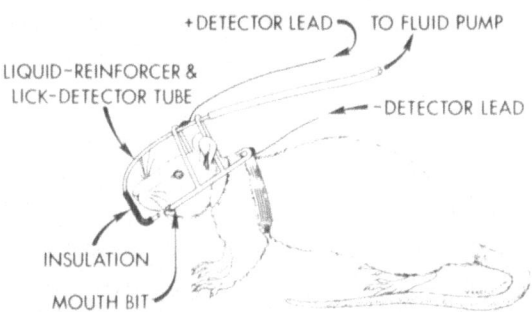

Figure 2.   Lick-detection and lick-reinforcing apparatus by which a discrete quantity of water as an unconditional stimulus for licking can be presented to a rat—and unavoidably so. Licks are electronically detected and are respondently reinforced via a small aluminum tube that addresses the animal's mouth. (Redrawn, with permission, from Deaux and Patten, 1964.)

polarization or electrolysis because of the rapid make-and-break of tongue contact during licking, but the findings that low intensities of ac and dc are detectable and that both are positively reinforcing raise an issue that would delight the disciples of Werner Heisenberg: *the flow of electrons (ions) that is used to detect licking may modify the licking response by acting as a perceptible or otherwise perturbing stimulus.* While preliminary studies by Weijnen (1972) have revealed that rats do not express a preference between electrified and nonelectrified water (i.e., water with and without a lick-sensing current), it is possible that the two are discriminable—and more, that thresholds of taste discriminability of weak solutions may be lowered or contaminated by sensing currents, ac or dc (see Martonyi and Valenstein, 1971).

2.1.2. Acoustic Detection

To combat potential behavioral artifact from electrosensing by the tongue, alternate modes of lick detection have been developed. A rather ingenious device that employs acoustically transduced detection of the lick—when, specifically, the metal ball in a balltype waterspout, after displacement by the tongue, returns to its seat with a "click—was used by Schaeffer and David (1973b) in a study of rabbits. The authors attached a small microphone to the drinking spout that converted acoustic signals to electrical pulses. A subsequent study of gerbils by the same authors (1973a) was performed with a conventional waterspout and electrosensing, which may indicate that the reduced mass and thrust of the small animal's tongue may not lend its licking response to the ball-displacement technique of acoustic detection. Another potential problem would be adventitious triggering of the acoustic transducer by ambient noise. Potential problems notwithstanding, further development of the acoustic sensor—perhaps by auscultation via a miniature microphone on the animal's throat—may provide resolution of problems inherent in and peculiar to electrosensing.

2.1.3. Photodetection

Another alternate to electrosensing is photodetection. King *et al.* developed (1970) and extensively tested (Justesen and King, 1970; King *et al.*, 1971; Justesen *et al.*, 1971) a device in which the breaking of a tiny beam of light by the tongue at a small aperture provided for detection and for fluid reinforcement of operant and respondent licks (see Figure 3). Martonyi and Valenstein (1971) later independently developed essentially the same technique of detection.

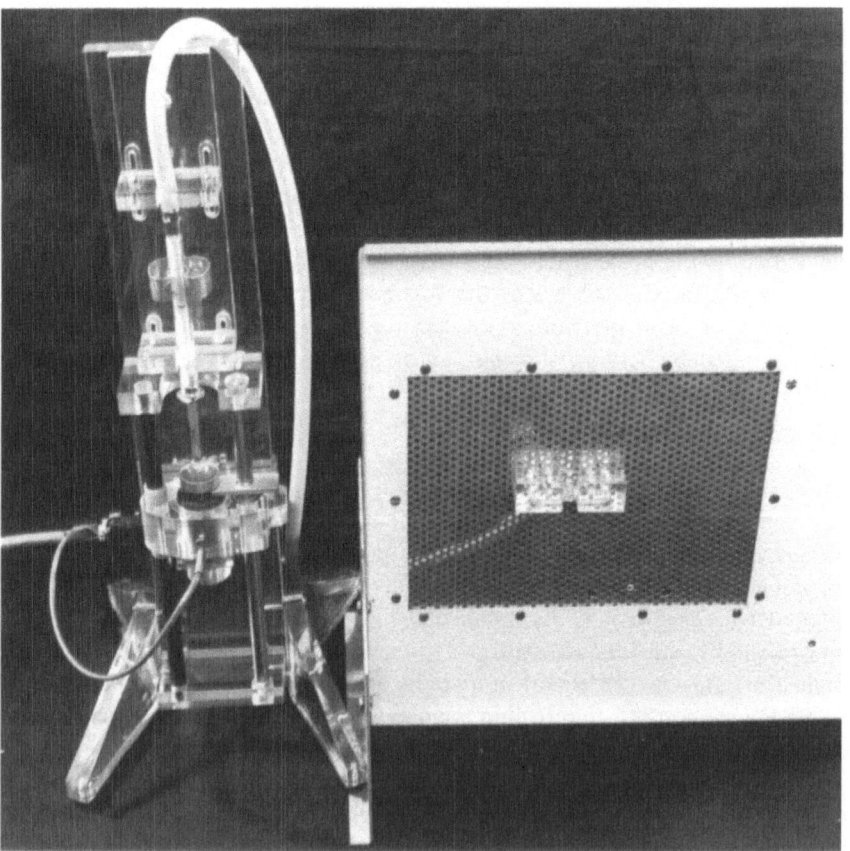

Figure 3A.   Photograph of a photo-operandum, liquid-feeder ensemble. (Adapted from that of King *et al.*, 1970).

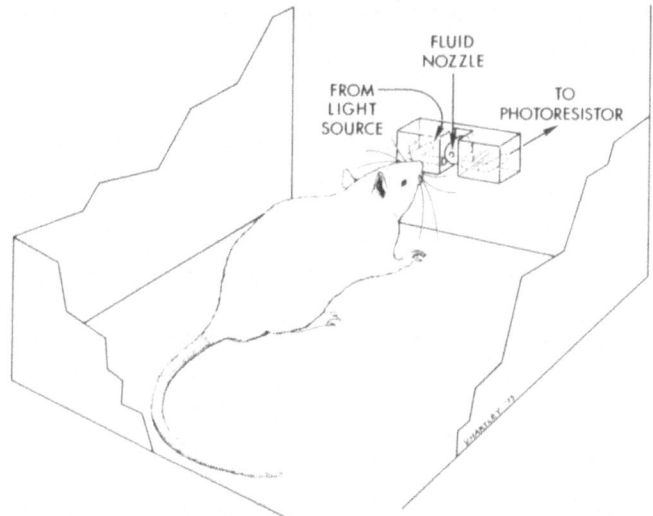

Figure 3B.   Artist's rendering of detail of the photo operandum.

Photodetection of the tongue lick offers demonstrable promise as a sensing technique but does have a potential drawback. The rat and many other licking animals are photophobic, and several accounts have been reported of inhibition of licking by ambient illumination (cf. Cone, 1974, and Cone and Cone, 1973, with Fitzsimons, 1957, and Weijnen, 1972). To the extent that light from a photodetector illuminates the animal, global rates of responding (especially of operant licking) may be reduced. The remedies are efficient shielding of the source of light, the minimum of luminous flux necessary to provide efficient lick-sensing, and probably most efficacious, the use of an infrared source of illumination to which the rat is insensitive both visually and thermally (see Hutcheson and Mills, 1974).

2.1.4. Licking-Evoked Potentials

To my knowledge, neither myogenic nor neurogenic electrical potentials as evoked by licking or associated with its efferent input have been reported in behavioral studies of the tongue lick. Because of the ever-present probability of undetected licks or of spurious contacts of the tongue with a sensor (e.g., "double-tonguing"—detection of two contacts in a single cycle of licking as the tongue touches first one and then the other margin of a waterspout), it would be useful to develop a technique of detecting licks from the lingual musculature, from its innervating nerves, or from nuclei in which the nerves originate. Because a muscle sensor on or in the tongue may interfere with articulation of licking movements (shades again of Heisenberg!), it would appear more feasible to attempt through stereotaxic surgery to place chronically indwelling electrodes in proximity to intracranial structures from which "licking signals" can be obtained. The toil and bother of producing such a preparation would probably militate against its use as a methodological standard for the study of licking behaviors, but it would be useful in *calibrating* the accuracy of sensing techniques that are standard. One could, for example, observe for local and global rates of licking via a conventional detector in each of a number of rats, then implant each animal with chronic recording electrodes to viable target areas (to, e.g., the hypoglossal nucleus or nerve), and then, following postoperative behavioral tests to ensure equivalence with preoperatively measured rates of licking, proceed to determine accuracy of the conventional systems of detection. The achieving of some sort of *intrinsic neurogenic signal* that varies one-to-one with the licking cycle would not only resolve the question of accuracy of the conventional sensor but may be crucial to resolution of the controversy over the invariance of the rate of the licking response, a controversy I shall address in a later section of the chapter.

2.2. Reinforcing Techniques

2.2.1. The Cup

Only a few investigators have utilized the cup (including the dipper [Wyckoff, Sidowski, and Chambliss, 1958] and spoon [Braud and Prytula, 1969]) in studies in which licking has been detected by contact sensing with the fluid contents of the cup. Schaeffer and Huff (1965) used electrical sensing of licks by kittens and cats that lapped milk or water from a cup and then from a spout. Rates of licking ranged from 3 to 5/sec but did not differ reliably as a function of source of liquid. Schaeffer and Premack (1961) found that neonatal rats during initial exposure to cup drinking licked at a local rate of 5.3 to 11.4 licks/sec and averaged 7.9 licks/sec. The possibility of spurious contacts (e.g., of the rats' noses with the water) is suggested by the highest rates of licking but may also reflect "microbehavioral" adjustments (cf. Allison, 1968, 1971) by an immature and experientially naïve animal. If, for example, the animal initially thrust its nose close to the surface of the water, thereby producing a larger volume of water intake per lick, the burst rate would be quite high in keeping with the observations of Hulse (1966) and of Hulse and Suter (1968, 1970). Subsequently, if the animal withdrew slightly from the cup, both volume of water per lick and rate of bursting would decrease. The apparent comparability of the cat's rate of licking from a cup and at a spout cannot of course be generalized to the rat, but it would be of interest to determine whether neonatal rats lick from the cup and the waterspout at comparable rates. I would predict that rates of cup licking would be higher, as would volume of water per lick, since the relatively small orifice of the spout would be likely to reduce volume of intake per lick.

2.2.2. The Waterspout

The waterspout—typically of a hollow glass or stainless steel cylinder that intubates a water bottle—provided the first means (Stellar and Hill, 1952) and remains the predominant means of reinforcing the licking response in the laboratory. The use of a glass spout or an externally insulated metal spout (see, e.g., Pierson and Schaeffer, 1975) would help control for "double-tonguing" of the electrical contact-sensor but creates the possibility that an air bubble or the retreat of meniscus in the orifice of the spout may sporadically insulate the tongue electrically during a licking movement, thereby precluding detection of the lick.

There are several additional factors that could and probably do affect the rate of licking at a conventional drinking spout. Hulse and Suter (1970) found by a special "one-drop" technique that as the volume of a drop of water that contacts the rat's tongue is increased, the number of elicited licks and the burst rate are both increased (see, too, Snyder and Hulse, 1961). If the rate and the duration of a burst of licks increase as a function of volume of the aqueous reinforcer, then any factor that produces variation of volume should increase the variance of both global and local rates of respondent licking. One would strongly suspect, then, that the diameter of the spout and especially of the orifice would be controlling factors, a suspicion originally stated by Stellar and Hill (1952) and more recently voiced by Cone (1974). While a majority of investigators do not specify dimensions of spouts, the minority has reported diameters of the orifice of the oft-mentioned "standard drinking spout" that range from 1.0 to 4.0 mm. Another factor that controls volume of fluid per lick is the height of the (closed) column of liquid in the reservoir, the two variables being inversely related (Allison, 1971). Other conditions being equal, as the viscosity of a fluid reinforcer increases, the volume per lick decreases (Allison, 1971), and one may conjecture from the data on airlicking (cf. Hendry and Rasche, 1961, and Oatley and Dickinson, 1970, with Fosset and Treichler, 1971) that the pressure or velocity with which the fluid reinforcer impacts upon the tongue is a controlling variable. Lending indirect support to this conjecture is Davenport's observation (1961) of an inverse relation between burst rate and the mass of the rat's tongue.

The spout of the water bottle that is filled with fluid is of necessity a device for integrant licking. Virtually all licks are wet licks, and all licks are

reinforced—unless an air bubble at the orifice prevents contact and ingestion of liquid or unless the animal moves from the spout, which may prevent
elicited licks from being detected. Because residual fluid may gather at the
orifice and be electrically continuous with fluid inside the spout, contacts by
an animal's nose or forepaws may be registered as licks even with glass or
insulated metallic spouts. Adventitious contacts unrelated to licking can
largely be controlled if the spout is recessed behind a shield with an aperture
(see, e.g., Hulse, 1960; Alvord *et al.*, 1971), but the investigator who wishes
to gain precise control over the volume of fluid reinforcement or who wishes
to study effects of intermittent or delayed reinforcement must turn to other
devices. The remainder of this section is devoted to such devices—those that
subsume as reinforcers of adjunct licking.

2.2.3. Needle Dispensers

Weisman (1965; see also Boice and Denny, 1965) has fashioned a
heavy-gauge (No. 11) hypodermic needle into a water dispenser, the tip of
which is ground and polished until flat and smooth. A small ring of copper
is positioned around and near the end of the needle in such a way that
contacts by a rat's tongue with the ring complete an electrosensing circuit.
A solenoid-operated valve under a head of pressure can be programmed to
eject a submilliliter volume of water from the tip of the needle. Boice (1967,
1968) has employed a variation of Weisman's device by which rats lick
directly at a 17-gauge hypodermic needle that is flattened and smoothed.
The system dispenses 200-μl quantities of water. In a personal communication from Professor Boice in February of 1976, I learned that he had never
observed signs of wear and tear on his animal's tongues, even after
prolonged scheduling with the needle dispensers.

2.2.4. Drop Dispensers

While Weisman's needle-reinforcer system is clearly of the adjunct
sort, he was not the first to develop a device that permits detection of licking independently of programming the presentation of precise volumes of
fluid reinforcement. The original device was developed several years before
by Hulse (1960). Hulse's fluid dispenser (Figure 4) consists of a small channel through the center of a small cylinder of Plexiglas. Just below the channel opening of the cylinder, the lower aspect of which is tapered, is a brass
contact-sensor that is flush with the tapered end. An infusion pump that
drives a remote hypodermic syringe can be programmed to provide a
discrete volume of reinforcing fluid that ranges upward from a microliter.
An important component of Hulse's system is a solenoid-driven lever that
engages and squeezes the rubber tubing that conveys fluid from the syringe

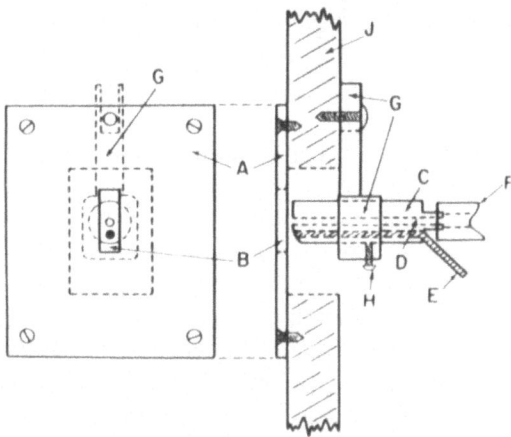

Figure 4. The "one-drop" adjunct detector-reinforcer system of S. Hulse. A = Plexiglas shield. B = aperture through which animal licks at drinking tube. C = Plexiglas drinking tube. D = fluid channel. E = brass rod for making contact. F = pressure tubing to liquid pump. G = adjustable holder and sleeve for drinking tube. H = set screw. J = wall of experimental chamber. (Reproduced with permission of the author and of the *Journal of the Experimental Analysis of Behavior (JEAB)*. From Hulse, 1960. Copyright 1960, the Society of the Experimental Analysis of Behavior, Inc.)

to the Plexiglas channel. The infusion pump and the lever actuate simultaneously, the latter providing by its return stroke from the tubing a slight negative pressure that causes fluid within the channel near the opening to be retracted, thereby preventing residual fluid from contacting an animal's tongue. With his system, Hulse was to achieve a means of presenting a precise quantum of fluid (i.e., water or aqueous solution in the form of a discrete packet or drop) that is posited directly on the animal's tongue so long as the animal maintains a working posture at the detector-reinforcer tube.

Variations on Hulse's theme have been reported by Justesen *et al.* (1967), who used a solenoid valve in place of an infusion pump, and by King *et al.* (1970), who developed a special manifold system with an array of combinatorially operated syringes by which a preselected quantum of fluid through a range of specified volumes can be programmed to reinforce any given lick in an ongoing schedule of reinforcement.

2.2.5. Intra- and Perioral Dispensers

The chronically implanted intraoral dispenser of DeBold *et al.* (1965) was briefly addressed in Section 2.1.1 and is mentioned again as a device— the first to appear in the literature—by which fluids are directly injected into the oral cavity of the experimental animal and licking is detected *in situ.* A major advantage of DeBold *et al.*'s dispenser is that complete control over programming of fluid reinforcers is vested in the investigator, a necessary factor in studies of classical conditioning of lingual behavior.

To avoid the outlay of time and effort and to preclude the possibility of physiological artifact that can occur after any traumatic surgical maneuver, the investigator of licking behaviors has recourse to a dispenser that provides most of the amenities of DeBold *et al.*'s system without the need for surgical intervention. Reference is made to the removable detector-reinforcer system of Deaux and Patten (1964), which is illustrated in Figure 2. A small brass rod serves as a mouth bit that holds the apparatus in place in conjunction with flexible elastic belts that encircle the animal's body, just anterior to and posterior to the forelimbs. Fastened to a wire frame that is continuous with the bit and belts—and forms part of a removable harness— is an aluminum tube (2.4 mm internal diameter), the end of which is located in line with and 3.2 mm from the rat's mouth. Water is expelled from the tube into the mouth and licks are electrically detected by contact of the tongue with the end of the tube. Of necessity, the animal is tethered to the wire that leads to an electronic sensor and by the tubing that leads to the source of fluid, but restraint of the animal in an immobilizing holder—apparently necessary with DeBold *et al.*'s dispenser—is effectively circumvented.

3. Classical and Instrumental Conditioning of Licking Behavior

3.1. Respondent Aspects

3.1.1. Properties of the Unconditional Stimulus

The apparatus of DeBold *et al.* (1965) and of Deaux and Patten (1964), which were discussed in the preceding section, were explicitly designed for classical conditioning of the rat's tongue-lick, that is, as means of dispensing water onto the tongue independently of the experimental animal's disposition to drink. A major question arises in connection with the forced-reinforcement methodology that has bearing on the specification of the physical properties of lick-eliciting stimulation. Is the neurophysiological "oscillator" or "pacemaker" that controls respondent licking so programmed that fluid does not elicit licking responses unless the tongue is extended from the mouth? Since DeBold *et al.* selectively elicited licks *in situ* by direct intraoral administration of water, the answer to the question is in the negative and it becomes apparent that *fluidity* as such is not an adequate stimulus for licking—otherwise the resident fluid of the oral cavity, saliva, would evoke a continuous bout of lingual activity.

Evidence from several lines of research suggests that a necessary if not sufficient property of the lick-eliciting stimulus is *thermal* in nature (cf., e.g., Mendelson and Chillag, 1970a,b, with Riccio, Hamilton, and Treichler, 1966; Carr, Levin, and Dissinger, 1968; and Mendelson and Zec, 1972). Specifically, cooling of the tongue or other oral tissues by a gas, a fluid, or a solid that is 15–20° below the body temperature does act as an operant reinforcer. Because several investigators have successfully brought about both operant and respondent conditioning of the tongue lick in the same experimental subjects and by the same reinforcers (cf., e.g., Weisman, 1965, with DeBold *et al.*, 1965) it is safe to presume that *some* of the stimulus properties that promote instrumental licking are associated with respondent licking. It has been well established, for example, that operant licking to puffs of cool air and to a cold object as well as to water is associated with thirst, that is, that licking is enhanced by a prior period of water deprivation. The only stimulus properties that are common to these reinforcing agents as typically presented in the laboratory are "coolness" and tactility. Tactile stimulation *per se* would not be sufficient—or contact of the tongue intraorally with the palate would trigger licking—but may be necessary along with cooling to evoke the tongue lick.

The observations by Slangen and Weijnen (1972) and by Weijnen (1972) of operant licking by rats to electrical stimulation of the tongue may be taken as evidence that cooling is not an exclusive reinforcing property in the Pavlovian sense or, conversely, that electrical stimulation activates (hypo)thermal receptors. Since Harrington and Linder (1962; cf. with Hartlep and Bertsch, 1974) have found that rats will work for mild electrical shocks to the feet, which are bereft of gustatory receptors, the possibility that positively reinforcing electrical stimulation of the tongue lacks gustatory or thermal equivalence must be entertained. One notes, however, that licking to electrical stimulation of the tongue is thirst-dependent, at least in the rat (Weijnen, 1972). It is unfortunate that the mechanism of gustatory reception of electrical stimuli is still without satisfactory explanation (Bujas, 1971), but it is of more than passing interest that Ogawa, Sato, and Yamashita (1968) found that "a majority of the *chorda tympani* nerve fibres [of the rat and hamster] respond . . . to gustatory as well as thermal stimulations of the tongue" (pp. 238–239). Also of interest are correlations noted by Ogawa *et al.* in neural activity evoked from fibers of the chorda tympani by thermal and chemical stimuli. Cooling of the tongue and applications of quinine and HCl solutions were highly likely to evoke responses from one set of fibers, while other fibers selectively responded to warming and to application of a sucrose solution, thereby providing evidence of a marked degree of stimulus equivalence for two classes of receptors.

On the basis of evidence that at best is spotty, I would hazard the guess that the lingual or oral receptors that are sensitive to *hypo*thermal stimulation are those that participate in the triggering of elicited licking. Granted that sucrose solutions are confirmed as strongly reinforcing of respondent as well as operant licking, such solutions are generally presented at the ambient temperature of the experimental surround, i.e., 15°C or more below mammalian body temperature. By my line of reasoning, it is the activation by cooling of one set of fibers (and not activation of another set by warming or by a sugar or saccharin solution) that, through a neural pacemaker, initiates the respondent burst. Comporting with this notion are findings reported by Hulse (1966, 1967) and Hulse and Suter (1968), who observed that rates of respondent licking by rats to differing concentrations of a given volume of saccharin, sucrose, and saline are essentially invariant. Davis and Keehn (1959) independently confirmed Hulse's findings in studies in which rats were tested with differing concentrations of saccharin, saline, and sucrose. Goodrich (1960) and Schaeffer and David (1973a) also reported data indicating that the concentration of the sucrose solution does not affect the rates of integrant licking of, respectively, rats and gerbils.

Cone, Wells, Goodson, and Cone (1975) have published the only data that appear to contravene the proposition that the local integrant rate is independent of the sucrose concentration of the reinforcing solution. These authors found that the averaged local rate of licking (by rats from a waterspout) increased from ~6.25 to ~7.25 licks/sec as the concentration of sucrose in an aqueous solution increased from 0% to 8%. At 16%, licking rates fell to an average near 6.85. Perhaps, as Cone *et al.* suggested, the data of Davis and Keehn (1959) and of Schaeffer and David (1973a) would have reflected concentration dependencies if appropriate statistical analyses had been performed on licking rates. Hulse's work, however, was performed without at least two potentially confounding sets of factors that have been present in Cone *et al.*'s study and indeed in all studies of integrant licking. First is the absence of control over the volume and frequency of reinforcement—both have powerful controlling influences on lick rate—and second is the absence of control over the temperature of the reinforcing solution. The rat that is licking continuously at a contact sensor may impart to it a considerable amount of caloric energy (see Weijnen, 1972); a rise of spout temperature, in keeping with the notion of thermal dependency, would reduce the potential for eliciting respondent licks, which have a higher local rate than their operant counterparts. By using a dispenser of low enthalpy (low thermal capacity) and by deploying a system that withdraws fluid from the end of the delivery channel after the reinforcer is expelled, thereby minimizing transfer of thermal energy from animal to the reinforcer solution, Hulse (1960) knowingly or unknowingly controlled to some extent for thermal artifact.

It could be argued that the concentration-dependent changes in rates of licking observed by Cone *et al.* are vindicated by the Law of Error—that there is no reason to suspect that either a volumetric or a thermal artifact is systematically related to the concentration of the reinforcer. The counterargument is that the volume of the reinforcer solution exerts strong control over the operant rate of licking, as revealed, for example, in studies by Hulse (1967) and by Hulse and Suter (1968). Postural attitude and distance from the waterspout are components of the total instrumental act, and both may change as the stimulus quality of a reinforcer is varied. If a reinforcer of higher palability resulted in a closer stance to a waterspout by a rat, a larger volume per lick would result with a commensurately higher rate of elicited licks (see Wilson and Barboriak, 1970). Too, a greater volume per lick would lessen the amount of warming of the spout by the licking animal because of an increased rate of flow of water through the drinking tube.

The prescription for resolving the question of a possible concentration dependency of the respondent lick-rate and, indeed, for clarifying much more that is not known about the unconditional stimulus for licking does not lie, of course, in speculation but in adequate instrumentation, in pervasive experimental control, and in systematic investigation. The work during the decade of the 1960s by Stewart Hulse of the Johns Hopkins University provides a rather singular model from this threefold perspective of investigative adequacy, but only one hard conclusion can be drawn from Hulse's data about properties of the unconditional stimulus for respondent licking. As mentioned several times previously, the volume of the reinforcer is positively related to the burst rate (Hulse, 1966, 1967; Hulse and Suter, 1968, 1970; and Snyder and Hulse, 1961). Figure 5, which is reproduced from Hulse and Suter (1970), presents data on durations of lick contacts and on interlick intervals for respondent bursts of 3–8 licks by rats as a function of "single-drop" reinforcers of water of 4-, 12-, or 20-μl volume. The burst rate ranges between 7 licks/sec to a 4-μl reinforcer and 8 licks/sec to a 20-μl reinforcer. While the range of rates of bursting is small (less than 15%), so is the range of the volumes tested by Hulse and Suter. The total range of effectively reinforcing volumes reported in the literature on adjunct licking extends from 1 μl (Hulse and Firestone, 1964) to 200 μl (e.g., Deaux and Patten, 1964). It is reasonable to assume that the rate of bursting would range more widely in consequence of more than two orders of magnitude in difference of reinforcer volume, but the upper and lower limits have not been established systematically in a given experimental setting.

The stimulus properties of aqueous reinforcers that control or are

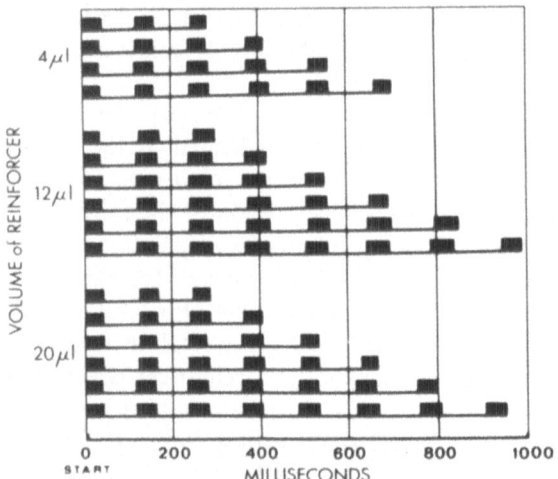

Figure 5. Interlick intervals (ILIs) and lick-contact durations by rats. Licking was elicited by one of three different volumes of water, each presented as a discrete, "one-drop" reinforcer. Careful scrutiny reveals that while ILIs decrease, contact periods *increase* as a function of the increasing volume of the reinforcer. The increase may not be a reflection of increased duration of contact *per se* between tongue and sensor but of a larger volume of water that bridges more space between tongue and sensor. The lick rate and the median number of licks in an elicited burst also increase as a function of reinforcer volume. (Reproduced, with permission, from Hulse and Sutter, 1970.)

suspected of controlling the local rate of licking have been addressed frequently in this chapter and can be summarized as:

a. *Volume*. A positively monotonic relation exists between the volume of the aqueous reinforcer and the burst rate within systematically tested limits.

b. *Concentration*. The concentration of sucrose, saccharin, or salts in solution has essentially no effect on burst rate in tests in which a single reinforcer is used to elicit a burst of licking; however, the lick rate may be concentration-dependent when the rat is continuously reinforced, at least in the case of integrant licking.

c. *Temperature*. By inference from studies of cold- and airlicking, the burst rate may be quantal or a continuous function but is probably inversely related within limits to the temperature of the oral cavity.

d. *Viscosity*. The burst rate appears to be inversely related to the viscosity of the reinforcer solution.

e. *Tactility*. Since application of *any* reinforcer to the tongue— whether solid, liquid, or gas—has a high probability of acting as a tactile stimulus, the question of tactility as a necessary property may be experimentally moot; the question of sufficiency is testable (see Section 3.2.1) but has not been resolved experimentally.

Other physical properties of the reinforcer solution (e.g., specific gravity, specific heat, conductivity, and pH) as it impacts upon the tongue and other intraoral tissues may undoubtedly influence the lingual burst rate, but these have not been evaluated by studies that have appeared in the literature.

3.1.2. The Unconditional Response

The unconditional response (UR) in traditional laboratory practice is defined as some *reflexive* element of autonomic or skeletal behavior that is easily and regularly evoked by application of the appropriate unconditional stimulus (US). That the licking of rats and some other small mammals is reflexive is apparent from three converging lines of evidence. First, direct application of an appropriate US such as water to the tongue evokes a burst of licks even in animals that are satiated with water (DeBold *et al.*, 1965). Second, the neonatal animal exhibits bursts of licking when first given access to water (cf. Schaeffer and Premack, 1961, with Schaeffer and David, 1973a, and Pierson and Schaeffer, 1975). And third, the temporal character of respondent or "wet" licking as compared with operant or "dry" licking differs considerably as revealed by studies of intermittent schedules of aqueous reinforcement (e.g., see King *et al.*, 1970) or by studies with the "one-drop," single-reinforcer technique (e.g., see Hulse and Suter, 1968). Not only is the rate of elicited licking higher, but the expected decrease in the interlick interval, which accounts for 60–70% of the total operant-licking cycle in Hulse's procedure, is initially associated with a modest *increase* in the time that the tongue remains in contact with the source of fluid (Hulse and Suter, 1968). By careful scrutiny, the reader can visualize in Figure 5 these alterations in the time course of elicited bursts.

Hulse and Suter have made more of the datum of *burst length* than I do—they note, for example, that a larger volume of aqueous reinforcement begets a higher median number of *recorded* elicited licks per burst than a smaller volume—but their response-detection system is insensitive to licks that may occur after an animal backs away or raises its head from the dispenser. Given a device that unobstrusively counted all licks, one would almost be certain to find within limits that the averaged number of respondent licks to a single reinforcer is a monotonic function of the reinforcer volume. My reservation about Hulse and Suter's burst-size datum is solely its quantitative accuracy.

When treated as an operant, the elicited burst of licking of the rat preserves its respondent character even after prolonged periods of training;

that is, elicited licking appears to be immune to rate modification by scheduling. Pierson and Schaeffer (1973) were successful in training rats to lick water from a spout in order to gain access to an activity wheel. However, differential reinforcement of higher rates of licking—attempts to increase the average rate of integrant licking through contingent application of an established, higher-order operant reinforcer—were futile. Perhaps related to this intransigence are data that have been reported by several investigators. The local rate of elicited licks exhibits a small but reliable decay within discrete bursts; too, subsequent bursts within a session exhibit small successive declines in rate. These decay phenomena occur in adjunct licking (Snyder and Hulse, 1961; Hulse and Suter, 1970) as well as integrant licking (Allison and Castellan, 1970; Cone, 1974) and may be related to microbehavioral adjustments by the licking animal (Allison, 1971) or to physical factors such as increasing temperature of the reinforcing solutions. It is common practice to draw fresh water from a tap just before the commencement of an experimental session or to store sucrose or saccharin solutions in a refrigerator until time of use. In either case, the temperatures of reinforcing solutions would be rising 10–20°C or more across time from the start of a session until the ambient level of the surround were reached. Whatever the source(s) of decay of burst rate across time, such a decay within an experimental session would continuously reduce the probability of a higher rate of licking that is available for differential reinforcement.

Apparent success in limiting rats to a single integrant lick was reported by Teitlebaum and Derks (1958), who made escape or avoidance of painful foot-shock contingent upon a single licking response. Not clear from the report, however, is whether there were single licks as such or merely single *recorded* licks; that is, their rats while still licking may have withdrawn their heads from the waterspout after initial contact.

3.1.3. Acquisition, Extinction, and Spontaneous Recovery of Conditional Licking Responses

Classical conditioning of licking has been demonstrated by several investigators, for example, Weisman (1965), Boice and Denny (1965), DeBold *et al.* (1965), Miller and DeBold (1965), and Patten and Deaux (1966). The pioneering study by Weisman is illustrative. Illumination of a 10-W incandescent lamp served as the conditional stimulus (CS). The CS was presented for 5 sec on each trial and overlapped presentation of the water US from a needle dispenser by 2 sec. Aperiodic intertrial intervals were used to control for temporal conditioning. After 13 daily sessions of pretraining and tests for sensitization and pseudoconditioning (see Figure 6), a total of 8 formal sessions of classical conditioning (40 CS–US pairings per session) was conducted, followed by 1 session of experimental extinc-

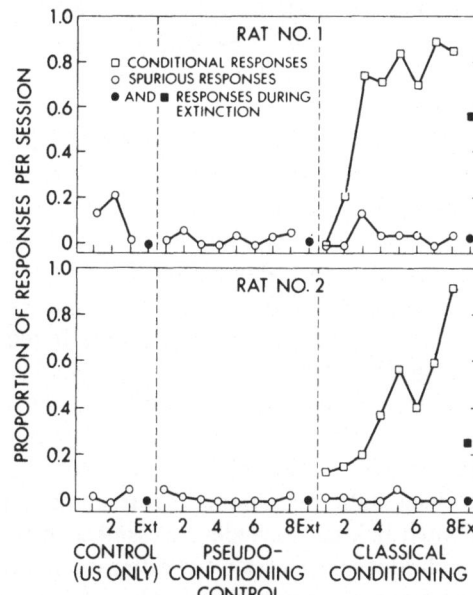

Figure 6. Data on classical conditioning of the licking response for each of two rats. A photic conditional stimulus was paired with a water unconditional stimulus during conditioning trials. (Redrawn, with the author's permission, from Weisman, 1965.)

tion. The criterion of a conditional response (CR) on a given trial was the occurrence of licking during the first 3 sec of CS illumination, provided there was no licking during the 3-sec period just before presentation of the CS. The individual data of two rats are shown in Figure 6. While both rats required more than 160 CS–US pairings before exhibiting CRs on more than 80% of the trials of a daily session, classical conditioning was unequivocally demonstrated. Resistance to extinction, considering the lengthy course of acquisition, was relatively modest. The percentage of CRs approximated 40% during the single session in which the US was withheld.

Data on spontaneous recovery were not reported by Weisman but were in a study by Patten and Deaux (1966). These authors employed a 500-Hz auditory signal as a CS in conjunction with the illumination of a 15-W incandescent lamp. The detector-reinforcer harness shown in Figure 2 was used to present 75-μl of water as the US. The compound CS was presented for 5 sec during each trial and overlapped presentation of the US during the final 2 sec. The criterion of a CR on each trial for each of a group of nine rats was similar to that used by Weisman. Conditional responding was demonstrated, asymptotic performances being attained after approximately 70 trials. Resistance to extinction after a total of 120 conditioning trials was modest, falling to zero level within 15 trials. The CRs in a test for spontaneous recovery about 24 hr after extinction were even more modest in number but were evident for at least 10 consecutive trials.

The relatively lengthy courses of acquisition of conditional licking and the modest levels of resistance to extinction that were found by Weisman and by Patten and Deaux may be taken as evidence that conditionally elicited licking is somewhat anomalous, but I suspect that the use of light as a CS was a contaminating source of variation. In Weisman's study, which showed slower rates of acquisition, the incandescent lamp that was used to present the CS was in closer proximity to the water dispenser and thus to the animal. As mentioned in Section 2.1.3, the rat is photophobic and licking may demonstrably diminish in the presence of light.

Through 1976, all published studies of classical conditioning of the tongue lick in my possession were based on the rat, and none contained fine-grained data on lick rates. It is highly likely that the conditionability of licking is general for other species, but it will be of interest to determine whether the lick response of a preferentially "sucking" animal such as the guinea pig is amenable to classical conditioning. The absence of fine-grained data on numbers, rates, and interlick intervals of conditionally elicited licking precludes assessment in terms of the Pavlovian dictum that conditional responses often differ quantitatively and qualitatively from their unconditional counterparts. One gathers, however, in reading between the lines of published reports that the lick CR is diminished in amplitude and is spare in the number of licks that is elicited by a given CS.

3.2. Instrumental Conditioning of the Licking Response

3.2.1. The Undiscriminated versus the Discriminated Operant

Hulse and Suter (1968) suspected that the rat's tongue-lick—except for the first in a free-drinking situation—is a tactually elicited reflex; if the suspicion were pressed to an assumption of exclusive sufficiency, one could make an argument that there is no free or "voluntary" licking—or what in the older Skinnerian lexicon is called an "undiscriminated operant." However, the rat's own oral cavity, its frequent lingual grooming of its pelt and paws, and its contacts with solid food—all frequent sources of tactile stimulation of the tongue—ostensibly do not elicit continuous bouts of repetitive licking. Tactile stimulation, as mentioned previously, may be a necessary, but is probably not a sufficient, condition for bursting.

Except for inference and conjecture, there are no data on the freely or spontaneously emitted licks of rats or of other animals that have been observed prior to the establishment of reinforcing contingencies. DeBold *et al.* (1965) did record spontaneous, intraorally detected licks of rats after subjecting them to classical conditioning; rates between 0.25 and 1.50 licks/ sec were observed. To what extent these extraordinarily high rates—for an undiscriminated operant—are a reflection of prior conditioning, of artifactual (nonlicking) movements of tongue or mandible, and of spontaneous licking *per se* is a question that defies reckoning. From a pragmatic standpoint, it is unlikely that an uncontaminated spontaneous level of licking will ever be measured, since in principle one would have to rear animals from birth until time of testing, all the while preventing them from licking for nutriment or hydration. A more relaxed and topographically specialized measure (e.g., rate of spontaneous licks by experimentally naïve animals at a perioral sensor akin to that of Deaux and Pattern, 1964) is achievable, but the yield from such an approach has yet to be reported in the literature.

Once having learned that licking provides for consumption of water and other liquids (i.e., once the licking operant is discriminated), the thirsty rat will readily approach a source of water, be it the stream in the wild or the waterspout in the laboratory cage. In accord with Hulse (1967, p. 483), I assume that the *first* lick by a rat that has approached an established source of water is a discriminated operant. Given the integrant case—for example, licking at the conventional spout from a water bottle—all subsequent licks in an uninterrupted series are reinforced in the Pavlovian sense by water and are therefore elicited. *The next instrumental act by the licking animal is presumably that of bodily withdrawal from the source of water.* While it has never been reported in the literature on integrant licking, it is highly likely that as the animal backs away or moves its head from the source of water, the final elicited burst continues its course—and without terminal licks being detected.

Inherent in the study of licking is a captious dilemma. On the one hand, integrant licking is that rat's *natural* way of drinking water. Nature does not often employ fixed-ratio schedules of reinforcement. On the other hand, continuous reinforcement of each lick by water transcends the technical case of continuous reinforcement (CRF) because it is the burst of several licks, not the individual lick, that is elicited. Increasing the ratio of licks required for reinforcement so that the burst runs its course before a subsequent reinforcer is presented does permit one experimentally to isolate operant and respondent licks, but as a result the natural case has been distorted for the sake of experimental analysis. The crux of the dilemma is that reinforcement in the integrant case is *compounded:* How much of the velocity or amplitude of lingual movement of any given lick is due to the immediately preceding reinforcer and how much to all prior reinforcers of each lick in an uninterrupted series of licks?

There is no path at present through the horns of the dilemma, only a caveat: except for the initial lick, integrant licking preserves a special and inescapably respondent character that can differ by millimeters or by kilometers from adjunct licking. With scheduling of CRF, where adjunct licking has its closest parallel to integrant licking, there is still an unavoidable difference in delay of reinforcement. All devices for adjunct licking depend upon electromechanical pumps or valves to provide a reinforcer of fluid after detection of a lick. Electromechanical devices are inertial devices, and the time required to pump a bolus of water into an animal's mouth is on the order of 20–50 msec. If conventional relays are used in lick-detection circuitry, an additional 15–50 msec may be imposed on delay of reinforcement. Given a licking rate of 7/sec, a single cycle of licking requires about 140 msec of time. Problems of phasing may arise, since the bolus of water may reach the tongue well into or even after the intake stroke of licking. Comparisons of lick rates, of tongue-contact durations, and of interlick intervals across the cases of integrant and adjunct CRF would likely reveal differences, but the differences would be confounded by the factor of delay. Indeed, comparisons within the adjunct case may also be confounded because of variability from laboratory to laboratory in operating characteristics of lick-detection and lick-reinforcing apparatus.

A discussion of effects of intermittent scheduling on the operant licking response perforce eliminates consideration of the literature on integrant licking. A further restriction is imposed since much of the extant literature on adjunct licking contains data of studies in which the operant demand placed upon the experimental animal was too low to permit a clear-cut separation of respondent and operant behaviors.

3.2.2. *Control of Licking by Intermittent Schedules of Reinforcement*

Hulse, Synder, and Bacon (1960) were the first investigators to institute fixed-ratio (FR) schedules of reinforcing the tongue lick, but the highest demand made upon their rat subjects was through a nominal FR-6 schedule. Hulse later reported (1967) use of FR-5, -10, and -15 schedules. In 1958, Wyckoff *et al.* reported the effects of variable-interval 14-sec (VI-14s) schedules, and later, in 1965, Stricker and Miller reported the effects of VI-12s and VI-60s schedules and of fixed-interval 60-sec (FI-60s) schedules of reinforcement. The ratio requirements imposed by Hulse and his associates are within the range of the number of consecutive licks that may be elicited by the aqueous reinforcer, hence the FR-5 to FR-15 contingencies they used are not *a priori* different from nominal CRF schedules. The VI and FI contingencies employed by Wyckoff *et al.* and by Stricker and Miller do not necessarily place much operant demand on the scheduled animal. Global licking rates only were reported by Stricker and Miller, the highest of which (2.07 licks/sec) suggests that the animal was responding less than a third of the time during 30-min sessions of testing. Of the reports cited, only in Wyckoff *et al.*'s study did the authors publish a cumulative record, the graphic *sine qua non* of operant performance that is the delight of orthodox Skinnerians.

Justesen *et al.* (1967) obtained cumulative records on 10 albino rats that were assigned to work on nominal FR-50, FI-30s, FI-60s, or VI-16s schedules of reinforcement in which the reinforcer was a 50-μl volume of 10% sucrose in water. During the first session of scheduling and after 21–23 hr of food and water deprivation, each rat was given 30 continuous reinforcements and then, after a 60-sec "time-out" (illumination in the test chamber was extinguished), was shifted directly to one of the four aforementioned schedules. Each rat continued to work until 30 additional reinforcers were delivered or until 20 min had elapsed. On subsequent days, each rat worked until 60 reinforcers were delivered or until 120 min had elapsed. Two rats each made successful transitions to FI-30s, FI-60s, and VI-16s schedules during the first session in the sense that all six of them subsequently worked for the total of 60 reinforcers. Their performances during earlier sessions were ragged, and several daily sessions of training were required before stable patterns of responding developed; however, each of the six animals obtained 60 reinforcers during most of the 25–43 sessions in which they were scheduled. In contrast, all four rats that were deployed on the nominal FR-50 schedule extinguished during the first session. During the second session, demand was reduced to that imposed by a nominal FR-10 schedule; stable responding by all four rats occurred during this session. Subsequently, the ratio requirement was increased from session to session until by the 10th session stable responding to nominal FR-50 was achieved. Comporting with the thesis that FR-10 is veridically related to CRF is the finding that all four FR rats responded consistently during their first session with the (nominal) FR-10 schedule.

A subsequent study by Alvord (1968; see also Alvord, Cheney, and Daley, 1971), who employed the modified Hulse-type detector–reinforcer of Justesen *et al.* (1967), revealed that rat and guinea pig generate highly comparable cumulative records when scheduled on nominal FR-50 to 50μl reinforcers of water (Figure 7). Global licking rates were the same for two rats (one studied by Alvord, the other by Justesen *et al.*) and averaged 3.9 licks/sec. The averaged global rate of the guinea pig was 3.1 licks/sec. From inspection of the records in Figure 7, it is apparent that the guinea pig took fewer time-outs in nominal FR-50 than did the rat.

The available data on the lick response as an operant in relatively high-demand, appetitive schedules are virtually limited to those provided by Alvord and by Justesen *et al.* (1967). One obviously does not make a case for generality on such a limited base, but two observations of a qualitative sort appear to be justified. One is that the character of tongue-lick responding to intermittent schedules is more comparable to that of the pecking pigeon than to that of the lever-pressing rat (see, e.g., Ferster and Skinner, 1957). Consummatory time-out is virtually absent with the rat's licking of water and is quite short for a pigeon pecking a grain of wheat, while the

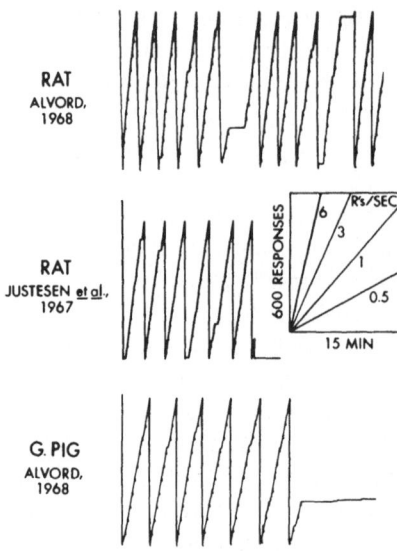

Figure 7. Segments of cumulative records of licking responses of two rats and of the guinea pig. All three subjects were reinforced by a 50-μl volume of sugar water through a fixed-ratio schedule in which every 50th lick was reinforced (*nominal* FR-50). (Redrawn, with permission, from Alvord *et al.*, 1971.)

rat's ingestion (e.g., of a Noyes food pellet) requires relatively much more time. One obvious difference, therefore, is the proportion of time during a session that the water-licking animal expends *exclusively* in consummatory activity—very little.

The other observation is not readily apparent in the cumulative records of licking animals and falls more appropriately under the rubric of anecdote. Actuation of the solenoids used in the lick detection and reinforcing apparatus of Justesen *et al.* (1967) and of King *et al.* (1970) was clearly audible to the experimenters. When a rat had achieved a high degree of stability in responding to nominal FR-40 or FR-50 schedules, one could readily differentiate the rate of licks in the elicited burst (faster and highly regular) and that of subsequent operant licking (slower, but still highly regular). I toyed with the idea of installing a rate-sensitive discriminator that could be used to block all but operant responses from being displayed on the cumulative record. The idea never materialized but presages a prototype by which to make quantitative comparisons of conventional operant behaviors with the licking response. Only when a purely operant signature is achieved for the tongue lick will it be possible to make fine-grained comparisons with lever-pressing and key-pecking behaviors.

3.2.3. Discriminative-Stimulus Control of Licking

Alvord (1968) presented data that indicate that the guinea pig's licking response is highly amenable to control by a discriminative stimulus. A complex schedule was used in which nominal FR-5 for 10 sec regularly alternated with a minimum time-out of 15 sec during which reinforcement was withheld (i.e., FR-5/Ext-15s). A white light on an intelligence panel was illuminated during the nominal FR-5 component; two red lights, during extinction. Fifteen sec of no-responding by the animal in the extinction component was required before an ensuing FR-5 component was reinitiated. An impressive degree of stimulus control is apparent (Figure 8) in the cumulative record, and this degree of control was maintained across 70 one-hour sessions.

Because licking is not a prepotent behavior of the guinea pig and because of the possibility of temporal conditioning in the complex schedule used by Alvord, the question of stimulus control of the rat's licking behavior under a temporally nonconfounded schedule is raised. Cumulative records in the report by King *et al.* (1970) reveal that acoustic cuing of rats in a complex (nominal FR-40/Ext) schedule provides excellent control of operant licking (Figure 9). Variable periods (60-, 120-, or 180-sec) of acoustic signaling by a 525-Hz signal alternated with varying periods (60-, 120-, or 180-sec) of no-signaling; reinforcement of every 40th lick by 35 μl of 16% dextrose in water was only available during the periods of acoustic signaling. Inspection of the cumulative record in Figure 9 reveals that licking by a rat was virtually continuous in the presence of the discriminative stimulus and rarely occurred during periods of extinction, except for runoff of a burst of elicited licks.

A different form of discriminative control, that which involves the mix

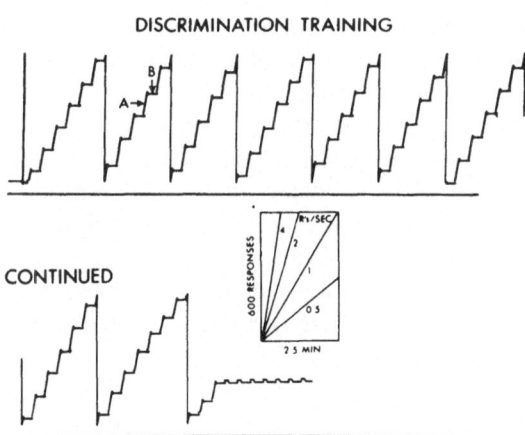

Figure 8. Cumulative record illustrating photically cued discriminative control of licking by the guinea pig. A complex schedule alternated A, a 10-sec period of reinforcement (FR-5, 50-μl sugar water) with B, a 15-sec period of no reinforcement. The alternations continued until the animal was satiated. (Redrawn, with permission, from Alvord, 1968.)

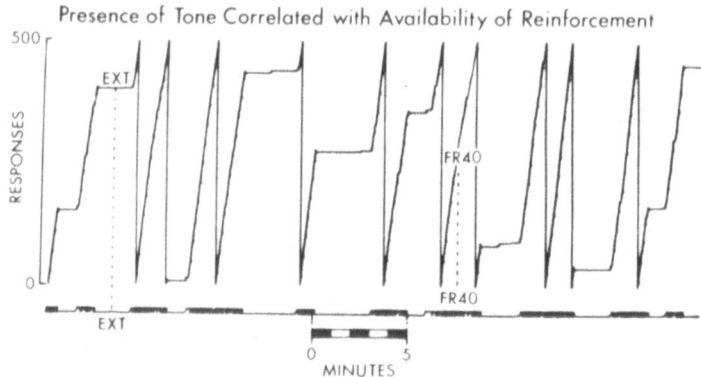

Figure 9.   Cumulative record of operant licking by a rat for which a 35-μl reinforcer of sugar water was available on a (nominal) FR-40 schedule in the presence of a 525-Hz auditory cue. Reinforcement was not available in the absence of the acoustic stimulus. (Redrawn, with permission, from King *et al.*, 1970.)

of classical and instrumental conditioning known as *conditional suppression* (see Estes and Skinner, 1941), has also been demonstrated for the licking behaviors of rats. Justesen *et al.* (1971) and King, Justesen, and Clark (1971) utilized 525-Hz acoustic stimulation or a small dose of microwave energy as a conditional stimulus and paired the CS with a US of aversive foot-shock. The CS–US pairings were superimposed upon ongoing, moderately low-demand aperiodic schedules of sugar-water reinforcement of the tongue lick of rats. Licking in the presence of either CS was quickly but selectively suppressed. Resistance to extinction of the suppressed response was high, being virtually absolute in the presence of acoustic stimulation during a 30-min session of testing.

The literature on discriminative control of operant licking is meager but does not forestall the conclusion that the lick response can be a highly utile behavior in, particularly, determinations of sensory thresholds. For example, in the study by King *et al.* (1971), thresholds of suppression of the rat's licking response by microwave irradiation were nearly an order of magnitude below other behaviorally determined thresholds to nonionizing electromagnetic radiation (*cf.* King *et al.* with Justesen and King, 1970).

4. A Controversy: The Doctrine of Invariance

In a series of reports by D. Cone and her associates (Cone, 1974; Cone and Cone, 1973; Cone, Cone, Golden, and Sanders, 1973; Cone, Wells,

Goodson, and Cone, 1975; Wells and Cone, 1975), the venerated doctrine of the invariance of the local rate of integrant licking has been challenged. The classical paper by Stellar and Hill (1952) gave first voice to the doctrine: "Under all conditions, the rat drinks at a constant rate or it does not drink at all. The tongue always laps water at the rate of six to seven times a second, and with each lap, rats get between [4 and 5 μl] of water. Thus, whenever the rat drinks, it drinks at a rate of about [30 μl] per second" (p. 102).

A large number of confirmatory reports and several reviews in support of the assumption of constancy have subsequently appeared in the biopsychological literature (see, e.g., Corbit and Luschei, 1969; Davis and Keehn, 1959; Keehn and Arnold, 1960; Pierson and Schaeffer, 1973; Premack, 1965; Schaeffer and David, 1973a,b; Schaeffer and Huff, 1965; Schaeffer and Premack, 1961). Cone's contention is that many investigators of licking behavior have failed to perform adequate statistical analysis of their data. When for example, Cone *et al.* (1975) performed an analysis of variance on averaged local rates of integrant licking that approximated 6.3, 6.8, 6.9, and 7.2 licks/sec for four groups of adult male albino rats that had been deprived of water for, respectively, 12, 24, 36, and 48 hr, they obtained an *F*-ratio of 8.03; at 3 and 54 degrees of freedom, the probability that the averages only differ from each other by chance is quite remote ($P < 0.01$). Similar analyses by Cone and her colleagues have yielded highly reliable differences between or among averages of local lick rates as a function of several variables: sex (females lick faster than males); circadian period (rats and opossums lick faster at night); saccharin and sucrose concentration of reinforcer solution (rats generally lick faster as concentration is increased); and age (older rats generally lick faster). While not denying the *reliability* of these differences, the statistically sophisticated scientist is likely to raise a fundamental (if contentious) question: When is a *statistically reliable* difference a *phenomenally significant* difference? I recall two preceptors from my days in graduate school, Lyle Bourne and Paul Porter, trying to drive the import of this Jamesian question home with the observation that even slightly effective sources of variation can yield impressively small *P*-values when large numbers of observations are thrown into the statistical hopper. The integrant tongue-lick is a high-frequency event of high durability, and so the contentious question can be furthered sharpened: *What constitutes invariance with respect to local rate of licking?* It is clear that invariance for Stellar and Hill means local rates of 6–7 licks/sec. It is also clear that averaged local rates that range from 6.3 to 7.2/sec constitute a marked departure from invariance for Cone *et al.* (1975).

Is the challenge issued by Cone and her colleagues merely a matter of statistical semantics? The interpretive and experimental context in which Stellar and Hill (1952) obtained their data certainly has bearing on the answer. They clearly recognized individual differences: "The amount [of water taken] per tongue lap varies slightly from rat to rat" (p. 98). They recognized that the volume of intake per lick" varies with the size of the opening of the drinking tube" (p. 98). And finally, they recognized that volume of intake per lick is reduced by "the amount the rat has to stretch its tongue to reach the water in the tube" (p. 98). Moreover, Stellar and Hill did *not* employ different solutions or vary concentrations or volumes of the reinforcer. They did *not* study female or infant rats or rats of different strain. And they did *not* study opposums, gerbils, or cats. In short, Stellar and Hill, accepted a ~15% range of local lick-rates under a restricted set of parameters as a tolerable spectrum of biological noise that is not incompatible with the interpretation of constancy.

A pivotal point of the controversy lies in the initial phrase of Stellar and Hill's statement of invariance, "*Under all conditions,* the rat drinks at a constant rate . . ." (italics mine). If the italicized phrase is taken as an all-inclusive claim, its resounding rebuttal in the light of present knowledge would vindicate the challenge of Cone and her colleagues. I think it more likely, however, that Stellar and Hill, given their holistic physiological bent, were extrapolating to the natural integrant case—to the feral rat that drinks unencumbered by laboratory paraphernalia from the pond or the stream of water. If so, the paucity of data on behavior directly analogous to that of feral licking would force the conclusion that the thesis of constancy is still a conjecture in search of confirmation.

If Cone *et al.* have a legitimate challenge, their gauntlet is more justly thrown at the feet of investigators who followed Stellar and Hill—those who did introduce several potentially influential variables and who in affirming constancy perhaps failed to analyze their data with due statistical acumen. To a great extent, my sympathies would lie with the accused followers. In view of the myriad factors that could conspire to inject variation even into the adult rat's licking behavior—physical, chemical, physiological, and behavioral, as well as the variables incumbent with widely differing means for reinforcing and detecting the licking response—it is something of a wonder that so many investigators have reported rates and ranges of local licking that are in accord with a liberal interpretation of constancy.

If forced to render a decision, I would reject the doctrine of invariance but on grounds that differ somewhat from those of Cone *et al.* First is the recognition that what often transpires in the laboratory is a marked departure from activities in the wild. Second is the demonstration that licking is two different kinds of behavior, emitted and elicited. In the rat, the former is generally slower, less time locked, and more amenable to modification by training. The latter is faster and more stereotyped, and its rate should be less sensitive to all but its physiologically adequate stimuli. Third is the thesis that the temperature of the eliciting stimulus is a major controlling influence with respect to the local rate of elicited licking. Except for the emphasis that I place upon thermal prepotency of the eliciting stimulus, my thinking about the invariance of the licking response is closely aligned with that of Hulse and Suter (1970):

> our analysis of the behavioral components of fluid ingestion in rats generates a cautionary tale for those who insist that licking in rats must be a fixed, all-or-nothing affair, that rats lick or they don't but when they do, the tongue moves at a constant rate. It is true that the tongue can be remarkably consistent under some conditions. What seem to be required are constant tactile and proprioceptive stimuli in the form of constant drop sizes from lick to lick . . . and a constant mechanical arrangement of the rat with respect to the supply of fluid. (p. 314)

5. Suggestions for Development and Research

5.1. Instrumentation and Measurement

5.1.1. Standardization versus Specification

Experimental psychologists are notoriously resistant to standardization of instrumentation, not because they want to be but because they have to be. Almost any new program of research that is to be based on the intact, freely behaving animal imposes a demand for novel instrumentation. There is no resort to the mass-produced armamentarium of analytical tools to which, for example, the biochemists have access. Consider the behavioral investigator who is faced with the apparently simple task of determining the effect of differing diameters of the orifices of otherwise uniform waterspouts on, say, rate of licking by rats. No extant manufacturer offers an assortment of spouts the orificial diameters of which are graduated in neat millimetric steps from one through five. No extant manufacturer, moreover, is going to tool up for such an enterprise without exacting formidable payment. The behavioral researcher will either make his own spouts or, if lucky enough to be in an institution where there are instrumentation shops and creative specialists, he or she can acquire custom-made (but uncustomary!) devices. Whichever, the product is likely to differ from that made somewhere else by someone else. The end result may be a significant but unrecognized source of variation. There is little to be gained by making the traditional plea for standardization of behavioral instrumentation. On economic grounds alone, that will likely never happen on a grand scale. What *is* necessary and what is sadly lacking in much of the current literature on licking is *specification* of demonstrably or potentially important variables.

The investigator of integrant spout-licking should be concerned with and should specify the composition and dimensions of the waterspout; the internal dimensions and fill level of the reservoir; the angle and location of the drinking spout in relation to the drinking animal; the spatial arrangement and dimensions of the aperture shield that may (and should) be used to reduce artifactual contact between animal and sensor; and the chemical composition and physical properties of the reinforcing solution. It is not enough to say "tap water," since the pH and other conductivity-incrementing contaminants in untreated water may differ from tap to tap by orders of magnitude. While it would be desirable to specify pH, conductivity, surface tension, viscosity, etc., of reinforcer solutions, the accoutrements for determining these are beyond the means of many investigators. By using pure, distilled water—perhaps with reagent-quality solutes such as sugars, saccharine, salts, quinine, etc.—and carefully cleaned spouts, tubing, and reservoirs, the investigator of modest means can provide almost all of the information that is necessary for someone else to duplicate solutions or to determine physical properties.

The one physical property that is not contained in a carefully scripted formulary is the situational variable of temperature. Minimally, the investigator should permit reinforcer solutions to reach the ambient temperature of the laboratory before using them. Implied here is the need to measure and to report the ambient temperature of the laboratory, particularly that of the immediate surround within which the animal is drinking. Since the temperature of the solution probably exerts a controlling influence on licking in virtue of its relation to the temperature of the drinking animal, it may prove of value to determine rectal, colonic, oral, or even hypothalamic temperatures.

The same regard for specification should be exercised by investigators of adjunct licking, upon whom also rests the requirement to specify volume and dispensing velocity of the reinforcer and the delay between time of contact of the tongue with a sensor and delivery of the reinforcer. Incumbent upon all investigators of licking is specification of the means of lick detection. If electrosensing is used, an effort should be made to measure and report not only the nominal dc or rms current but the momentary peak levels of current that may be much higher because of distributive capacitive buildup of voltage at the sensor during the intercontact interval (see Chapter 2).

Finally, an effort should be made to determine with precision the volume of fluid intake of individual animals during a given session. Not all of the water that leaves a spout gets into the licking animal. A carefully designed drinking well that is detachable from the main cage and into which the animal inserts its head to drink from a dispenser would permit relatively precise determinations of consumed volumes of fluid. The investigator could use a balance to determine the mass of the detached cage, with the animal inside, before and after a drinking session. (It is assumed that noningested reinforcer fluids would be contained within the detachable drinking well.) The difference of mass in grams, given a reinforcing fluid of established specific gravity, could readily be extrapolated into total volume of consumed fluid. Averaged volume per record lick could then be determined, losses of body fluid by respiration and evaporation being the only sources of error. Also, by determining the lost mass of fluid in the detachable reservoir of reinforcing solution, one could obtain the ratio of ingested to noningested fluid during a session, thereby achieving an index of the efficiency of the fluid reinforcing system. The utility of data on reinforcing efficiency for assessment of licking or consummatory constancy should be apparent.

5.1.2. Perfection of Measures of Lick Detection

In Section 5.1.1., I began a discussion of the need for instrumentation by which to *calibrate* the accuracy of detectors of the licking response. In amplifying on the earlier remarks, I shall begin by noting that all reported devices save two that are used to detect licking are vulnerable to the animals' withdrawal from the contact sensor; unavoidably, some elicited licks will not be detected. The exceptions (DeBold, Miller, and Jensen, 1965; Deaux and Pattern, 1964) involve intra- or perioral contact sensors that depend upon flow of ions through the tongue of the subject for detection at licking. Critical to a putatively uncontaminated measure of licking is the development of a means of detecting lingual movements that is literally incorporated—*is on or in the animal.* Auscultation of licking via a microphone that is mounted directly on the throat is suggested by the work of Schaeffer and David (1973b) and should be investigated to determine whether accurate and unobtrusive acoustic detection of licking movements is feasible. If licking and swallowing (peristalsis of the esophagus) are coextensive—there are no data bearing on the correlation between operant licking and esophageal activity—perhaps plethysmographic detection of licking is feasible.

A potentially ideal means for gaining a behaviorally and instrumentally unconfounded measure of licking and one that I discussed earlier—via neurogenic potentials that are generated during licking—is yet to be reported in the literature. I recently learned from Professor Halpern, who authored the first chapter of this text, that neurogenic electrical responses associated with licking have been observed by S. Wiesenfeld at Cornell University. Wiesenfeld implanted chronically indwelling recording electrodes in each of several rats and subsequently observed in approximately 30% of her animals—those with placements in the hypoglossal nucleus—regular successions of electrical responses that selectively appeared while the animals were licking water.

The promise of neurogenic potentials as a means of calibrating the efficiency of conventional sensors of licking should be carefully and systematically explored. The eventual use of such potentials (or of auscultative or plethysmographic signals) as a calibrating technique will, however, give rise to the vexing question of criterional adequacy: How is the investigator to calibrate his calibrator? or what, so to speak, will be the platinum rod, the ultimate standard of lick detection? The answer probably lies in the cinematographic analysis of film or videotape. An infrared camera with a split-lens system that simultaneously records the licking animal and the signals generated by the detector–reinforcer device under calibration can yield graphic data that would have to be scrutinized and validated, frame by frame, by independent observers. The onerous nature of cinematographic analysis forbids common use as a working calibrator, but the need for rigorous consensual validation probably precludes any other technique as a final criterion (see Marowitz and Halpern, 1973).

If neurogenic potentials associated with licking prove to be reliable and accurate means of detecting licks, their utility would certainly transcend that of a working calibrator. One application of particular interest to the experimental neuropsychologist would be deployment of the neurogenic, lick-associated signal as an operant response. Perhaps the hypoglossal signal that was observed by Wiesenfeld occurs only under elicited or under operant licking—or perhaps it is common to both. Some leverage may be gained, that is, on central nervous electrical correlates that differentiate emitted from elicited licking.

5.2. Unresolved Issues

The reader has arrived at two conclusions about the licking response if he or she has been able to follow my admittedly tortuous threading through conjecture and data: (1) the field is noisy, but 2) it is lush with opportunity for interesting experimentation. My gut reaction to the state of the licking art is something of a poignant *déjà vu*. Not since the early 1950s, when I was observing the waning game of miniature models—Clark Hull was the coach—have I felt the exhilaration that attends those brief glimpses of behavior as lawful and predictable in consequence of simple physical manipulations. While there is every reason to suspect that a lobster thrown into a pot of boiling water will quickly cease functioning as a behavioral entity, I have long been jaded by the realization that the emergent properties incumbent in the complex nervous systems of mammals forestall safe bets about the effect of small variations of physical stimuli on "molar" behavior. But there may exist in the licking behavior of the rat the stuff of which Clark Hull's vision was made.

Consider, once again, the evidence that the temperature of the reinforcer of the licking response is a primary controlling influence. If a rat is properly dehydrated, it will lick for cool water, for cool air, or even at a cool object that it cannot ingest. Since an elicited burst of licks follows reinforcement of the tongue by water at ambient temperatures, it follows that, if hypothermal stimulation of lingual or oral receptors is an adequate physiological stimulus, the elicited response will extinguish as the temperature of the reinforcer is increased, presumably to or near the biological isotherm. If the hypothesized extinction of elicited licking takes place, there is no guarantee that the operant component will not follow suit. On the other hand, a solution of sugar at the biological isotherm may be palatable—may serve selectively as an operant reinforcer—which would permit studies of purely operant licking even with integrant devices.

A variation of the thermal experiment would use air, not water, as a reinforcer. Puffs of air at 100% relative humidity would be presented aperiodically by an intra- or perioral device, and ascending or descending series of air puffs at changing temperature would be used. If there were extinction of evoked licking at temperatures near the isotherm, one would have evidence not only for thermal control but against tactility as a sufficient property of the physiologically adequate stimulus for licking.

Two other response variables that deserve further systematic investigation are those of burst rate and number of licks in a burst as a function of volume of a single fluid reinforcer. The unresolved question is whether rate or number over a feasible range of reinforcer volumes—say, $0.1-100 \mu l$—is a continuous monotonic function of volume or whether there is some threshold volume of fluid too small to evoke respondent licking that is nonetheless of sufficient size to maintain operant licking.

More toward the ethological end of the behavioral spectrum are unresolved issues related to integrant licking. How much water is *ingested* by a licking animal from a cup or a natural source, and what are the resulting means and variances of global and local rates of licking that is unconfounded by conventional means of detection? Could parameters of licking by different strains or species of animal be used as taxonomic markers, as has been suggested, for example, by Boice (cf. 1967, 1968)? Similarly, in the adult animal, is elicited or preferential operant licking a behavior that drops out at a relatively low rung of the mammalian scale? Comparative studies may reveal that elicited licking in the mature animal is largely confined to lower mammals, but would operant licking follow suit or be more nearly a function of geographical or anatomical variables?

Finally, there is a host of "busy-work" studies that could provide a veritable library of master's theses: on the relation of licking to the chemical composition and physical properties of reinforcer solutions; on the interaction between chemical and physical variables and electrosensing currents of differing polarity and magnitude; on relative preferences of an animal for reinforcer solutions and their correlations, if any, with thresholds of operant and respondent licking; and on the relation between the minimum quantity of reinforcement that is required to maintain licking and that required for adequate hydration or nutrition.

Whether someone will eventually successfully generalize the data on licking into a hypotheticodeductive minitheory is unanswerable at present, but the challenge and the fun of finding out should make the requisite effort worthwhile. To be able to contain adaptive mammalian behavior—especially that of the individual animal—tightly within a coherent frame of physical predicates and predictors has proved elusive, but the biological system may be at hand for licking the problem.

6. Summary

Viewed from the perspective of the laboratory rat, drinking of fluids comprises a genetic unity of three behaviors: reflexive, instrumental, and consummatory. The unconditionally elicited response is a phasic reflex—a highly regular succession or burst of licks—that occurs when a single bolus of water contacts the tongue. Both the rate of licking and the number of licks in the reflexive burst are positively correlated with the volume of the reinforcer. Pavlovian conditioning—conditionally elicited licking to photic or acoustic stimulation—has been demonstrated in several laboratories.

Instrumental or operant licking must be interpreted in consequence of the three differing procedures that have been used to reinforce the lick response: *integrant,* simple *adjunct,* and compound *adjunctive.* Integrant licking is that in which fluid is taken from a cup or a waterspout and is characterized by direct consummation of fluid by each lap of the tongue. Because of the unconditionally eliciting properties of water and other aqueous solutions, all licks but the first in a sustained series of integrant licking preserve a respondent character. The ostensibly exclusive instrumental components of integrant licking are limited to those of the rat's approach to, its first lick at, and its withdrawal from the spout or the cup. The simple adjunct case of licking is that in which licks are detected by one means and reinforced by another. Such "dry" licks can be reinforced at the detector or remotely, positively or negatively, continuously or intermittently, with liquids or solids, but always with some element of delay of reinforcement. The compound adjunctive case is that of the combination of integrant or "wet" licking with adjunctive presentations of food or water. Schedule-induced polydipsia, in which an animal may lick at and continuously consumes water from a spout but also receives, say, a food pellet at one-minute intervals, is an exemplar of the compound adjunctive case.

Essentially "pure" operant conditioning of licking has been observed only in the adjunct case, which, however, may also involve elicited licking. For example, Stewart Hulse devised a small Plexiglas cylinder that terminates in a contact sensor that is situated just below a channel opening from which a bolus of water can be expelled into the rat's mouth by a remotely programmed pump. By imposing a fixed-ratio schedule that requires a greater number of licks than that in an elicited burst, one can obtain a series of operant licks; operant licking may continue until the next water reinforcer, after which another elicited burst occurs.

The local rate of operant licking is not only lower than that of elicited licking but is more sensitive to parameters of scheduling. The licking operant is easily brought under the control of discriminative stimuli and is impressively amenable to conditional suppression.

The doctrine of invariance of the rat's rate of licking, which was originally voiced in the classical study of integrant licking by Eliot Stellar and J. H. Hill, has recently been challenged by Donna Cone. Examination of the studies cited by Cone and her colleagues reveals three sources of variation: (1) the "noise" introduced procedurally by differences in detector–reinforcer instrumentation; (2) the semantic problem that inheres in the definition of "invariance"; and (3) data from several laboratories that indicate that the lick rate is reliably different as a function, for example, of volume of discrete reinforcers, age and sex of the subject, and period of testing within the circadian clock. Since Stellar and Hill may have intended to limit their statement of constancy to the narrower set of conditions of the *natural* integrant case (i.e., to the feral rat that drinks from a pond or a stream of water), the paucity of data on such licking precludes assessment of the validity of the doctrine. Considered as an all-embracing statement of constancy, one must in the light of present knowledge reject the validity of the doctrine. The datum of rate is clearly different for the elicited and the operant components of licking. Indeed, the two components of licking appear to be controlled to some extent by different processes; S. Hulse and his associates have shown that the rate functions of one may increase while those of the other are decreasing under certain conditions of scheduling.

A major complicating factor arises from the preponderant use of the integrant technique in most laboratories. A burst of 3–10 or more licks will occur to a single reinforcer of water, yet each lick after the first in the integrant case is reinforced unless the subject withdraws from the spout or the cup. To determine with precision how such variables as volume, density, pH, viscosity, impact velocity, and surface tension of the reinforcer differentially affect the elicited lick and the operant lick, more investigative emphasis will have to be focused on the adjunct, "one-drop" technique that was pioneered by Hulse.

The physical and chemical properties of the stimulus for elicited licking are currently in doubt. "Fluidity" and "aqueousness" can largely be ruled out. Hulse has proposed tactility as an influential if not sufficient stimulus property, but his own data tend to preclude tactual stimulation as the critical variable. The rat that responds consistently to the Hulse-type detector–dispenser is receiving tactual stimulation of the tongue on each lick. Yet if a water reinforcer is presented intermittently—say, after every 50th lick—there emerges each time water contacts the tongue a clear-cut series of (elicited) licks at a higher rate than the subsequent series of (operant) licks. Some physical or chemical property of the reinforcer, either in isolation or in combination with tactility, is responsible for triggering the burst of reflexive licks. Converging evidence from studies of cold, water, and air licking suggests that *temperature* of the reinforcer may be a critical variable. If this notion is valid, elicited licking, but not operant licking, should extinguish at some point as a rat drinks a palatable fluid from a source that is rising in temperature from the ambient of the surround toward the biological isotherm.

ACKNOWLEDGMENTS

The U.S. Veterans Administration, the Surgeon General of the U.S. Army, and the Federal Food and Drug Administration provided support during the period in which this chapter was written. The author thanks Virginia Hartley and Harold Bowen for art work, Ann Howie and Lynn Bruetsch for archival research and for editorial assistance, and E. L. Wike, C. L. Sheridan, D. M. Levinson, and R. L. Clarke as well as the editors for technical advice and criticism.

References

Allison, J., 1968, Individual differences in eating and drinking in the rat, *Psychon. Sci.* **13**:31–32.

Allison, J., 1971, Microbehavioral features of nutritive and nonnutritive drinking in rats. *J. Comp. Physiol. Psychol.* **76**:408–417.

Allison, J., and Castellan, N. J., 1970, Temporal characteristics of nutritive drinking in rats and humans, *J. Comp. Physiol. Psychol.* **70**:116–125.

Alvord, J. R. Development and control of licking behavior in the guinea pig, unpublished master's thesis, Utah State University, 1968.

Alvord, J., Cheney, C., and Daley, M., 1971, Development and control of licking in the guinea pig (*Cavia porcellus*), *Behav. Res. Methods Instrum.* **3**:14–15.

Barbach, L. G., 1975, "For Yourself: The Fulfillment of Female Sexuality," Doubleday, New York.

Boice, R., 1967, Lick rates and topographies as taxonomic criteria in Southwestern rodents, *Psychon. Sci.* **9**:431–432.

Boice, R., 1968, Operant components of conditioned licking in desert rodents, *Psychol. Rep.* **22:**1161-1167.

Boice, R., and Denny, M. R., 1965, The conditioned licking response in rats as a function of the CS-UCS interval, *Psychon. Sci.* **3:**93-94.

Braud, W. G., and Prytula, R. E., 1969, Licking suppression and recovery as functions of contingency, locus, and frequency of punishment, *Psychon. Sci.* **14:**16-18.

Bujas, Z., 1971, Electrical taste, *in* "Handbook of Sensory Physiology (Vol. 4, Chemical Senses, 2) (L. M. Beidler, ed.), Springer, Berlin, pp. 180-199.

Carr, J., Levin, B. H., and Dissinger, M. L., 1968, Water drinking and air drinking: Some physiological determinants, *Psychon. Sci.* **13:**23-24.

Cone, A. L., and Cone, D. M., 1973, Variability in the burst lick rate of albino rats as a function of sex, time of day, and exposure to the test situation, *Bull. Psychon. Soc.* **2:**283-284.

Cone, D. M., 1974, Do mammals lick at a constant rate? A critical review of the literature, *Psychol. Record* **24:**353-364.

Cone, D. M., Cone, A. L., Golden, A. J., and Sanders, S. L., 1973, Differential lick rates in opossum: A challenge to the invariance hypothesis, *Psycho. Rec.* **23:**343-347.

Cone, A. L., Wells, R., Goodson, L., and Cone, D. M., 1975, Changing lick rate of rats by manipulating deprivation and type of solution, *Psychol. Rec.* **25:**491-498.

Corbit, J. D., and Luschei, E. S., 1969, Invariance of the rat's rate of drinking, *J. Comp. Physiol. Psychol.* **69:**119-125.

Crawford, M. L. J., 1970, Shock-avoidance and shock-escape drinking in rats: Rate of licking. *Psychon. Sci.* **21:**304-305.

Davenport, R. K., Jr., 1961, An investigation of the drinking behavior of the white rat, unpublished doctoral dissertation, University of Kentucky. (University Microfilms, Ann Arbor, Mich. No. 61-294)

Davis, J. D., and Keehn, J. D., 1959, Magnitude of reinforcement and consummatory behavior, *Science* **130:**269-271.

Deaux, E. B., and Patten, R. L., 1964, Measurement of the anticipatory goal response in instrumental runway conditioning, *Psychon. Sci.* **1:**357-358.

DeBold, R. C., Miller, N. E., and Jensen, D. D., 1965, Effect of strength of drive determined by a new technique for appetitive classical conditioning of rats, *J. Comp. Physiol. Psychol.* **59:**102-108.

Eibl-Eibesfeldt, I., 1970, "Ethology: The Biology of Behavior," Holt, Rinehart & Winston, New York.

Estes, W. K., and Skinner, B. F., 1941, Some quantitative properties of anxiety, *J. Exp. Psychol.* **45:**218-224.

Ferster, C. B., and Skinner, B. F., 1957, "Schedules of Reinforcement" Appleton-Century-Crofts, New York.

Fitzsimons, J. T., 1957, Normal drinking in rats, *Society* 19-20:39.

Fossett, C. K., Jr., and Treichler, R. F., 1971, Air drinking by partially ageusic rats, *Physiol. Behav.* **7:**759-761.

Goodrich, K. P., 1960, Running speed and drinking rate as functions of sucrose concentration and amount of consummatory activity, *J. Comp. Physiol. Psychol.* **53:**245-250.

Harrington, G. M., and Linder, W. K., 1962, A positive reinforcing effect of electrical stimulation, *J. Comp. Physiol. Psychol.* **55:**1014-1015.

Hartlep, K., and Bertsch, G., 1974, Facilitation of licking by response-contingent electric shock, *Anim. Learn. Behav.* **2:**196-198.

Hendry, D. P., and Rasche, R. H., 1961, Analysis of a new non-nutritive reinforcer based on thirst, *J. Comp. Physiol. Psychol.* **54:**477-483.

Hulse, S., 1966, Stimulus intensity and the magnitude of the licking reflex in rats, *Psychon. Sci.* **6:**33-34.

Hulse, S. H., 1967, Licking behavior of rats in relation to saccharin concentration and shifts in fixed-ratio reinforcement, *J. Comp. Physiol. Psychol.* **64**:478-484.

Hulse, S. H., Jr., 1960, A precision liquid feeding system controlled by licking behavior, *J. Exp. Anal. Behav.* **3**:1-3.

Hulse, S. H., and Firestone, R. J., 1964, Mean amount of reinforcement and instrumental response strength, *J. Exp. Psychol.* **67**:417-422.

Hulse, S. H., Snyder, H. L., and Bacon, W. E., 1960, Instrumental licking behavior as a function of schedule, volume, and concentration of a saccharine reinforcer, *J. Exp. Psychol.* **60**:359-364.

Hulse, S., and Suter, S., 1968, One-drop licking in rats. *J. Comp. Physiol. Psychol.* **66**:536-539.

Hulse, S. H., and Suter, S., 1970, Emitted and elicited behavior: An analysis of some learning mechanisms associated with fluid intake in rats, *Learn. Motiv.* **1**:304-315.

Hutcheson, J. S., and Mills, K. C., 1974, A compact and inexpensive drinkometer for use with small animals,*Physiol. Behav.* **13**:179-181.

Justesen, D. R., and King, N. W., 1970, Behavioral effects of low-level microwave irradiation in the closed space situation, *in* "Biological Effects and Health Implications of Microwave Radiation—Symposium Proceedings" (U.S. Public Health Service Publication No. BRH/DBE 70-2) (S. F. Cleary, ed.), U.S. Government Printing Office, Washington, D.C.

Justesen, D. R., King, N. W., and Clarke, R. L., 1971, Unavoidable gridshock without scrambling circuitry from a faradic source of low-radio-frequency current, *Behav. Res. Methods Instrum.* **3**:131-135.

Justesen, D. R., Levinson, D. M., and Daley, M., 1967, A modification of the Hulse licker-andum: Preliminary studies and suggested application, *Psycho. Sci.* **7**:111-112.

Keehn, J. D., and Arnold, E. M. M., 1960, Licking rates of albino rats, *Science* **132**:739-741.

King, N. W., Justesen, D. R., and Clarke, R. L., 1971, Behavioral sensitivity to microwave irradiation, *Science* **172**:398-401.

King, N. W., Justesen, D. R., and Simpson, A. D., 1970, The photo-lickerandum: A device for detecting the licking response, with capability for near-instantaneous programming of variable quantum reinforcement, *Behav. Res. Methods Instrum.* **2**:125-129.

MacDonald, J. W., 1972, The spout lick as a viable operant response, unpublished M.S. thesis, University of Wyoming, Laramie.

Marowitz, L. A., and Halpern, B. P., 1973, The effects of environmental constraints upon licking patterns, *Physiol. Behav.* **11**:259-263.

Martonyi, B., and Valenstein, E. S., 1971, On drinkometers: Problems and an inexpensive photocell solution, *Physiol. Behav.* **7**:913-914.

Masters, W. H., and Johnson, V. E., 1966, "Human Sexual Response," Little, Brown, Boston.

Masters, W. H., and Johnson, V. E., 1970, "Human Sexual Inadequacy," Little, Brown, Boston.

Mendelson, J., and Chillag, D., 1970a, Schedule-induced air licking in rats, *Physiol. Behav.* **5**:535-537.

Mendelson, J., and Chillag, D., 1970b, Tongue cooling: A new reward for thirsty rodents, *Science* **170**:1418-1421.

Mendelson, J., and Zec, R., 1972, Effects of lingual denervation and desalivation on air licking in the rat, *Physiol. Behav.* **8**:711-714.

Miller, N. E., and DeBold, R. C., 1965, Classically conditioned tongue-licking and operant bar pressing recorded simultaneously in the rat, *J. Comp. Physiol. Psychol.* **59**:109-111.

Oatley, K., and Dickinson, A., 1970, Air-drinking and the measurement of thirst, *Anim. Behav.* **18**:259-265.

Ogawa, H., Sato, M., and Yamashita, S., 1968, Multiple sensitivity of *chorda tympani* fibres of the rat and hamster to gustatory and thermal stimuli,*J. Physiol.* **199**:223-240.

Patten, R. L., and Deaux, E. B., 1966, Classical conditioning and extinction of the licking response in rats, *Psychon. Sci.* **4**:21–22.

Pierson, S. C., and Schaeffer, R. W., 1973, Differential reinforcement of specific lick rates in the rat, *Bull. Psychon. Soc.* **2**:31–34.

Pierson, S. C., and Schaeffer, R. W., 1975, Lick rate development in infant mongolian gerbils, *Bull. Psychon. Soc.* **5**:47–48.

Premack, D. 1965, Reinforcement theory, *in* "Nebraska Symposium on Motivation" (D. Levine, ed.), University of Nebraska Press, Lincoln.

Riccio, D. C., Hamilton, D. M., and Treichler, R., 1966, Effects of age and ingestive experience on air-drinking behavior in the rat, *Psychon. Sci.* **7**:295–296.

Schaeffer, R. W., and David, M., 1973a, Lick rates in gerbils, *Bull. Psychon. Soc.* **2**:257–260.

Schaeffer, R. W., and David, M., 1973b, Lick rates in New Zealand white rabbits, *Bull. Psychon. Soc.* **2**:43–44.

Schaeffer, R. W., and Huff, R., 1965, Lick rates in cats, *Psychon. Sci.* **3**:377–378.

Schaeffer, R. W., and Premack, D., 1961, Licking rates in infant albino rats, *Science* **134**:1980–1981.

Slangen, J. L., and Weijnen, J. A. W. M., 1972, The reinforcing effect of electrical stimulation of the tongue in thirsty rats, *Physiol. Behav.* **8**:565–568.

Snyder, H. L., and Hulse, S. H., 1961, Effect of volume of reinforcement and number of consummatory responses on licking and running behavior, *J. Exp. Psychol.* **61**:474–479.

Stellar, E., and Hill, J. H., 1952, The rat's rate of drinking as a function of water deprivation, *J. Comp. Physiol. Psychol.* **45**:96–102.

Stricker, E. M., and Miller, N. E., 1965, Thirst measured by licking and reinforced on interval schedules: Effects of prewatering and of bacterial endotoxin, *J. Comp. Physiol. Psychol.* **59**:112–115.

Teitlebaum, P., and Derks, P., 1958, The effect of amphetamine on forced drinking in rats, *J. Comp. Physiol. Psychol.* **51**:801–810.

Weijnen, J. R., 1972, "Lick-Contingent Electrical Stimulation of the Tongue: Its Reinforcing Properties under Dipsogenic Conditions," Drukkerij Elinkwijk, Utrecht, Netherlands.

Weisman, R. G., 1965, Experimental comparison of classical and instrumental appetitive conditioning, *Am. J. Psychol.* **75**:423–431.

Wells, R. N., and Cone, A. L., 1975, Changes in the burst lick rate of albino rats as functions of age, sex, and drinking experience, *Bull. Psychon. Soc.* **6**:605–607.

Williams, D. R., and Teitelbaum, P., 1956, Control of drinking behavior by means of an operant-conditioning technique, *Science* **124**:1294–1296.

Wilson, A. S., and Barboriak, J. J., 1970, Lick volume determined by food schedules in rats, *Psychon. Sci.* **20**:271–272.

Wyckoff, L. B., Sidowski, J., and Chambliss, D. J., 1958, An experimental study of the relationship between secondary reinforcing and cue effects of a stimulus, *J. Comp. Physiol. Psychol.* **51**:103–109.

4

Airlicking and Cold Licking in Rodents

Joseph Mendelson

1. Airlicking

Hendry and Rasche (1961) accidentally discovered that when water-deprived rats are exposed to a stream of air, they will lick the stream persistently (Figure 1). To the naïve observer, airlicking (AL) looks very much like water drinking, and both behaviors generate similar interlick-interval histograms (Oatley and Dickinson, 1970). Since the discovery of AL, a number of experiments have been published concerning its reinforcing and satiating effects and the organismic and airstream parameters that govern its occurrence.

1.1. Measurement Problems

Airlicking is usually induced by the exposure of water-deprived rats to an airstream that is being pumped through a standard drinking tube. Attempts to detect AL by the use of standard drinkometers (see Chapters 2 and 3) reveal that rats do not have to contact the drinking tube in order to

Joseph Mendelson ● Department of Psychology, University of Kansas, Lawrence, Kansas. The author's research reported herein was supported in part by grants to the author from the National Institute of Mental Health, U.S. Public Health Service, U.S. Department of Health, Education, and Welfare (MH-14410 and MH-21955), the National Science Foundation (GB7370), and the University of Kansas General Research Fund (3080-5038 and 3582-5038) and by a Biomedical Sciences Support Grant (RR-07037) to the University of Kansas from the U.S. Public Health Service.

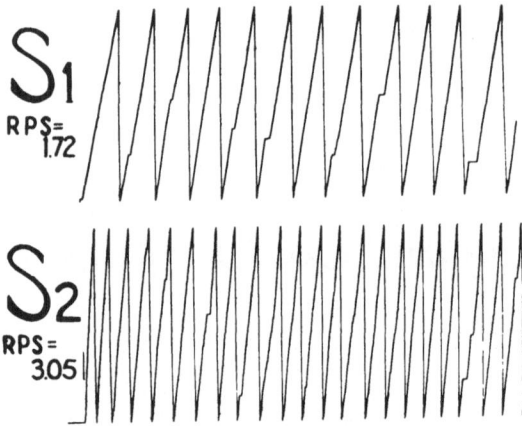

Figure 1. Cumulative response records for a 1-hr session of airlicking for the slowest (S1, 1.72 licks/sec) and the fastest (S2, 3.05 licks/sec) water-deprived rat on a particular test day. The pen reset after every 500 licks. (From Hendry and Rasche, 1961. Copyright 1961 by the American Psychological Association. Reprinted by permission.)

airlick; much of their AL is directed at the airstream several millimeters away from its point of exit from the drinking tube. There are two handy solutions to this problem. One is to place in front of the drinking tube a plate bearing a hole (Mendelson, Zielke, Werner, and L. M. Freed, 1973). A photocell beam is positioned behind the plate in such a way that it is interrupted whenever the rat's tongue interrupts the airstream. Interruption of the photobeam activates a recording device. Another way that is more frequently used is to make the airstream lick-contingent by passing it through a valve whose activation is controlled by drinkometer-detected contacts with the drinking tube (Hendry and Rasche, 1961). A timer is set so that each lick opens the valve for, for example, 0.1 sec. Since a continuous airstream with unrestricted access seems to have mildly aversive effects (Hendry and Rasche, 1961), this procedure has the additional advantage of reducing this aversion; the animals now approach the drinking tube without hesitation. However, extreme caution must be exercised if one is interested in using a drinkometer to record the number of licks emitted during an AL session. We have noticed that some animals lick the drinking tube in such a way as to make two discrete tongue contacts with each lick, one while the tongue is advancing and the other while it is being retracted. In our laboratory, this has happened only with AL and not with drinking (but cf. Chapter 3). Presumably, this form of tongue contact occurs because the AL tube is dry, thereby increasing the likelihood of a contact break during a lick. Since the rate of AL never exceeds 8 licks/second (Oatley and Dickinson, 1970), this recording problem can be solved by the electronic filtering-out of any lick that follows the immediately preceding lick by less than 0.1 sec.

1.2. Ontogeny

Riccio, Hamilton, and Treichler (1967) tested different groups of rats for AL as soon as they reached the ages of 18, 23, 30, and 40 days. Prior to testing, the body weights of the older animals were reduced by 10% by deprivation of water, while the weights of the 18-day-old animals, which had not been weaned, were reduced by maternal deprivation. The authors presented data for the first two 30-min AL sessions. Individual animals of the two older groups generated about 10,000 licks, while individuals of the two younger groups licked only about 3000 (23 days old) and 200 (18 days old) times per session. The authors concluded that in water-deprived rats, AL is an increasing function of age, reaching an asymptote near 30 days of age. There are at least two possible explanations for the deficits exhibited by the rats of the two younger groups.

1. The 18-day animals were deprived of both food and water, hence their body-weight deficit was not brought about solely by water deprivation; consequently, dehydration may have been less severe than for the other groups. Furthermore, when water-deprived rats do not eat they do not become severely dehydrated (Kutscher, 1971). Perhaps a better way to study the ontogeny of AL would be to determine for each age group the dose of hypertonic saline required to generate equal amounts of water ingestion (ml of water/ kg body weight) and then to use these doses to study AL.
2. An airstream in the vicinity of the face seems to be somewhat aversive even to adult rats and perhaps much more so to very young rats, since a greater proportion of their smaller faces is stimulated by the airstream. In any event, it seems clear that the ontogeny of AL is a research area that needs further exploration.

1.3. Acquisition

When water-deprived rats are exposed to an airstream, they do not begin immediately to exhibit high rates of AL. Rats with 1-4 hr of preexposure to a continuous airstream licked at a rate of 45-65 licks/min during a subsequent 1-hr session of lick-contingent AL, while rats with no prior exposure to an airstream licked at a rate of only 23 licks/min in their first 1-hr session (Treichler and Weinstein, 1967).

Prior experience in drinking water is not necessary for the manifestation of AL. Normal levels of AL were shown by rats reared without water and tested at 18 or 23 days of age (Riccio, Hamilton, and Treichler, 1967). Thus, AL must be considered a primary reinforcer rather than a secondary reinforcer whose efficacy is based on prior drinking experience. This finding seems to controvert the notion that "the typical response of a rat to an air tube represents a simple failure to discriminate between the immediate sensory effects of drinking air and those of drinking water" (Hendry and Rasche, 1961, p. 482). Furthermore, rats can discriminate between "the immediate sensory effects of drinking air and those of drinking water." When confronted with a choice in a T-maze between an airstream and water, they consistently select the side of the maze leading to the water, and this preference persists during a series of position reversals of the airstream and the water. Even a mere 10-sec period of exposure to water is preferred to 300-sec access to an airstream (Ostroot and Mendelson, 1974).

1.4. Organismic Determinants

1.4.1. Deprivation State

With one exception (see Section 1.4.5), every organismic state that induces water drinking and that has been tested for AL has also been found to induce AL. These states include water deprivation, systemic injections of hypertonic saline solution (Carr, Levin, and Dissinger, 1968; Mendelson, Zec, and Chillag, 1972), and food deprivation followed by the intermittent presentation of small morsels of food (schedule-induced AL, Mendelson and Chillag, 1968, 1970a; see Section 1.4.2). Furthermore, AL is an increasing function of the length of prior water deprivation (12–48 hr, Oatley and Dickinson, 1970), and schedule-induced AL is an increasing function of food-deprivation–induced body-weight deficit (Chillag and Mendelson, 1971), as is schedule-induced polydipsia (Falk, 1971). Food deprivation, by itself, fails to induce AL (Treichler and Hamilton, 1967).

Treichler and Hamilton (1967) provided continuous 24-hr access to an airstream for one-half of a group of rats that was deprived of water. The rats with an airstream available ate less food and lost weight more quickly than the rats that were denied access to an airstream. Hendry and Rasche (1961) also noted that AL induces a loss of body weight, which they attributed to evaporation of saliva by the airstream. The mean weight-loss attributable to 1 hr of AL was 1.16 gm, which was calculated to represent 0.69% of total body weight (Hendry and Rasche, 1961). Thus, the daily loss of saliva from the mouths of the Treichler and Hamilton airlickers (which were given 24-hr sessions) must have constituted a considerable portion of the total loss of body water from which the animals were suffering as a consequence of terminal water deprivation. This additional AL-induced dehydration was probably responsible for the extreme depression of intake of dry food reported for the animals (cf. Section 1.4.3 and Mendelson, Zec, and Chillag, 1972).

1.4.2. Food Deprivation: Schedule-Induced Airlicking

When food-deprived rats are fed small amounts of food every minute or every few minutes, they drink exceedingly large quantities of water, sometimes as much as half of their body weight within about a 3-hr period (Falk, 1969, 1971). The volume of water consumed is a function of the schedule of food delivery to the rat; therefore, the excessive drinking has been called *schedule-induced polydipsia* (SIP). Since the rats have not been found to be under the influence of any physiological water deficit (Stricker and Adair, 1966), other motivating factors have been sought to account for the polydipsia (see Chapter 9).

One theory that had been proposed as an explanation of this phenomenon is that the rat, after eating each dry food pellet, drinks in order to moisten its mouth (Teitelbaum, 1966). We tested this hypothesis by replacing the water in a drinking tube with a stream of air. All the rats rapidly developed schedule-induced AL (Mendelson and Chillag, 1968, 1970a). Schedule-induced AL was found to be even more pervasive than SIP; under comparable testing conditions, the rats would spend twice as much time airlicking as they would spend drinking (Figures 2 and 3). Since the rats receive no moisture during schedule-induced AL, we can conclude that orolingual moistening is not necessary for the maintenance of schedule-induced licking behavior.

Some rats emitted over 25,000 licks in the 90-min test period. Their AL was so intense that frequently it successfully competed with feeding. Sometimes starving rats would not stop airlicking in order to eat the pellet that was automatically dispensed every minute, despite the loud click of the pellet dispenser that was located beside the test cage. In contrast, starving rats that are licking water always stop immediately in order to eat each pellet as soon as it is delivered (Mendelson and Chillag, 1970a).

Having discovered schedule-induced AL, we then undertook two parametric studies to see if it is subject to the same influences as is SIP:

1. The amount of AL between feedings was found to be invariant with "meal" size (one or three 45-mg pellets) in half of the rats tested and to decrease with the larger meals in the other half (Figures 4 and 5; Mendelson, Zec, and Chillag, 1971). This finding contrasts with reports indicating that SIP increases when meal size is

Figure 2. Lick records of two food-deprived rats receiving 1 food pellet/min (indicated by a vertical line between the records) during a typical 90-min session of asymptotic performance of schedule-induced licking of water (W) or an airstream (A). (From Mendelson and Chillag, 1970a. Copyright 1970 by Pergamon Press, Inc. Reprinted by permission.)

increased from one to two or three 45-mg pellets (Hawkins, Schrot, Githens, and Everett, 1972; Keehn, 1972).

2. AL sharply decreased as the level of starvation was decreased (Figure 6; Chillag and Mendelson, 1971). This finding is in essential agreement with Falk's data for SIP, except that as his animals' body weights were permitted to increase beyond 80% of their *ad libitum* levels, polydipsia did not begin to decrease substantially until the rats had recovered to 95%, whereas our rats showed a substantial decrement in AL by the time their weights had recovered to 90% of normal. These studies demonstrate that both SIP and schedule-induced AL are sensitive to the same experimental manipulations but that they do not respond to them in an identical fashion.

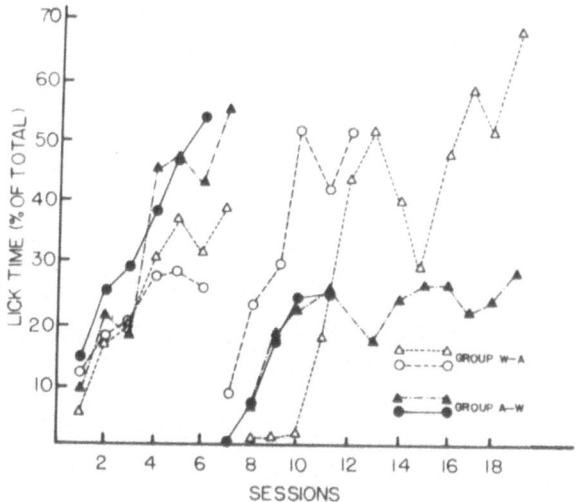

Figure 3. Percentage of total session time spent licking water or an airstream by each of four food-deprived rats receiving 1 food pellet/min during daily 90-min sessions. The breaks in the curves occur where an airstream was substituted for water (Group W-A) or vice versa (Group A-W). (From Mendelson and Chillag, 1970a. Copyright 1970 by Pergamon Press, Inc. Reprinted by permission.)

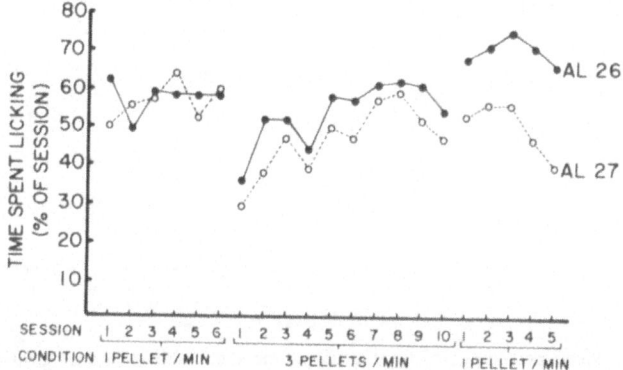

Figure 4. Percentage of total session time spent airlicking by each of two food-deprived rats during 90-min sessions as a function of the number of 45-mg food pellets delivered at the end of each minute. (From Mendelson, Zec, and Chillag, 1971. Copyright 1971 by Pergamon Press, Inc. Reprinted by permission.)

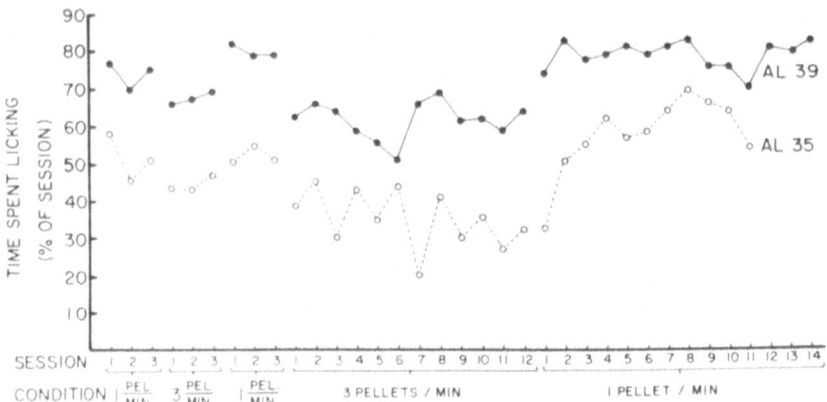

Figure 5.   See legend for Figure 4.

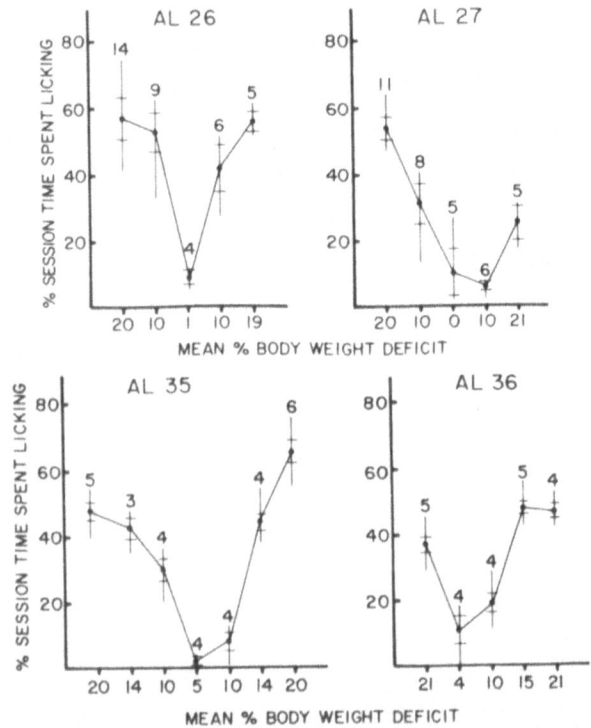

Figure 6.   Mean percentage of total session time spent airlicking during 90-min sessions by each of four rats as a function of body-weight deficit brought about by food deprivation. Each rat was receiving one 45-mg food pellet/min throughout each session. The number of sessions at each level of deprivation is indicated above each point on the graphs. The vertical lines give the ranges, and the horizontal lines intersecting the vertical ones indicate the standard error of the mean. (From Chillag and Mendelson, 1971. Copyright 1971 by Pergamon Press, Inc. Reprinted by permission.)

1.4.3. Desalivation

In the course of our studies on the interaction of feeding and drinking, we decided to test the hypothesis that the desalivated rat (DS) engages in prandial (feeding-associated) drinking (Kissileff, 1969) in order to moisten its oral cavity and/or its throat after eating dry food. Kissileff (1969) found that if DSs are maintained on a diet of 45-mg Noyes food pellets, they gradually develop a tendency to drink water after consuming each pellet; this behavior results in polydipsia. If DSs engage in prandial drinking only in order to moisten their mouths, then they should not engage in prandial AL. We had already shown that schedule-induced polydipsia (which is a form of prandial drinking) is not maintained by the moistening property of water. What about desalivation-induced polydipsia?

To answer this question, we desalivated four rats. Our plan was first to allow prandial drinking to develop and then to test for prandial AL. Several weeks after desalivation, the rats were food-deprived and trained to bar-press on a continuous-reinforcement schedule for 45-mg pellets in a Skinner box with water available from a drinking tube. All of them rapidly developed a tendency to drink after consuming each pellet. When we then replaced the water with an airstream, licking of the drinking tube rapidly extinguished (Mendelson, Zec, and Chillag, 1970). However, this is not unusual, even in normal water-deprived rats; when the water in a drinking tube is first replaced by an airstream emerging from the tube, many water-deprived rats extinguish their tendency to lick the tube. But within one or two sessions, persistent AL becomes firmly established. However, our food-deprived DSs never reestablished their prandial licking behavior when the water in the drinking tube was replaced by an airstream. We therefore decided to train them to airlick under water deprivation and then to reintroduce them into the situation designed to demonstrate prandial AL. Much to our surprise, despite many sessions of exposure of the water-deprived DSs to the airstream, none of them developed persistent AL (Chillag and Mendelson, 1969; Mendelson, Zec, and Chillag, 1972). Although there was some AL during the first session, it almost completely extinguished in subsequent sessions (Figure 7).

We at first conjectured that the failure of the DSs to airlick might be a consequence of a reduction in the amount of fluid in the oral cavity that is available for evaporative cooling by the airstream (Mendelson and Chillag, 1970a; see Section 1.5.3 for a detailed discussion of the role of orolingual cooling in sustaining AL). However, an alternative interpretation arose from the observation that desalivation also reduces the amount of water that is consumed after water deprivation (Epstein, Spector, Samman, and Goldblum, 1964; Vance, 1965). It could be that the reduced amount of AL and drinking exhibited by water-deprived DSs is due to their diet of dry food pellets. Under such conditions, the following three factors could

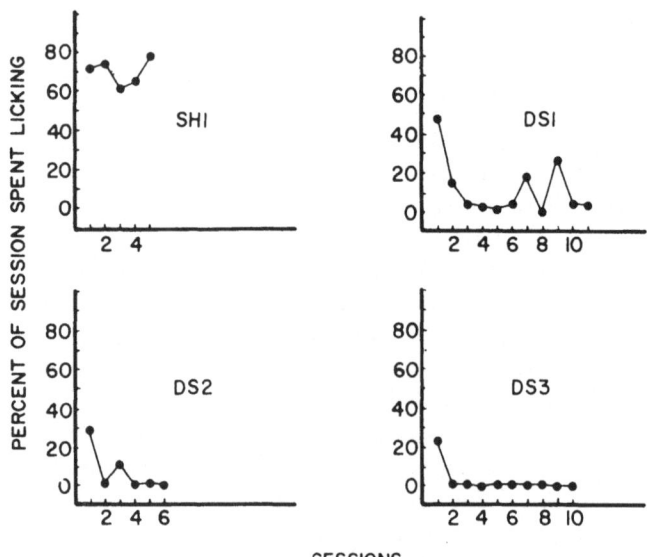

Figure 7.   Percentage of total session time spent airlicking during successive 30-min sessions by one sham operate (SH1) and three desalivated rats water-deprived to 85% of their initial *ad libitum* body weights. (From Mendelson, Zec, and Chillag, 1972. Copyright 1972 by the American Psychological Association. Reprinted by permission.)

operate to generate the observed AL and drinking deficits. (1) During water deprivation DSs eat much less dry food than do control rats (Vance, 1965). (2) The resulting combination of water deprivation and food deprivation fails to induce the increase in blood-plasma osmolarity that is characteristic of water deprivation (Kutscher, 1971). (3) Probably as a result of the latter factor, the combination of water deprivation and food deprivation induces much less drinking than does water deprivation alone (Lal and Zabik, 1970).

In view of these considerations, we evaluated the possibility that water-deprived DSs maintained on *ad libitum* dry food get less thirsty than controls because the DSs are essentially self-deprived of food. We found that when our water-deprived rats were maintained on moist mash, which they could and did eat, subsequent drinking and AL were about the same for desalivated and sham-operated animals. We also found that after combined food and water deprivation, animals of both groups airlicked at greatly reduced rates. A further indication that desalivation does not directly induce any deficit in AL comes from our finding that subcutaneous injections of hypertonic saline induce large and equal amounts of AL in both desalivated and sham-operated animals (Figures 8 and 9; Mendelson, Zec, and Chillag, 1972). We concluded from this series of studies that desalivation does not produce any specific decrement in either drinking or AL

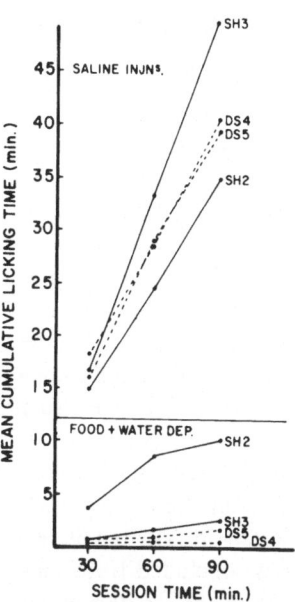

Figure 8. Mean cumulative airlicking time for each of four rats given six sessions after they were deprived of food and water until they lost 15% of their body weight (lower section of figure) and 3 hr following injections of hypertonic saline administered after a 24-hr period of food and water deprivation (upper section). DS4 and DS5 had been surgically desalivated; SH2 and SH3 had been subjected to sham desalivation operations. (From Mendelson, Zec, and Chillag, 1972. Copyright 1972 by the American Psychological Association. Reprinted by permission.)

induced by water deprivation and that only after DSs have been maintained under water-deprivation conditions that engender subsequent deficits in drinking, will they also manifest AL deficits. Thus, it is unnecessary and probably incorrect to postulate (Vance, 1965) that water-deprivation-induced drinking is less rewarding for DSs because they lack certain chemical constituents of saliva that, in normal animals, modify the response of tongue receptors to water so as to increase the reward obtained while drinking water.

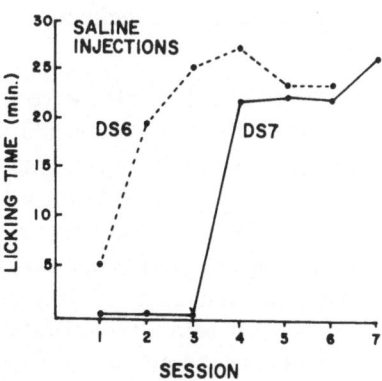

Figure 9. Mean airlicking time during 30-min sessions for each of two desalivated rats that had been injected with hypertonic saline 75 min prior to the beginning of each session. (From Mendelson, Zec, and Chillag, 1972. Copyright 1972 by the American Psychological Association. Reprinted by permission.)

1.4.4. Lingual Denervation

Is AL maintained by some kind of sensory feedback from the tongue? If so, which tongue nerves convey the essential feedback to the central nervous system? As an initial approach to this problem we attempted to denervate sensorily as much of the rat tongue as possible without debilitating our animals to such an extent that it would be necessary to keep them alive by intubation of water or nutrients into the stomach (Mendelson and Zec, 1972). We found that bilateral section of the lingual and chorda tympani nerves (which innervate the anterior two-thirds of the tongue) and of the lateral branch of the glossopharyngeal nerve (which innervates the posterior third of the tongue) could be done without our encountering survival problems. However, surprisingly, such denervated rats showed no deficits in already-acquired AL; they continued to airlick for about 80% of the duration of each 30-min session, even after they had bitten off the anterior portions of their tongues. Only one of the four rats showed a postdenervation deficit in AL, but it recovered to its preoperative level within six sessions. We concluded from these results that sensory stimulation from the anterior two-thirds of the tongue is not necessary to sustain AL and that the essential sensory feedback is derived from the posterior tongue, the nonlingual oral tissues, or both. Weijnen (1976) has confirmed and extended these observations to rats with total sensory denervation of the tongue; combined bilateral section of the chorda tympani, the lingual and glossopharyngeal nerves, and the pharyngeal branch of the vagus nerve failed to suppress AL behavior.

These results contrast sharply with those reported by Fossett and Treichler (1971). They found that bilateral section of the glossopharyngeal nerve was without effect on AL but that bilateral section of the lingual and chorda tympani nerves induced large AL deficits. There are at least two ways to account for the discrepancy.

1. The animals of Fossett and Treichler were given only two post-denervation AL sessions. Perhaps they would have recovered completely during subsequent sessions, just as one of our animals took six sessions to recover its preoperative baseline level of AL.
2. The apparent deficit may have been due to a recording artifact. The authors recorded only the number of contacts of the tongue with the drinking tube, assuming that each contact represented one lick. However, it is possible that before the operation, many of the licks had been generating double contacts with the drinking tube but that the operation induced a change in licking style (perhaps because the animals bit off parts of the denervated portions of their tongues), so that only one contact was made by each lick. This change in style would have been reflected by fewer tube contacts, although the

amount of time spent airlicking may have stayed the same. But the authors did not report the amount of time spent airlicking. This explanation of the AL decrement also is consistent with the lack of a drinking decrement, since double-contact licks are less likely to be recorded while rats are drinking water from a drinking tube.

1.4.5. Hypothalamic Stimulation

There is one organismic state that is believed to induce water drinking, but not AL. Wayner, Cott, and Greenberg (1970) tested three rats for stimulation-induced nitrogen licking. All three bore electrodes in areas of the lateral hypothalamus whose stimulation reliably evoked water drinking. In two of these animals almost no nitrogen licking could be evoked by the brain stimulation. But this finding is not surprising, since neither of them showed substantial amounts of nitrogen licking when tested under 23.5 hr of water deprivation, despite ten 30-min periods of exposure to the nitrogen stream or lick-contingent nitrogen puffs. Under water deprivation, one rat emitted less than 200 licks per session (equivalent to 30–100 sec of licking), and the other one licked only up to 1000 times on its better sessions (150–500 sec). This amount of AL contrasts sharply with reports from most other laboratories that seasoned airlickers tested after 23.5 hr of water deprivation will generate 3000–4000 licks in a 30-min period (see, e.g., Hendry and Rasche, 1961; Oatley and Dickinson, 1970). If these animals were poor nitrogen lickers under water deprivation, one might not expect them to be good lickers under brain stimulation.

The third rat of the Wayner group did learn to nitrogen-lick while water-deprived; it consistently generated over 7000 licks/30-min session, and, curiously it was the only one that showed any promising signs of becoming a stimulation-induced nitrogen licker. But even in this animal, stimulation-induced nitrogen licking was not nearly as strong as stimulation-induced drinking; in fact, during most of its sessions, stimulation-induced nitrogen licking failed to occur at all. The authors have suggested that it is difficult to demonstrate stimulation-induced nitrogen licking because lateral-hypothalamic stimulation increases the aversive properties of a nitrogen stream flowing in the vicinity of the face. Surely this experiment bears repeating with the use of a stream of nitrogen or air whose parameters have been shown to generate strong licking behavior in water-deprived rats.

1.4.6. Hypothalamic Lesions

Since lateral hypothalamic lesions severely depress drinking (Epstein and Teitelbaum, 1964), it would be reasonable to suspect that they might depress AL as well. We therefore attempted to compare the effects of such lesions on drinking and AL (L. M. Freed and J. Mendelson, unpublished experiment). Thirteen rats were reduced to 80% body weight and then were given a number of 30-min drinking and AL sessions. After both behaviors stabilized, they were returned to an *ad libitum* diet and bilaterally implanted with electrodes in the lateral hypothalamic area. Then they were again deprived to 80% body weight and were retested daily for drinking and AL until both behaviors restabilized. At this point, we began to inflict a number of lesions of increasing size upon each animal. If a particular lesion induced a deficit in either drinking or AL, no further lesions were made until stable levels of each behavior were reestablished. The first lesion was made by the passing of 1 mA dc through each electrode for 3 or 6 sec. On each subsequent occasion, the size of the lesion was increased by the addition of 3 sec to the duration of the lesion just previously used, until the duration was 21 or 24 sec. Then the duration was increased in increments of 5 sec until it reached 51 or 49 sec. At this point, the intensity of the current was raised to 2 mA and its duration was reduced to 5 sec. Subsequent lesions were increased by 5 sec each time. If a lesion induced a decrement in drinking or AL, the same parameters were repeated after the behavior(s) restabilized; if not, they were raised according to the foregoing schedule. When a lesion failed to induce a deficit, the parameters were increased on the following day. All rats were tested until they died or lost their electrodes.

The data were analyzed by expression of the decrements in amount of water consumed or time spent airlicking in terms of percentage of decrease from the score obtained on the session prior to the most recent lesion. The total effect of each lesion was quantified in terms of the sum of these decrements over days. All postlesion days were included until the decrement failed to exceed 20% or until testing was terminated. Decrements of less than 20% were ignored, since they were within the range of normal variability. This analysis revealed that in nine rats there were 21 lesions that dissociated AL deficits from drinking deficits. In 17 of these cases, the lesions produced a greater deficit in AL, and in 4 cases, the drinking deficit was greater. These data suggest that some lateral hypothalamic lesions may reduce licking behavior by attenuating the reinforcing potency of the sensory feedback derived from it. If the postingestional reinforcing consequences of drinking are less severely affected by such lesions, then we would expect lesion-induced drinking deficits to be less severe than lesion-induced AL deficits.

1.4.7. Hypothalamic Cooling

Cooling of the oral tissues or of the posterior portion of the tongue appears to be necessary for the maintenance of AL (see Sections 1.4.4 and

1.5.4). Since the hypothalamus lies just above the roof of the mouth, it is reasonable to suspect that the hypothalamus might be subject to conductive and convective cooling during AL. Carlisle and Laudenslager (1975) found that this, in fact, is the case. Ten minutes of continuous AL (about 3400 licks) may lower the temperature of the preoptic area of the hypothalamus by as much as 0.25°C. The authors did not attempt to test the hypothesis that this cooling is essential for the maintenance of AL behavior. They could have done this by clamping the hypothalamus at a fixed temperature via their chronically implanted water-perfused thermodes in order to prevent the hypothalamic temperature from dropping during AL. If hypothalamic cooling is necessary for AL, it is easy to understand why anterior-tongue cooling is not sufficient to maintain AL (see Section 1.5.4); such cooling is not likely to be transferred to the hypothalamus.

Although the authors did not attempt to prevent the AL-induced hypothalamic cooling, they did study the effects on AL of cooling the preoptic area of the hypothalamus via the thermodes. Such cooling prevented the acquisition of AL by naïve rats and reduced AL by about 50% in experienced airlickers. There are at least two ways to account for this decrement. (1) Preoptic cooling reduces drinking in water-deprived rats (Banet and Seguin, 1970), presumably by inhibiting a central thirst mechanism; therefore it would be expected to inhibit AL as well. (2) The authors reported that preoptic cooling induced a rise in core temperature, presumably by causing the hypothalamus to signal that the core temperature was too low. Therefore the orolingual cooling effect of AL would work against the brain's perceived need to increase the core temperature. Thus, the rats reduced their AL, or did not acquire it, because of their perceived need to defend against a decreasing core temperature.

1.4.8. Comparative Aspects

There have been no reports in the literature of attempts to demonstrate AL in species other than the rat. We have examined the AL behavior of water-deprived guinea pigs, hamsters, and gerbils (W. J. Freed, L. M. Freed, and J. Mendelson, unpublished experiments).

Two guinea pigs were given small daily rations of water until their body weights dropped to 85% of their initial values. Then they were given daily 30-min periods of exposure to a lick-contingent airstream. One hour after each session they were given a ration of water that was calculated to be insufficient to maintain their new body weights. One animal acquired robust AL after 6 sessions, by which time its body weight had fallen to 70%. During 11 sessions at 70% body weight, it airlicked for a mean of 22.7 min. It was then given an additional 10 sessions at 80% body weight, and its performance dropped to a mean of 11.6 min. The other animal (Figure 10)

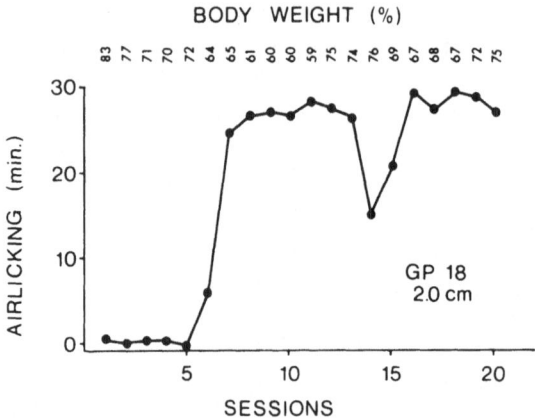

Figure 10. Amount of time spent airlicking by a guinea pig during successive 30-min sessions as a function of the guinea pig's body weight expressed as a percentage of its initial body weight before water rationing was instituted (W. J. Freed and J. Mendelson, unpublished experiment).

did not start to airlick until its weight had fallen to 64% and the airstream was allowed to flow continuously. Thereafter it licked for at least 25 min per session on 11 of 14 consecutive sessions (mean = 26.1 min), as its body weight was allowed to vary between 59% and 76%. During 4 sessions at 74–76%, it airlicked for a mean of 24.3 min per session.

Six hamsters were given fifteen 30-min sessions of AL training under various degrees of water deprivation (70–90% of normal body weight or 24-hr to 48-hr water deprivation). The airstream pressure was varied from 1 to 2 cm of water; the first two sessions were conducted with a continuous airstream, while a lick-contingent airstream was used for the remaining sessions. There was considerable intra- and intersubject variability in the amount of session time spent airlicking. Five hamsters showed substantial amounts of AL on several sessions; each animal's best performance equaled at least 8 min of AL, and only one animal ever licked for more than half of a session. However, there was no clear relation between the level of deprivation in a particular session and the amount of AL generated in that session. These data are in contrast to those for rats whose AL behavior generally shows low levels of variability (in terms of the amount of time per session spent airlicking) and is correlated with deprivation level (Oatley and Dickinson, 1970).

Four gerbils were water-deprived to 85–90% of their normal body weights and given ten 30-min sessions of AL training, eight sessions with a lick-contingent airstream at a pressure of 2-cm water for the first four sessions and 1 cm for the next four, and two sessions with a noncontingent airstream at a pressure of 2 cm. In the first session, the best animal licked for only 53 sec, and thereafter no animal licked for more than 32 sec/ session.

On the basis of these data, we tentatively conclude that guinea pigs airlick about as persistently as rats, although water deprivation may have to be more severe for the guinea pig. Most hamsters airlick, but their AL behavior is much weaker than that of rats and guinea pigs. Gerbils do not airlick after their body weight has been reduced by 10–15% by water deprivation. We are unable to offer a satisfactory explanation for these species differences, but we would like to suggest that the strength of the AL tendency of a species of rodent may be directly related to the amount of water found in the species' natural habitat.

1.5. Airstream Parameters

1.5.1. Airstream Pressure

Airlicking is usually demonstrated by the pumping of air through a standard drinking tube. The air pressure used varies considerably from laboratory to laboratory. We therefore ran a study to determine the minimum pressure required to develop and sustain AL in water-deprived rats (maintained at 85% of their original *ad libitum* weights). Air was pumped through a stainless-steel drinking tube whose orifice measured 2.6 mm. We measured the pressure with a water-filled U-tube manometer by attaching a hypodermic needle to one of its ends and inserting the needle into the rubber air hose leading to the drinking tube. Four animals were given a mean of eighty-five 30-min sessions, beginning with a pressure of 1.0 cm of water ($= 0.014$ psi $= 980$ dn/cm^2; 1 cm refers to the difference between the heights of the columns of water in the two arms of the U-tube manometer). One animal learned to airlick within four sessions. The other animals did not begin to airlick persistently until the pressure was raised to 1.5 or 2.0 cm. After the animals had acquired persistent AL behavior, the pressure was reduced slightly from session to session until AL extinguished. Then, at the beginning of each of the subsequent sessions, the pressure was slightly raised; these sessions continued until AL was reinstated. Data for two representative animals are shown in Figure 11. In different rats, AL extinguished at 0.2, 0.05, 0.025, and 0.005 cm and was reinstated at 0.4, 0.1, 0.1, and 0.01 cm, respectively. The results of this study indicated that the airstream pressures that we have used in our studies (2–4 cm) are much greater than those required to support maximal AL behavior. Of course, it is possible that the reinforcing effects of AL are not accurately reflected by the amount of time spent airlicking during a 30-min session and that in fact

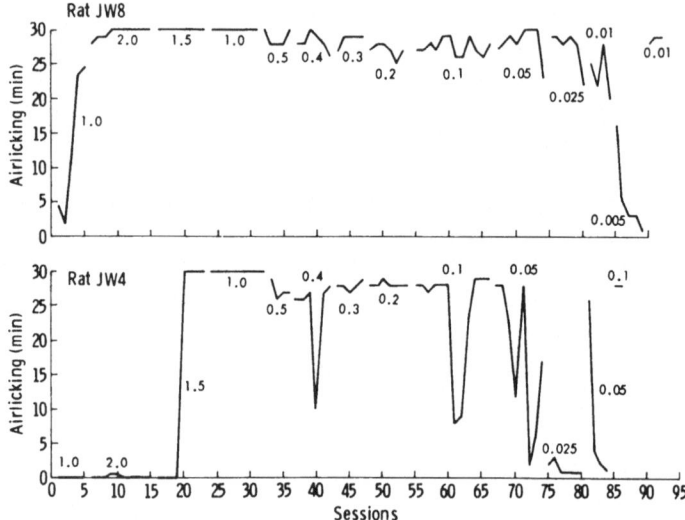

Figure 11. Amount of time spent airlicking as a function of airstream pressure during successive 30-min sessions administered to each of two rats maintained by water deprivation at 85% of their initial *ad libitum* body weights. The pressure (in centimeters of water) is indicated by the numbers next to the segments of the graphs (J. S. Werner and J. Mendelson, unpublished experiment).

higher pressures are more reinforcing. We therefore designed a preference test to determine whether rats prefer to airlick at high- or low-pressure airstreams.

Three rats were water-deprived to 80% of their *ad libitum* body weights and then tested for 30 min/day in chambers in which lick-contingent air puffs at pressures of 0.5 and 2.5 cm of water were simultaneously available from two drinking tubes. For the first 11 sessions, the positions of the tubes were changed according to an alternating sequence, and for the next 8 days on an ABBA sequence.

Two of the rats showed an initial clear preference for the low-pressure air puffs but then switched their preference to the high-pressure puffs, while the third rat initially manifested no stable preference but eventually also developed a strong and persistent preference for the high-pressure puffs (Figure 12). The data suggest that the high-pressure puffs are more rewarding; perhaps they are also more aversive and thus more discouraging to naïve rats that have not yet learned how to position themselves with respect to the drinking tubes in such a way as to minimize facial exposure to the air puffs. Once they have learned this, their preference switches to the high-pressure puffs.

1.5.2. Puff Duration

Treichler and Weinstein (1967) have investigated the effects of air-puff duration on the rate of bar pressing for air puffs by water-deprived rats. Each bar press yielded 0.05, 0.1, 0.2, or 0.4 sec of an airstream. For the three longer durations, there was no difference in the amount of time it took the rats to complete 300 bar presses; but at the shortest duration the animals took significantly longer to complete 300 bar presses.

However, there are two considerations that might adversely affect the meaningfulness of these observations: (1) The minimum duration of an air puff that is just sufficient to provide a reward is likely to be dependent upon

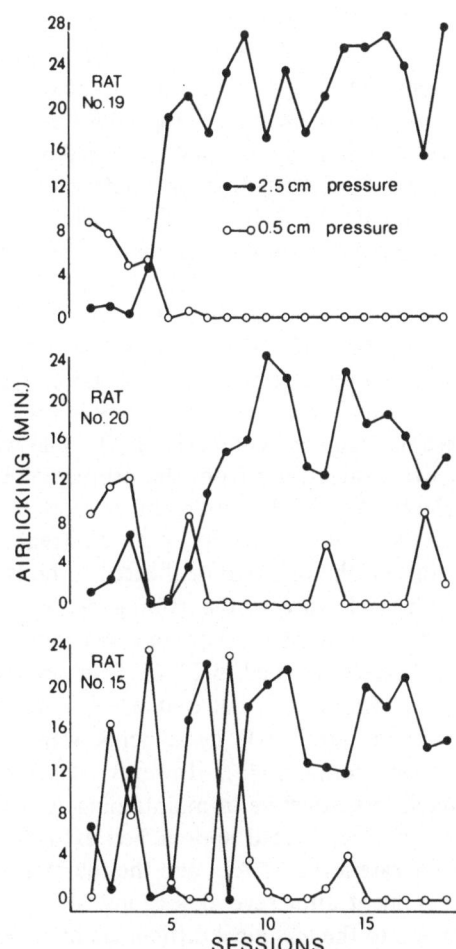

Figure 12. Amount of time spent airlicking during successive 30-min sessions by each of three rats simultaneously exposed to two airstreams with pressures of 0.5 and 2.5 cm of water (W. J. Freed and J. Mendelson, unpublished experiment).

the strength of the thirst of the animal and the velocity, shape, and size of the puff. (2) The 0.05-sec puff was so short that the animal may have frequently failed to be exposed to part or all of it while orientating to the drinking tube after each bar press, although the tube and the bar were separated by a vertical distance of only 1.9 cm. They could have avoided this problem by eliminating the bar-press requirement and simply making the air puffs lick-contingent. Under such conditions, the minimum effective air puff required to sustain AL could be determined as a function of water-deprivation level, airstream pressure, and the diameter of the drinking-tube orifice.

1.5.3. Airstream Temperature and Humidity

Hendry and Rasche (1961) originally demonstrated AL with the use of an airstream that had been cooled to about 60°F. However, they mention in a footnote that warming the airstream to about 100°F did not diminish its capacity to elicit AL reliably in water-deprived rats. The report of this finding is perhaps unfortunate, since it seems to have suppressed attempts to test the hypothesis that the airstream must have cooling properties in order to support AL behavior. Curiously, the authors themselves did not reject this possibility; in their discussion they suggested that "rapid evaporation of saliva probably cools the mouth. This may account for the reinforcing effect insofar as cooling the mouth is a normal consequence of drinking water" (p. 482). Despite this speculation, about 10 years passed before anyone attempted to test the "cooling hypothesis" of AL.

Warming dry air up to body temperature is, in fact, a very poor way to test the cooling hypothesis of AL. The warmer the air, the more effectively it evaporates fluid from the orolingual tissues, thus enhancing the cooling effectiveness of the airstream. A more efficient way of reducing or eliminating the cooling capacity of an airstream is to saturate it with water. If, in addition, the airstream is heated to body temperature or above, then it will not be able to cool the orolingual tissues either by contact or by evaporation. We attempted such an experiment and found that when the airstream was merely warmed to 23–45°C or brought up to about 100% humidity, AL was undiminished, but that when it was both warmed to 39–45°C and humidified, large AL decrements were induced (Mendelson, Chillag and Paramesvaran, 1973). However, although the warm, humid airstream was much less effective in maintaining AL, the licking behavior of only one of the four rats tested approached extinction, two of the rats still maintained moderate levels of AL, and the fourth rat completely recovered from its AL decrement after several sessions. We attributed the incomplete suppression of AL to the technical difficulties of (1) ensuring that the airstream was still sufficiently warm and humid by the time that it came into contact with the orolingual tissues, and (2) preventing condensation of water vapor within the drinking tube, resulting in the occasional emergence of a drop of water, which would provide powerful intermittent reinforcement for licking behavior.

In view of these difficulties, we attempted in a subsequent experiment to measure rats' preferences for licking a warm, dry airstream or a warm, humid airstream (Freed and Mendelson, 1974). Water-deprived rats were simultaneously confronted with two 42°C airstreams, only one of which was saturated with water. All of the eight animals tested developed a very strong and persistent preference for the relatively dry airstream and this preference was maintained throughout many position reversals (Figure 13). We concluded that evaporative cooling of the orolingual tissues makes an important contribution to the reinforcement obtained from AL.

The importance of tissue cooling for the maintenance of AL behavior suggested to us a reinterpretation of the Epstein self-injection phenomenon. Epstein (1960) discovered that rats can be trained to obtain water by pressing a lever to inject water directly into the stomach via a nasopharyngeal-esophageal fistula, so that no water comes into contact with the oropharyngeal tissues. He subsequently reported that hungry rats would similarly self-inject liquid food (Epstein and Teitelbaum, 1962). Holman (1968) raised the question of the possible role of tissue cooling in the maintenance of self-injection behavior. He reported that hungry rats would inject food into their stomachs via a nasopharyngeal–esophageal fistula only if the food was cold, so that it could cool the nasopharynx and esophagus as it passed through the fistula on its way to the stomach. Unfortunately, Holman did not report whether self-injection of water is also dependent upon cooling, and Epstein did not report the temperature of the water used in his experiment. Perhaps it was at room temperature, as in a subsequent self-injection experiment in Epstein's laboratory (Borer, 1968). If so, it would have had a cooling effect on its way to the stomach. It is too early to speculate as to whether this cooling effect is related to the AL phenomenon.

1.5.4. Airstream Accessibility

During the course of the humidity experiments it became clear to us that in order to conduct a convincing test of the cooling hypothesis of AL, it would be necessary for the experimenter (rather than the rat) to gain control over the temperature, pressure, and humidity of the airstream at the point where the rat's orolingual tissues make contact with it. In order to accomplish this, we initially restricted the rat's access to an airstream in such a way that the rat could airlick only by inserting its tongue into a circular hole (1.7 cm in diameter) in a wall of the test chamber to make contact with the airstream as it emerged from a drinking tube located just

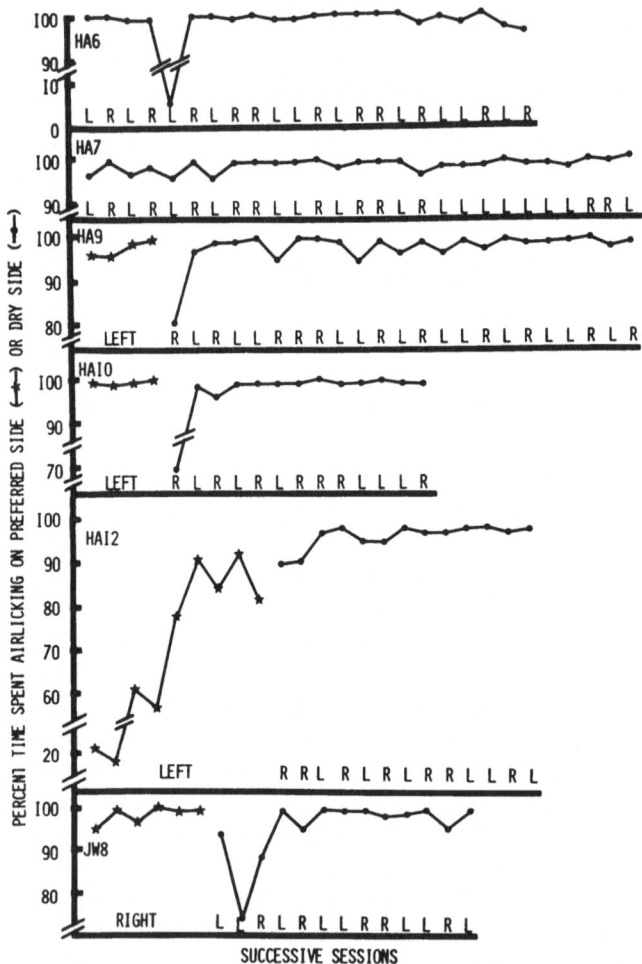

Figure 13. Water-deprived rats were simultaneously confronted with two warm airstreams, one of which was saturated with water. Each section of the figure shows for each session the percentage of a rat's airlicking that was directed at the dry airstream (filled circles). On each of the graphs (except for the upper two), the left-hand portion indicates the percentage of time spent airlicking at the left or right (as indicated) airstream (filled stars). In these sessions both airstreams were dry and at room temperature. *L* and *R* on the abscissa indicate the location of the dry airstream, left or right, respectively. (From W. J. Freed and Mendelson, 1974. Copyright 1974 by Pergamon Press. Reprinted by permission.)

behind the hole (Mendelson, Zielke, Werner, and L. M. Freed, 1973). However, after fourteen 30-min sessions of training on a dry, room-temperature airstream, only one of six rats had learned to airlick. But within two sessions of testing with an enlarged oval access hole (2.5 × 3.3 cm), all of the rats substantially increased their AL scores and four of them stabilized at fairly high levels. However, we noticed that none of the six rats adopted the usual licking posture assumed by rats that are drinking water; rather, all of them stuck their heads through the access hole and twisted them almost completely upside down, that is, rotated through an angle of about 140°. These observations convinced us that mere contact of the tongue with the airstream provides insufficient reinforcement to sustain AL and that rats will airlick only if they can position their heads in such a way that the airstream can penetrate well into their mouths. We therefore undertook systematic tests of this hypothesis (Mendelson, Zielke, Werner, and L. M. Freed, 1973).

In the first test, we varied the size of an access hole situated in front of the airstream as it emerged and descended from a vertically oriented drinking tube. After the rats had learned to airlick with unrestricted access, a plate bearing a circular 3.6-cm diameter hole was inserted into the test chamber, 1.1 cm in front of the drinking tube. From session to session, the access hole was diminished in small steps (of about 10% each), until AL almost completely extinguished. The mean extinction access-hole size was 2.2 cm (based on six rats; Figure 14). In order to be sure that each rat was

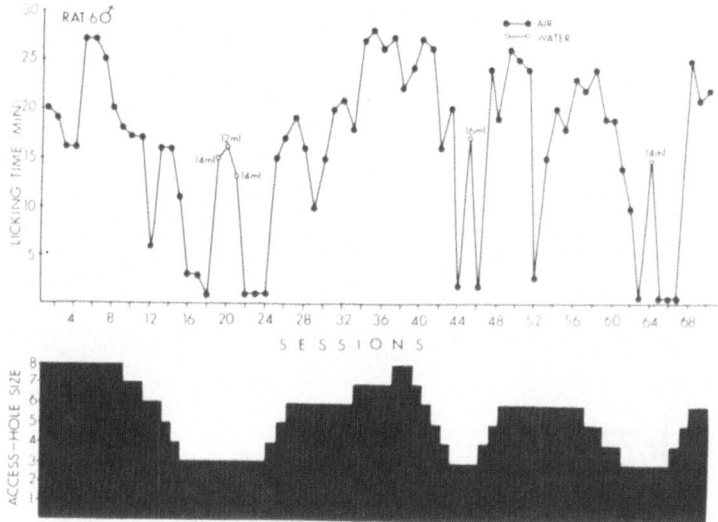

Figure 14. Amount of time spent airlicking or drinking by a rat in successive 30-min sessions is shown as a function of the size of the circular access hole, which restricted the rat's access to the drinking tube. The smallest hole size (2.0 cm) is indicated by the number 1 on the lower part of the ordinate; the largest hole size (3.6 cm) is indicated by 7. The number 8 indicates that the rat had unlimited access to the airstream; that is, no plate was inserted into the entrance of the compartment. The number of milliliters consumed is indicated over each water session. A tongue-shaped orifice was introduced on the 52nd session. (From Mendelson, Zielke, Werner, and L. M. Freed, 1973. Copyright 1973 by Pergamon Press, Inc. Reprinted by permission.)

capable of licking at the orifice of the drinking tube through its extinction-hole size, it was then tested with water during three sessions in which the same extinction-hole size was used. When it was found that the rats drank under these conditions, they were retested with an airstream, whereupon they promptly reextinguished.

In a second test, we restricted the rats' access to the airstream by placing a Plexiglas platform underneath the drinking tube, at a variable distance from it. As this distance diminished, the rats' AL extinguished, although they would still drink water from the tube (Fig. 15). The results of these two experiments strengthened our suspicion that rats will engage in persistent AL only if the airstream can penetrate into their mouths while they airlick.

An interesting observation was made during the course of the restricted-access experiment. As the airstream's accessibility was reduced, the rats seemed to become extremely agitated. Thus, one rat repeatedly clawed and bit the platform and the drinking tube along its entire length. Occasionally, the rat would interrupt its licking–clawing–biting activity and make a rapid 360° whirl. Another rat also bit the tube but confined its biting to the area close to the tip. The third rat initially bit the tube along its entire length, but as the platform was moved still closer to the tube's orifice,

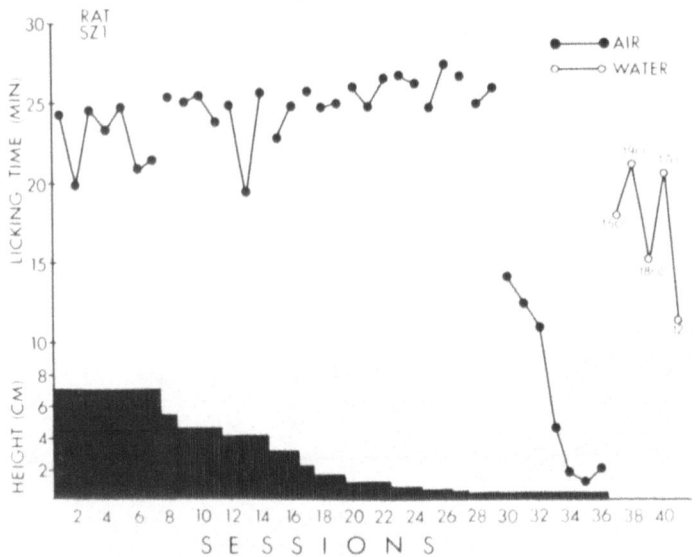

Figure 15. Amount of time spent airlicking or drinking by a rat in successive 30-min sessions as a function of the vertical distance (height) between the orifice of a drinking tube and a Plexiglas plate lying beneath it. The number of cubic centimeters consumed is indicated next to the point for each water session. (From Mendelson, Zielke, Werner, and W. J. Freed, 1973. Copyright 1973 by Pergamon Press, Inc. Reprinted by permission.)

it began to confine its biting to a point about 3 cm above the tip. These behaviors also occurred during an earlier experiment when we warmed and humidified an airstream without restricting access to it (Mendelson, Chillag, and Paramesvaran, 1973). In that experiment, stereotyped AL behavior tended to be supplanted by vigorous pawing and biting of the drinking tube. This behavior developed into stable forms of tube contact that could not be mistaken for "normal" styles of AL but that nevertheless activated the drinkometer and registered on the recording device. Accordingly, we were unable to obtain valid measurements of the amounts of "real" AL occurring under these conditions. We believe that these behaviors occurred as a consequence of the frustration arising from the fact that the animals were no longer able to position their heads in such a way as to permit the airstream to penetrate into their mouths so that it could cool the posterior portions of their tongues and/or their nonlingual oral tissues.

1.6. Electrophysiology

Although cooling of the anterior tongue is not sufficient to maintain AL behavior, there are no data that contradict the possibility that cooling of the posterior tongue may be sufficient for AL. In fact, there are some electrophysiological findings that suggest that the posterior tongue may be more important. Oakley (1967) reported that the rat glossopharyngeal nerve (which innervates the posterior third of the tongue) is much more responsive to cooling than are the rat chorda tympani and lingual nerves (which innervate the anterior two-thirds). The slope of the integrated response-versus-temperature curve was about 10 times greater for the glosso-pharyngeal nerve than for the chorda tympani nerve. Oakley gives no quantitative estimate of the magnitude of the responsivity difference between the glossopharyngeal and lingual nerves, but he does say that the lingual nerve is much more responsive to cooling than is the chorda tympani nerve. In any event, it is clear from Oakley's work that the posterior third of the tongue is much more responsive to cooling than are the anterior two-thirds.

The electrical responses of "cold fibers" in the cat lingual nerve were first described by Zotterman (1936, 1959):

> Leading off from a fine branch of the lingual nerve generally a number of small action potentials are seen, with spike heights of $\frac{1}{3}$–$\frac{1}{10}$ of that of the largest potentials elicited by touching the tongue.
> When the tongue is washed in warm water, above 30°C, these small action potentials disappear, to return after a while if the tongue is laid free in air of room temperature. A faint draught over the tongue increases the number of impulses, and a sudden fine stream of air from a syringe on the receptive field elicits a distinct volley of small

potentials. When the blow of air is harder, so as to make a noticeable deformation on the tongue, large action potentials appear amongst an increased number of small ones. If the air in the syringe is successively warmed, one comes to a point when the air blow does not elicit any small action potentials, while the large ones still appear as soon as the pressure is raised sufficiently to occasion a noticeable deformation on the tongue. After cooling the tongue by washing with water of 5°C the air blow failed to elicit any of the small potentials for several minutes, the large potentials still appearing in response to a strong air stream. The small potentials could be made to appear again in response to faint air blows by washing the cold tongue with warm water.

The same small action potentials are to be seen when a drop of cold water falls on the tongue. The moment of contact is signalled by a large action potential. It will thus be evident that these small action potentials are elicited from cold spots, the specific receptors being excited by the cooling of the tongue. (Zotterman, 1936, pp. 111–112)

The discharge pattern in single fibers of the lingual nerve produced by sudden sustained cooling of the tongue has two components. The immediate response is a short-lived phasic discharge at a very high frequency. This discharge rapidly declines to a sustained, low tonic frequency that may be as little as 7% of the initial frequency. If these findings can be generalized to the cold fibers that innervate the posterior tongue and the nonlingual tissues of rodents, then they have interesting implications for our understanding of airlicking and cold licking. That the phasic discharge tends to be so much greater than the tonic discharge may account for (1) the tendency of cold-licking guinea pigs (see below) to move their orolingual tissues continuously over a cold dry drinking tube, rather than to get a firm oral grip on the tube and hang on tightly, and (2) the tendency of rodents to lick an airstream at a high rate instead of merely positioning their orolingual tissues so as to keep them in constant, invariant contact with the airstream. In view of Weijnen's (1976) sensory denervation work, which indicates that total elimination of lingual sensations fails to eliminate AL behavior, it may be that airlicking rats are merely using their tongues as convenient tools for segmenting the airstream into puffs that repetitively make and break contact between the cooling airstream and their oral tissues; this behavior results in maximization of the occurrence of the high-frequency phasic response of the cold fibers innervating the oral tissues. If this is true, then if the experimenter provided air puffs instead of a continuous airstream, this would eliminate the need to lick at the puffs. However, the animals might still lick if the air puffs are adequate stimuli for elicitation of the lick reflex (see Chapter 3). Weijnen (1976) has reported that if the tongue is paralyzed by bilateral sectioning of the hypoglossal nerve, rats still display "AL" behavior by making "licking" movements with the lower jaw, which supports the tongue. Such jaw movements could serve the same function as tongue movements in that they cause the airstream to come into alternate contact with different portions of the receptive surfaces enclosing the oral cavity. According to the above hypothesis as to the function of tongue movements in normal animals, a pulsating airstream would probably eliminate the jaw movements of these denervated animals. It is rather unlikely that such movements would be reflexively elicited by the air puffs in the same way as licking movements might be.

1.7. Satiating Effects

Airlicking might be said to satiate thirst if it reduces subsequent AL or water drinking. The duration of the satiation effect would then be the post-AL interval during which subsequent AL or drinking is suppressed. Hendry and Rasche (1961) reported that when 1 hr of AL was immediately followed by the opportunity to drink, it reduced the volume of water ingested by 23-hr water-deprived rats by about 22%. Interposing a 1-hr delay between the end of the AL period and the presentation of water reduced the decrement to about 15%. A 2-hr delay resulted in only a 4% decrement, which was not statistically significant. A 1-hr bout of AL also induced significant decrements in the rate of subsequent bar pressing for water on a VI 1-min schedule.

Similarly, Williams, Treichler, and Thomas (1964) reported that a 2-hr bout of AL by water-deprived rats reduced their subsequent CRF bar pressing for 3-sec puffs of air from about 165 (no pre-AL) to about 40 bar presses in a 15-min test. When the bar-pressing test was delayed by at least 30 min from the end of the pre-AL period, very little decrement was obtained. The authors compared their results to the temporary hunger-satiating effects produced by the ingestion of nonnutritive saccharin solution by food-deprived rats (Miller, 1957).

The temporary thirst satiation induced by AL behavior may be a consequence of (1) the consummatory licking behavior itself, (2) the consequent sensory feedback, or (3) the feedback from stomach and small-intestine distension induced by swallowing air. A 1-hr AL session results in as much distension as that produced by an injection of 10 ml of air via stomach tube (Oatley and Dickinson, 1970).

1.8. Reinforcing Effects

Airlicking can serve as a reinforcer for the acquisition of new instrumental responses. Thus, water-deprived rats learn to traverse a straight alley for the opportunity to airlick for 20 sec; their running behavior extinguishes when they are satiated or when the airstream is turned off (Hendry and Rasche, 1961). Rats can also be trained to bar-press on a CRF schedule for 0.1-sec to 3.0-sec puffs of air (Carr *et al.,* 1968; Treichler and Weinstein, 1967; Williams *et al.,* 1964). Bar-pressing rates are comparable to those rewarded by 0.0125–0.1 cc of water (Treichler and Weinstein, 1967). No one has reported attempts to use partial reinforcement schedules with AL as the reinforcer; such schedules could be used to measure the reinforcement potency of AL under various organismic and airstream conditions.

2. Cold Licking

Some of the evidence presented above suggests that AL behavior is maintained by the airstream's capacity to cool the orolingual tissues. If this

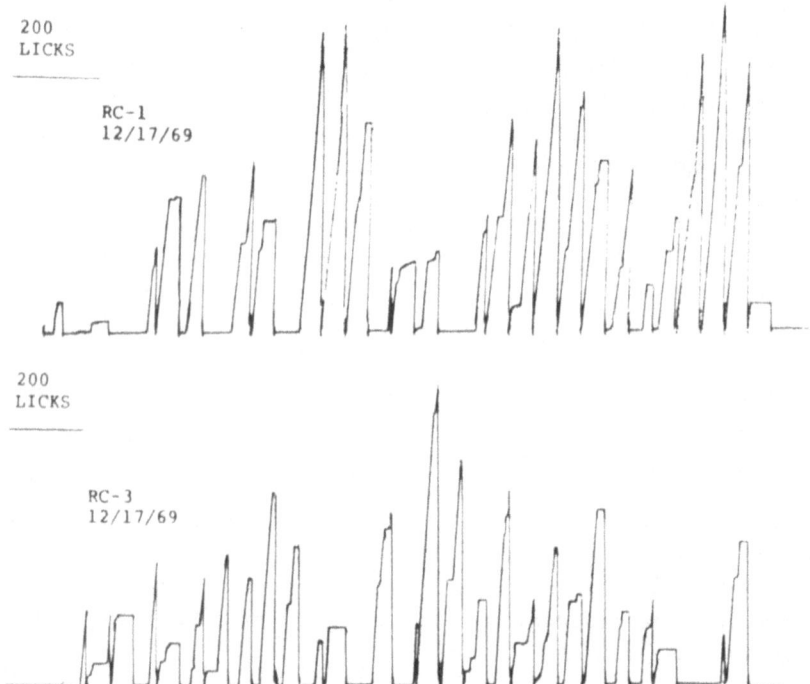

Figure 16.   Records of licking for two water-deprived rats in their first session of exposure to a cold metal tube. The record moves from left to right, each contact with the tube moving the tracing vertically. The height representing 200 contacts is indicated by a horizontal line. The pen automatically reset to zero at the end of each minute. (This type of recording underestimates the number of licks, since the animals may maintain their lip or paw in continuous contact with the tube while they are licking it. Under such conditions the pen would not move vertically.) (From Mendelson and Chillag, 1970b. Copyright 1970 by the American Association for the Advancement of Science. Reprinted by permission.)

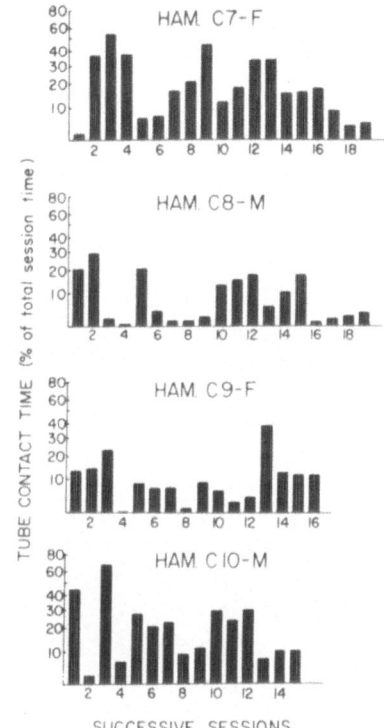

Figure 17. Percentage of total session time spent cold licking in successive 60-min sessions administered to four hamsters water-deprived to 80% of their initial *ad libitum* body weights. (From Chillag, 1970.)

is the case, then water-deprived rats should take advantage of other means of cooling these tissues. Accordingly, we presented them with a cold, dry metal drinking tube. They "cold-licked" (Fig. 16; Mendelson and Chillag, 1970b). So did mice, hamsters (Figure 17), and guinea pigs (Figure 18). However, in contrast to AL, which is usually very stable both between subjects and across sessions, inter- and intrasubject variability were very high for cold-licking rats and hamsters (Figure 17), while for guinea pigs, cold-licking is both very robust and very stable (Figure 18; Mendelson and Plotsky, 1974). Experienced guinea pigs, water-deprived to 80% of their *ad libitum* body weights, consistently spend over 80% of 30-min sessions orally palpating a dry stainless-steel drinking tube maintained at 12–32°C. Only when the temperature of the drinking tube approaches body temperature (36–38°C), does cold licking begin to extinguish. However, although 32°C is sufficient to maintain persistent cold licking in experienced cold lickers, colder temperatures (e.g., 15°C) are required for its acquisition (Figure 18; Mendelson and Plotsky, 1974). When offered a choice between a cold metal rod and a warm metal rod, water-deprived guinea pigs and hamsters prefer to lick the cold rod (Figures 19 and 20; Mendelson and Chillag, 1970b). Cold licking is highly reinforcing for guinea pigs; they can be trained to

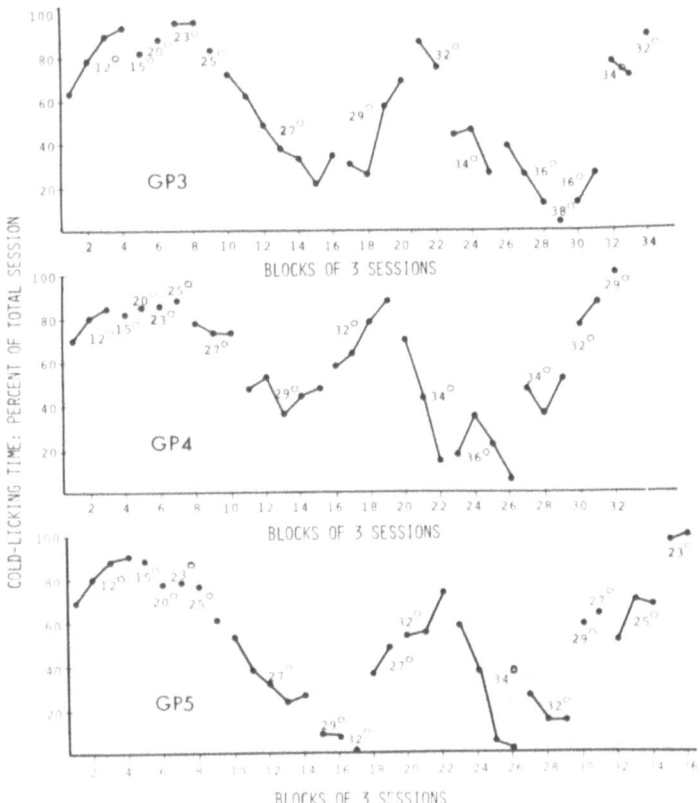

Figure 18. Mean percentage of total session time spent orally palpating a dry stainless-steel drinking tube as a function of tube temperature for successive blocks of three sessions administered to three water-deprived guinea pigs (GP3, GP4, and GP5). (From Mendelson and Plotsky, 1974. Copyright 1974 by Academic Press, Inc. Reprinted by permission.)

emit up to 380 lever presses on a progressive-ratio schedule in order to obtain 3-sec access to a cold, dry drinking tube (Figure 21). Cold licking appears to be a primary reinforcer, at least for hamsters; they will cold-lick even though they have never drunk water (Figure 20; Mendelson and Chillag, 1970b).

In all but one of the above experiments, the cold tubes or rods were maintained at temperatures above the dew point in order to prevent condensation of moisture that could reinforce licking. However, it has been correctly pointed out by Weijnen (1972) that when an animal approaches the tube or rod it brings with it a large increase in humidity. Thus, it is conceivable that the animal licks the cold metal only in order to obtain the moisture that has condensed onto it.

There are two reasons to believe that condensation cannot account for the cold-licking phenomenon:

1. Persistent cold licking was observed even at temperatures just 2–6°C below body temperature. It is highly unlikely that condensation occurred when the tubes were maintained within this range of temperatures. It is true that cold licking would not *develop* at these temperatures; only after it had developed at much lower temperatures (less than 20°C) would such temperatures support it. One may argue that orolingual cooling became a secondary reinforcer by association with the licking of moisture that had been condensing out at the lower temperatures used for training. But if cold licking were only secondarily reinforcing, then it should have extinguished at temperatures above room temperature, at which virtually no condensation would be expected. However, in well-practiced guinea pigs, such temperatures were capable of maintaining high levels of cold licking over many successive sessions.

2. Guinea pigs do not literally "lick" the cold tube as do other rodents; they take the tube well into their mouths and appear to move their

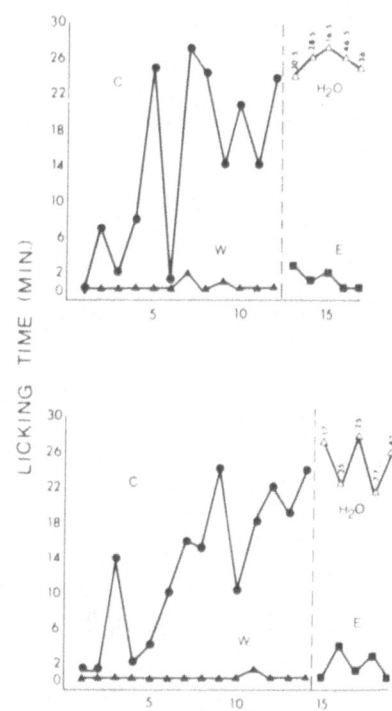

Figure 19. The amount of time during successive 30-min sessions spent orally palpating a cold (C, filled circles) and a warm (W, filled triangles) metal rod that were concurrently available (left portion of each graph), or a water-filled tube (H_2O, open triangles) and an empty tube (E, filled squares) that were concurrently available (right portion of each graph). The data are shown separately for each of two water-deprived guinea pigs. (From L. M. Freed, 1975.)

tongues back and forth over it and to pull and suck on it. The tube often stays in their mouths for several minutes at a time. Obviously, while the tube is in contact with the orolingual tissues, moisture cannot be condensing on it. Of course, the tube is moistened by the mouth, and it might be that it is the combination of wetness and coldness that is necessary to support cold licking. Desalivation would not be sufficient to test this possibility since mucous secretions enter the mouth via other glands in addition to the salivary glands.

2.1. Schedule-Induced Cold Licking

Schedule-induced polydipsia (SIP; see Chapter 9) has been observed in food-deprived rats (Falk, 1969), mice (Palfai, Kutscher, and Symons, 1971), pigeons (Shanab and Peterson, 1969), and monkeys (Schuster and Woods, 1966), but there are no reports on the guinea pig. Since guinea pigs are the best cold lickers after water deprivation (Mendelson and Plotsky, 1974), we

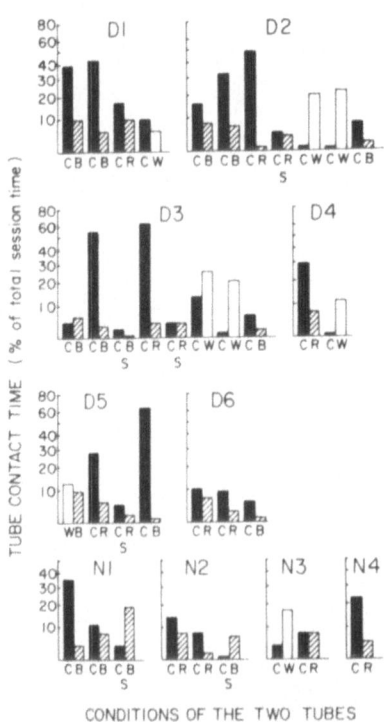

CONDITIONS OF THE TWO TUBES
IN SUCCESSIVE SESSIONS

Figure 20. Percentage of total session time spent in contact with each of two drinking tubes as a function of the conditions of the tubes in successive 20-min sessions (C = cold, dry; B = body temperature, dry; R = room temperature, dry; W = filled with room-temperature water). Hamsters D1–D6 had been raised from birth without the opportunity to drink water; hamsters N1–N4 had been raised in a normal environment. *S* indicates that the animals were satiated with water before testing; all other sessions were administered after 12–14 hr of water deprivation. (From Chillag, 1970.)

decided to determine if they would cold-lick under conditions that engender SIP. If we could demonstrate SIP in guinea pigs, would we also be able to demonstrate schedule-induced cold licking? In our first experiment (J. W. Werner and J. Mendelson, unpublished), eight guinea pigs were deprived of food until their weights dropped to 80% of normal. Then they were given daily 90-min sessions during which water was continuously available and one 45-mg Noyes food pellet was delivered every minute. Under similar conditions, almost all rats develop SIP within 4 sessions. However, none of our guinea pigs showed any signs of SIP even after 10 sessions.

Despite our failure to demonstrate SIP in the guinea pig, we decided to replace the drinking tube containing water with a metal rod maintained at 12–15°C. Seven out of the eight guinea pigs developed schedule-induced cold licking. After this behavior had developed, we replaced the cold rod with a water-filled drinking tube for several sessions, but there were still no signs of the emergence of SIP. In a second experiment (M. R. F. Wano and J. Mendelson, unpublished), eight guinea pigs were food-deprived to 80% of normal body weight and given 90-min sessions with both water and a cold rod available. All of the animals rapidly developed a strong tendency to engage in schedule-induced cold licking, but none of them showed even weak signs of SIP (Figure 22). In both of these experiments, the cold licking had the same postpellet characteristic as does SIP; that is, it tended to occur immediately after the consumption of each pellet. Appropriate control sessions revealed that for most guinea pigs schedule-induced rod licking

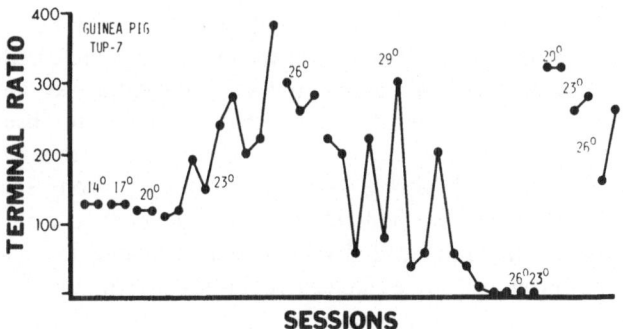

SESSIONS

Figure 21. A guinea pig was water-deprived to 80% of its *ad libitum* body weight and trained to bar-press for a 3-sec exposure to a cold, dry drinking tube. Then it was placed on a progressive-ratio schedule, with a ratio increment of 20 after each reinforcement. The terminal ratio was then studied as a function of the temperature of the tube. This ratio is shown for each session. The increment in performance at 23°C was probably due to the fact that the animal's performance had not yet reached asymptotic levels at the lower temperatures. Even when the tube was maintained at room temperature, 26°C, vigorous bar-pressing was observed. Apparently, even a slight amount of orolingual cooling is highly reinforcing for the thirsty guinea pig. (J. Mendelson, J. P. Grigsby, and W. J. Freed, unpublished experiment.)

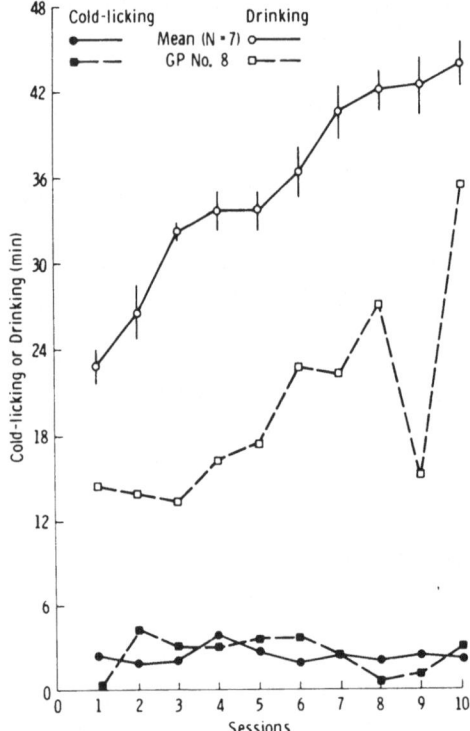

Figure 22. Eight food-deprived guinea pigs were given ten 90-min sessions of scheduled feeding during which they received one 45-mg Noyes pellet/minute. Both a cold rod and a water-filled drinking tube were available during each session. The figure shows the amounts of time spent drinking or cold licking for the worst animal (squares) and for the mean of the other seven animals (circles). (M.R.F. Wano and J. Mendelson, unpublished experiment).

was dependent upon the rod's being cold and the food's being delivered intermittently in small quantities. All of the guinea pigs preferred cold licking to warm licking when both a cold and a warm rod were available during the same session. We consider it unlikely that our failure to observe SIP in guinea pigs was due to the use of inappropriate parameters (e.g., wrong kind or quantity of food, wrong deprivation level, or wrong schedule; see Falk, 1969, 1971), since we had selected parameters that are optimal for demonstrating schedule-induced drinking and airlicking in rats and that generated powerful schedule-induced cold-licking in the guinea pigs themselves.

2.2. Stimulation-Induced Cold Licking

When confronted with a cold metal rod that is shaped like a drinking tube, stimulation-induced cold licking can be elicited from about 80% of guinea pigs whose electrodes are situated in areas of the lateral hypothalamus in which electrical stimulation induces drinking (Freed,

1975). However, the stimulation-induced orolingual palpation is not specific to the cold metal rod; its current-intensity threshold is no different from that obtained when the rod is warmed, and both thresholds are similar to the stimulation-induced drinking threshold (Figure 23). This lack of specificity is reminiscent of the persistent stimulation-induced licking of an empty drinking tube that can be elicited from some rats bearing stimulation-induced-drinking electrodes in the lateral-hypothalamic area (Valenstein, Kakolewski, and Cox, 1968).

When our guinea pigs were given the opportunity to stimulate their lateral hypothalamus electrically by orally palpating a cold or a warm rod, they self-stimulated at high rates but showed no reliable temperature preference. Furthermore, the self-selected durations of electrical current were unrelated to the temperature of the rod being palpated (Freed, 1975). Thus, once a stimulation had been self-initiated, orolingual cooling did not tend to prolong it. In addition, when confronted with a choice between self-stimulating on a water-filled or an empty drinking tube, only one of three guinea pigs preferred the tube containing water (its mean self-stimulation rate for each tube was 95 and 29 per half-hour session; Freed, 1975). These findings were quite unexpected in view of experiments that indicate that for

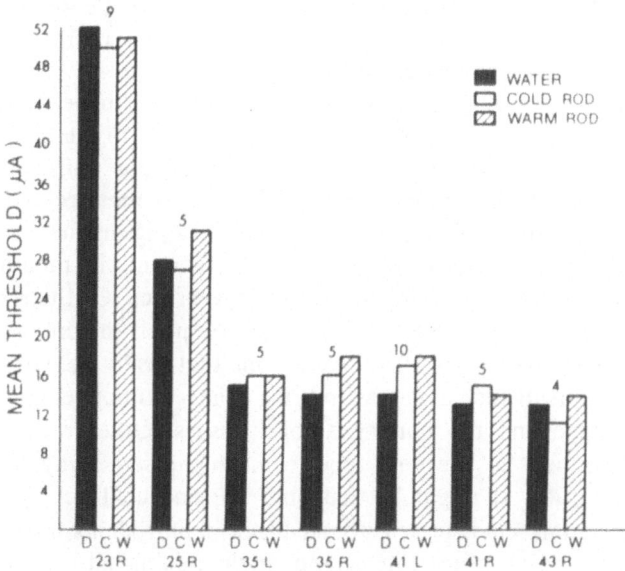

Figure 23. The mean thresholds for elicitation of drinking (D) or orolingual palpation of a cold (C) or warm (W) metal rod for each guinea pig's electrode(s), through which electrical stimulation elicited these behaviors. The numbers located above each triplet of bars indicate the numbers of sessions on which the means are based. (From L. M. Freed, 1975.)

rats bearing hypothalamic drinking-inducing electrodes, drinking increases the reinforcing effects of electrical self-stimulation (Mendelson, 1967, 1970; Morgenson and Morgan, 1967).

3. Conclusions

Both airlicking and cold licking are primary reinforcers for water-deprived rodents, in that they both occur in animals raised without the opportunity to drink water. The reinforcement obtained from AL is dependent upon the capacity of the airstream to cool the oral tissues or the posterior tongue. This thesis is supported by the findings that (1) large decrements in AL are induced by the warming and humidifying of an airstream, and (2) rats prefer to lick a warm, dry airstream rather than a warm, humid airstream. Cooling of the anterior tongue is neither necessary nor sufficient to sustain AL, since (1) rats whose tongues have been sensorily denervated continue to airlick, and (2) AL extinguishes in rats that can contact the airstream with only the anterior portion of the tongue. The posterior tongue or the tissues enclosing the oral cavity must be cooled if AL behavior is to be sustained. It may also be the case that these tissues have to cool the hypothalamus in order for AL to be rewarding. The hypothalamus is cooled during AL, but it is not known whether this cooling is essential for the maintenance of AL.

We believe that the fact that the guinea pig is a far-better cold licker than other rodents (rats, hamsters, or mice) is related to the fact that cooling of the anterior tongue is neither necessary nor sufficient for the maintenance of AL. The tongue of the guinea pig is more rooted than that of other rodents; this may explain why the guinea pig ingests water from a drinking tube not by licking but by taking the tube into its mouth and then sucking and/or swishing its tongue around the orifice. Of the rodents that we have so far tested for cold licking, the guinea pig is the only species that drinks in this fashion. It is also the only one that orally palpates the cold, dry drinking tube in such a way as to bring it into intimate contact with the posterior tongue and the nonlingual oral tissues. Other rodents tend to make discrete licks at the cold tube in much the same way as they lick water from a tube. It might be that both AL and cold licking are maintained by the same kinds of sensory feedback in all rodents, namely, by (1) the cooling of the posterior tongue and/or the nonlingual oral tissues, and (2) proprioceptive feedback from the movements involved in normal drinking behavior. This would explain why guinea pigs are the best cold lickers: they are the only rodents whose style of normal drinking behavior brings them into contact with the cold tube in such a way as to result in the cooling of the posterior tongue and the nonlingual oral tissues (cf. Chapter 5, end of Section 3.6.3.2). All other rodents merely lick the cold tube; while this gives them the appropriate proprioceptive feedback, it fails to generate cooling of the appropriate orolingual tissues. Therefore, their cold licking rarely reaches the high levels consistently maintained by guinea pigs. If the tongues of other rodents were sewn down to the underlying tissues and they learned to drink like guinea pigs, would they too become good cold lickers? We have done this with several rats, but they did not develop a tendency to drink like guinea pigs. Instead, they rotated their heads through an angle exceeding 90° and made repetitive lick-like movements with their lower jaws, so as to contact the water with their lower lips. The "lick" rate appeared to be the same as that for normal rats. In some of these rats, clipping of the incisors seemed to facilitate the emergence of this pattern of drinking (W. J. Freed and J. Mendelson, unpublished experiment).

It has been long known that there are fibers in the nerves innervating the tongue that are excited by cold stimuli placed on the tongue (see Appleberg, 1958, and Fishman, 1957, for data on the rat and the hamster). However, no one who has studied the responses of these fibers has ascribed any particular functional significance to them. It seems to have been assumed that since most of the skin is innervated by receptors responsive to cold stimuli, the tongue is similarly innervated. Our results suggest that the cold receptors in rodents' tongues and/or mouths might help to promote the rodents' survival. Feedback from these receptors to the central nervous system is rewarding to animals whose bodies have been depleted of water. Since, of all naturally occurring substances, water most efficiently cools body tissues, the fact that orolingual cooling is reinforcing for thirsty rodents is likely to lead to appropriate water-ingestion behavior and thus to promote survival. Thus, the orolingual cold receptors help thirsty rodents to detect water and increase the probability that they will drink it. That other rewards in addition to orolingual cooling are normally also derived from drinking water is suggested by experiments that have shown that rats will drink water that has been warmed to body temperature or above (see Chapter 6; Ramsauer, Mendelson, and Freed, 1974) and that rats can learn to regulate their water balance by pressing a lever to inject water directly into their stomachs or veins, thus bypassing the oropharynx (Epstein, 1960; Nicolaidis and Rowland, 1974). However, regardless of which rewards were operating in these experiments, our experiments show that these rewards are not essential for the maintenance of licking in thirsty rodents; orolingual cooling constitutes a sufficient reward for maintaining such behavior. Thus, for the water-deprived rodent, both water intake without the act of drinking and orolingual cooling without water intake are reinforcing events.

The scene is now set for the merger of two areas of research. The major thrust of research into thirst mechanisms has been directed at identifying the types of organismic states that predispose organisms to ingest water. In contrast, the aim of the research reviewed here and in Chapter 5 has been to identify the types of orolingual stimuli that are reinforcing to the rodent. Investigators concerned with the issue of organismic states have largely neglected the issue of reinforcement, and investigators concerned with the reinforcement issue have barely begun to study the reinforcing potency of various stimuli as a function of different types of thirst. Only when these two research areas have been fully integrated will we know which stimuli are reinforcing under which dipsogenic conditions. This information may give us new insights into the origins and workings of drinking mechanisms.

References

Appleberg, B., 1958, Species differences in the taste qualities mediated through the glossopharyngeal nerve, *Scand. J. Physiol.* **44**:129-137.

Banet, M., and Seguin, J. J., 1970, Effects of preoptic cooling in rats acclimated to 21 and 4°C, *J. App. Physiol.* **29**:385-388.

Borer, K. T., 1968, Disappearance of preferences and aversions for sapid solutions in rats ingesting untasted fluids, *J. Comp. Physiol. Psychol.* **65**:213-221.

Carlisle, H. J., and Laudenslager, M. L., 1975, Inhibition of airlicking in thirsty rats by cooling the preoptic area, *Nature* **225**:72-73.

Carr, W. J., Levin, B. H., and Dissinger, M. L., 1968, Water drinking and air drinking: Some physiological determinants, *Psychon. Sci.* **13**:23-24.

Chillag, D., 1970, Identification of a sensory input sufficient to maintain drinking in thirsty rodents, unpublished M.S. thesis, Rutgers University.

Chillag, D., and Mendelson, J., 1969, Air-licking in desalivate rats, *Am. Zool.* **9**:1059.

Chillag, D., and Mendelson, J., 1971, Schedule-induced air-licking as a function of body weight deficit in rats, *Physiol. Behav.* **6**:603-605.

Epstein, A. N., 1960, Water intake without the act of drinking, *Science* **131**:497-498.

Epstein, A. N., Spector, D., Samman, A., and Goldblum, C., 1964, Exaggerated prandial drinking in the rat without salivary glands, *Nature* **201**:1342-1343.

Epstein, A. N., and Teitelbaum, P., 1962, Regulating food intake in the absence of taste, smell and other oro-pharyngeal sensations, *J. Comp. Physiol. Psychol.* **55**:753-759.

Epstein, A. N., and Teitelbaum, P., 1964, Severe and persistent deficits in thirst produced by lateral hypothalamic damage, *in* "Thirst in the Regulation of Body Water" (M. J. Wayner, ed.), Pergamon, Oxford, pp. 394-410.

Falk, J. L., 1969, Conditions producing psychogenic polydipsia in animals, *Ann. N.Y. Acad. Sci.* **157**:569-589.

Falk, J. L., 1971, The nature and determinants of adjunctive behavior, *Physiol. Behav.* **6**:577-588.

Fishman, I. Y., 1957, Single fiber gustatory impulses in rat and hamster, *J. Cell. Comp. Physiol.* **49**:319-334.

Fossett, C. K., and Treichler, F. R., 1971, Air drinking by partially ageusic rats, *Physiol. Behav.* **7**:759-761.

Freed, L. M., 1975, Stimulus-bound metal rod licking and self-stimulation in the guinea pig, master's thesis, University of Kansas.

Freed, W. J., and Mendelson, J., 1974, Airlicking: Thirsty rats prefer a warm dry airstream to a warm humid airstream, *Physiol. Behav.* **12**:557-561.

Hawkins, T. D., Schrot, J. F., Githens, S. H., and Everett, P. B., 1972, Schedule-induced polydipsia: An analysis of water and alcohol ingestion, *in* "Schedule Effects: Drugs, Drinking, and Aggression" (R. M. Gilbert and J. D. Keehn, eds.), University of Toronto Press, Toronto, pp. 95-128.

Hendry, D. P., and Rasche, R. H., 1961, Analysis of a new nonnutritive positive reinforcer based on thirst, *J. Comp. Physiol. Psychol.* **54**:477-483.

Holman, G. L., 1968, Intragastric reinforcement effect, *J. Comp. Physiol. Psychol.* **69**:432-441.

Keehn, J. D., 1972, Schedule-dependence, schedule-induction, and the Law of Effect, *in* "Schedule Effects: Drugs, Drinking, and Aggression" (R. M. Gilbert and J. D. Keehn, eds.), University of Toronto Press, Toronto, pp. 65-94.

Kissileff, H. R., 1969, Food-associated drinking in the rat, *J. Comp. Physiol. Psychol.* **67**:234-300.

Kutscher, C. L., 1971, Hermatocrit, plasma osmolality, and plasma protein concentration as estimators of plasma volume in hooded rats during food and water deprivation, *Physiol. Behav.* **7**:283-285.

Lal, H., and Zabik, J., 1970, Increased food consumption in thirsty rats after water satiation: Inhibition by salts, *Psychon. Sci.* **20**:131-132.

Mendelson, J., 1967, Lateral hypothalamic stimulation in satiated rats: The rewarding effects of self-induced drinking, *Science* **157**:1077-1079.

Mendelson, J., 1970, Self-induced drinking in rats: The qualitative identity of drive and reward systems in the lateral hypothalamus, *Physiol. Behav.* **5**:925-930.

Mendelson, J., and Chillag, D., 1968, Schedule-induced air-licking in rats, *Amer. Zool.* **8**:744.

Mendelson, J., and Chillag, D., 1970a, Schedule-induced air licking in rats, *Physiol. Behav.* **5**:535-537.

Mendelson, J., and Chillag, D., 1970b, Tongue--cooling: A new reward for thirsty rodents, *Science* **170**:1418-1421.

Mendelson, J., Chillag, D., and Paramesvaran, M., 1973, Effects of airstream temperature and humidity on airlicking in the rat, *Behav. Biol.* **8**:357-365.

Mendelson, J., and Plotsky, P. M., 1974, Cold-licking in guinea pigs as a function of temperature, *Behav. Biol.* **10**:191-198.

Mendelson, J., and Zec, R., 1972, Effects of lingual denervation and desalivation on airlicking in the rat, *Physiol. Behav.* **8**:711-714.

Mendelson, J., Zec, R., and Chillag, D., 1970, Failure to demonstrate feeding-associated airlicking in desalivate rats, *Am. Zool.* **10**:473.

Mendelson, J., Zec, R., and Chillag, D., 1971, Schedule dependency of schedule-induced airlicking, *Physiol. Behav.* **7**:207-210.

Mendelson, J., Zec, R., and Chillag, D., 1972, Effects of desalivation on drinking and air-licking induced by water deprivation and hypertonic saline injections, *J. Comp. Physiol Psychol.* **80**:30-42.

Mendelson, J., Zielke, S., Werner, J. S., and Freed, L. M., 1973, Effects of airstream accessability on airlicking in the rat, *Physiol. Behav.* **11**:125-130.

Miller, N. E., 1957, Experiments on motivation, *Science* **126**:1271-1278.

Mogenson, G. J., and Morgan, C. W., 1967, Effects of induced drinking on self-stimulation of the lateral hypothalamus, *Exp. Brain Res.* **3**:111-116.

Nicolaidis, S., and Rowland, N., 1974, Long-term self-intravenous "drinking" in the rat, *J. Comp. Physiol. Phychol.* **87**:1-15.

Oakley, B., 1967, Altered temperature and taste responses from cross-regenerated sensory nerves in the rat's tongue, *J. Physiol. London* **188**:353-371.

Oatley, K., and Dickinson, A., 1970, Air drinking and the measurement of thirst, *Anim. Behav.* **18**:259-265.

Ostroot, D., and Mendelson, J., 1974, Airlicking vs. water drinking in a T-maze: Thirsty rats prefer water, *Physiol. Behav.* **13**:195-199.

Palfai, T., Kutscher, C. L., and Symons, J. P., 1971, Schedule-induced polydipsia in the mouse, *Physiol. Behav.* **6**:461-462.

Ramsauer, S., Mendelson, J., and Freed, W. J., 1974, Effects of water temperature on the reward value and satiating capacity of water in water-deprived rats, *Behav. Biol.* **11**:381-393.

Riccio, D. C., Hamilton, D. M., and Treichler, F. R., 1967, Effects of age and ingestive experience on air-drinking behavior in the rat, *Psychon. Sci.* **7**:295-296.

Schuster, C. R., and Woods, J. H., 1966, Schedule-induced polydipsia in the rhesus monkey, *Psychol. Rep.* **19**:823-828.

Shanab, M. E., and Peterson, J. L., 1969, Polydipsia in the pigeon, *Psychon. Sci.* **15**:51-52.

Stricker, E. M., and Adair, E. R., 1966, Body fluid balance, taste and postprandial factors in schedule-induced polydipsia, *J. Comp. Physiol. Psychol.* **62**:449-454.

Teitelbaum, P., 1966, The use of operant methods in the assessment and control of motivational states, *in* "Operant Behavior: Areas of Research and Application" (W. K. Honig, ed.), Appleton-Century-Crofts, New York, pp. 565-608.

Treichler, F. R., and Hamilton, D. M., 1967, Relationships between deprivation and air-drinking behavior, *J. Comp. Physiol. Psychol.* **63**:541-544.

Treichler, F. R., and Weinstein, A., 1967, Effects of adaptation and amount of reward on air drinking behavior, *Psychon. Sci.* **9**:525-526.

Valenstein, E. S., Kakolewski, J. W., and Cox, V. C., 1968, A comparison of stimulus-bound drinking and drinking induced by water deprivation, *Com. Behav. Biol.* **2**:227-233.

Vance, W. B., 1965, Observations on the role of salivary secretions in the regulation of food and fluid intake in the white rat, *Phychol. Monogr.* **79**:No. 598.

Wayner, M. J., Cott, A., and Greenberg, I., 1970, Stimulus bound nitrogen licking evoked during electrical stimulation of the lateral hypothalamus, *Physiol. Behav.* **5**:1455-1460.

Weijnen, J. A. W. M., 1972, Lick-contingent electrical stimulation of the tongue: Its reinforcing properties in rats under dipsogenic conditions, doctoral dissertation, University of Utrecht (University Microfilms, Order No. 76-6040).

Weijnen, J. A. W. M., 1976, Effects of denervation of the tongue on airlicking and current-licking behavior in the rat, paper presented at the annual meeting of the European Brain and Behaviour Society, Copenhagen, Denmark.

Williams, J. L., Treichler, F. R., and Thomas, D. R., 1964, Satiation and recovery of the "air-drinking" response in rats, *Psychon. Sci.* **1**:49-50.

Zotterman, Y., 1936, Specific action potentials in the lingual nerve of cat, *Skand. Arch. Physiol.* **75**:105-120.

Zotterman, Y., 1959, Thermal sensations, *in* "Handbook of Physiology (Section I: Neurophysiology, Vol. I)" (J. Field, ed.), American Physiological Society, Washington, D.C., pp. 431-458.

5

Current Licking: Lick-Contingent Electrical Stimulation of the Tongue

Jan A. W. M. Weijnen

Accidental discoveries can lead to extensive research programs. Two lines or research into the reinforcing properties of oral sensory stimulation have resulted directly from such serendipitous observations: the study of airlicking (Hendry and Rasche, 1961) and of current-licking behavior (Slangen and Weijnen, 1972).

The special electrical characteristics of a lick sensor (BRS-Foringer DO-101) made possible the discovery of current-licking behavior. Other types of lick sensors pass a very low current through an animal when its tongue touches a drinking tube. However, completion of the input circuit of the BRS-Foringer instrument results in an appreciable current flow through the animal. The reinforcing properties of lick-contingent electrical stimulation of the tongue were discovered in water-deprived rats, when this sensor was used for the registration of tongue contacts with a small stainless-steel ball, rather than for the recording of drinking behavior (Slangen and Weijnen, 1972).

In this chapter, an overview is presented of the present state of research on current-licking behavior. The text is more detailed than is usual for a review, as little has been published on this subject.

Jan A. W. M. Weijnen ● Department of Psychology, Physiological Psychology Section, Tilburg University, Tilburg, The Netherlands. Based on a doctoral dissertation submitted to the University of Utrecht, The Netherlands (1972). The present version is updated and incorporates results of subsequent studies of the phenomenon.

1. Current Licking: Some Basic Aspects of the Phenomenon

1.1. Reinforcing Effects of Peripheral Electrical Stimulation

Alimentary reinforcement of licking behavior under continuous-reinforcement conditions is an everyday experience of rats with *ad libitum* access to water or liquid food. In deprived rats, licking can be used as an operant response to obtain reinforcement. The licking response is easily executed, and high rates of responding without fatigue can be maintained over long periods if a suitable schedule of reinforcement is chosen (King, Justesen, and Simpson, 1970). Lick-contingent electrical stimulation of the tongue has no alimentary value. The existence *per se* of a nonalimentary reinforcer of licking behavior in water-deprived rats is not unprecedented. This has been well documented in the preceding chapter on airlicking and cold licking.

Reinforcing properties of peripheral electrical stimulation have been reported before, but never has the tongue been implicated as the site of the reinforcing stimulation.

In one study (Harrington and Linder, 1962), an increase in response rate occurred when rats received electrical stimulation of the feet contingent upon bar contact in a Skinner box. The authors reported "bursts of very rapid responses," but unfortunately, no description of the behavior of the rat during these bursts was given. The response rate declined from session to session.

Campbell (1968, 1972) reported peripheral electrical self-stimulation behavior in fishes of different species and in a crocodile. The results were obtained with spatial preference and with operant techniques under undeprived conditions. Satiation effects were clearly seen within sessions, and there was substantial recovery between sessions. Campbell suggested that peripheral sensory stimulation results in activation of limbic "pleasure areas," which can also be activated directly with intracranial self-stimulation techniques. This latter type of self-stimulation behavior, however, usually does not show satiation effects; it lacks the "naturalness" of peripheral stimulation (Campbell, 1972).

A spatial preference technique, in which rats were allowed to choose between the absence or the presence of foot shock, was also used by Campbell and Masterson (1969). They observed a temporary preference for low shock levels. The authors stated: "While these observations were casual at best, our conclusion is that no intensity of electrical stimulation is innately positively reinforcing and that any activity sustained in the presence of such stimulation is motivated by curiosity" (p. 17). From the results of these studies, performed with animals that were deprived neither of sensory stimulation nor of water, one would not predict persistent licking behavior in water-deprived rats when electrical tongue stimulation, but no water, is obtained for each lick. Research into the reinforcing properties of electrical tongue stimulation in rats under dipsogenic conditions was needed for a better understanding of the current-licking phenomenon.

1.2. The Study of Current-Licking Behavior: General Information

Some general information regarding the methods that were used to investigate current-licking behavior are given first.

1.2.1. Recording Licking Behavior

In Chapter 2 the technique of recording licking behavior is described, with special reference to current-licking behavior. Included in this chapter are descriptions of lick sensors and "tongue stimulators" that have been used in the study of current-licking behavior.

1.2.2. Test Chamber

Most of the experiments that are reported in this chapter employed a simple test chamber. This chamber, made of perspex, was equipped with a stainless-steel plate floor (Figure 1). Licking rod(s) and/or drinking tube(s) with glass sleeve(s) could be put in different positions to meet the experimental requirements (see Chapter 2, Figure 4). As a rule, the chamber was put in a dimly lit, sound-attenuated booth or housing; a ventilator supplied background noise. The capacitance of the input circuit of the lick sensor connected with the test chamber was approximately 0.01 nF in most cases.

1.2.3. Water-Deprivation Schedule

A schedule was chosen that allowed periods of five successive days of experimentation and involved a minimum of labor (Table 1). Food was

Table 1.   Water-Deprivation Schedule

Sat.	Sun.	Mon.	Tues.	Wed.	Thur.	Fri.	Sat.
←--------→			←-------------------→			←----→	
Water-deprivation living cage 48 hr			Water plus current individual watering cage 10 min/day			Water living cage *ad libitum*	

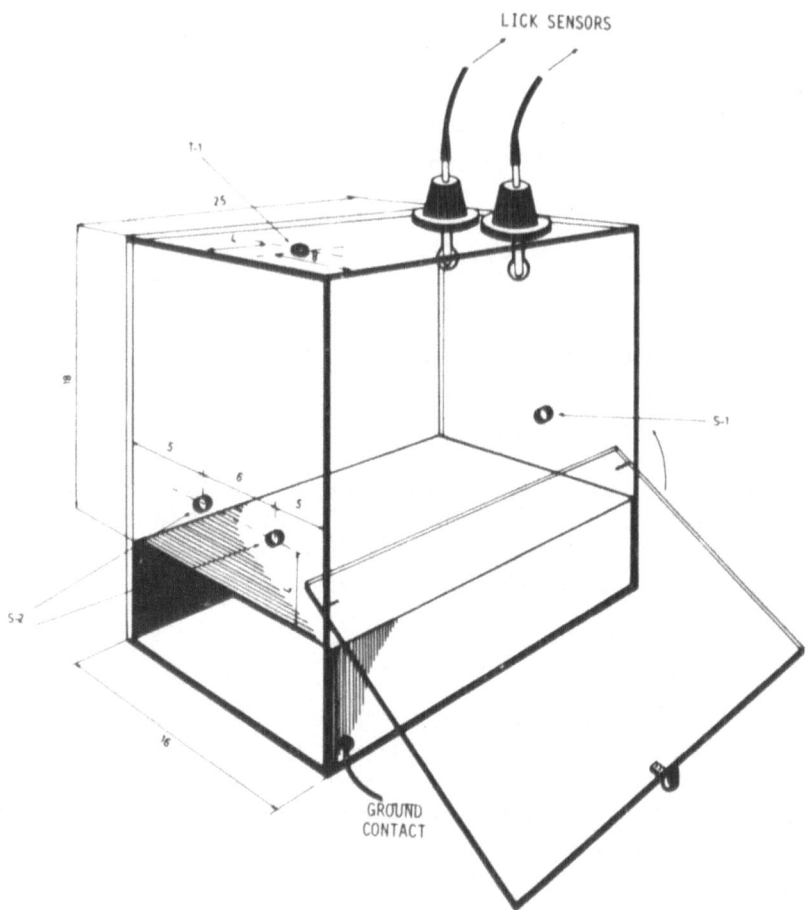

Figure 1. Test chamber that could be used for single-stimulus and two-choice test situations, for recording of licking at rods (or watering tubes) in top (T) or side (S) position. The top of the chamber was perforated to improve ventilation (not shown). Measurements are in centimeters. See also Figure 4 of Chapter 2.

present in the living cages only. Daily watering of the animals was carried out in a battery of individual cages. The ball-point drinking tube (fitting in a glass sleeve) of each cage was connected to the anode of the 30-V dc general power supply with a 465-kΩ series resistor (equivalent to the Type I lick-sensor circuit, see Chapter 2). The stainless-steel floors were grounded. This resulted in a current flow through the animal during licking behavior of approx. 40–50 μA, comparable with an estimate of the average current intensity in the input circuit of the BRS-Foringer sensor that was used in the original study (see Chapter 2, Figure 9). The animals seemed to bear these deprivation conditions very well (body weight fell to 85–90% of the *ad libitum* value). Good results were obtained, but it is of course not claimed that the schedule ensured a constant motivation level. Between different experiments with the same group of animals, rest periods were allowed under *ad libitum* conditions.

1.2.4. Standard Training

A substantial part of the data was obtained in 10-min sessions. Without prior exposure of the animals to the deprivation schedule and the test chamber, the first few 10-min test sessions might have resulted in a poor performance of the unadapted animal. For this reason, a standard training procedure was introduced, that lasted 1 week. On the first 2 days after the initial 48-hr deprivation period, the rats had access to water plus current for 20 and 15 min, respectively. From Day 3 to Day 5, sessions of 10 min were used. On the fifth day, the test chamber was used instead of the watering cages. *Ad libitum* conditions were then reinstated until the next day, followed by 48 hr of water deprivation. Then testing could begin.

1.2.5. Subjects

Most of the data in this chapter have been collected with Wistar albino rats. Frequently, female animals were employed, with a body weight of 150–170 g at the start of the experiment (their numerical availability happened to be favorable). The use of male rats or rats of other strains is indicated here by a comment. Other rodents have also been used: mice, gerbils, and hamsters. But data obtained with these animals are not included in this chapter.

1.3. Dependence of Current-Licking Behavior on Experience with Water plus Current

When current-licking behavior was observed for the first time, the animals that took part in the study had been exposed earlier to water intake with concomitant electrical stimulation of the tongue. This electrical stimulation was supplied by the lick sensor that was used to monitor licking behavior (Slangen and Weijnen, 1972).

The first step in the systematic study of current-licking behavior was to investigate whether superimposing electrical stimulation of the tongue on water licking *before* testing could fully account for the phenomenon. The effect of the training circumstances on subsequent current-licking behavior was investigated in tests with one licking rod (single stimulus) and with two rods (two-choice condition) available in the test chamber.

1.3.1. Single-Stimulus Condition

For this experiment, 40 naïve animals were used. One group (N = 20) received standard training with water *plus* current. The second group was submitted to the same procedure except that electrical stimulation of the tongue was not paired with water intake. Half of the animals of each group were tested for 10 min with an electrified licking rod. For the other rats, the rod was not electrified. In Figure 2 the performance of all four groups is shown.

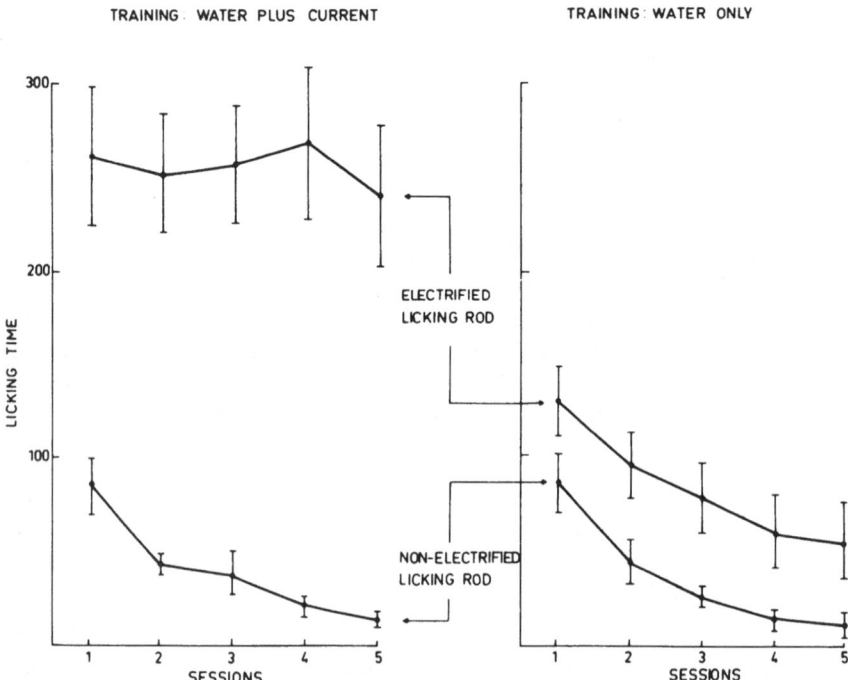

Figure 2.   Effects of prior training conditions—water with or without current—on subsequent current-licking behavior. Mean licking times (in seconds ± standard error) are given for four groups of 10 rats. There were five daily 10-min sessions. The electrified licking rod was connected to a 30-V dc supply via a 465-kΩ series resistor, resulting in tongue stimulation with approximately 40–50 μA (see Chapter 2, Figure 9). Time spent licking was measured with a manually operated keyboard connected to the running-time counters. This procedure was chosen as no other automatic method of recording licking behavior without the use of electric current was available. At the end of each of the five testing days, the animals had access to 8 ml water without current for 10 min.

The total licking time per animal, summated over 5 sessions, was used for statistical evaluation of the results.[1] For the animals trained with water *plus* current, the difference between the rats licking the electrified rod and the animals licking the nonelectrified rod was significant ($p < 0.002$). The results after training with water alone were also significant ($p < 0.02$). After both the training procedures, current licking could be demonstrated. However, contact with the electrified rod seemed to have the greatest reinforcing value for animals trained with water plus current ($p < 0.01$).

The animals of both training groups that were tested with the nonelectrified rod licked for approximately 15% of the session duration on Day 1; this percentage dropped to less than 2½% on the fifth day of the experiment. This licking behavior might be attributed to the fact that the rod was put in the position occupied by the drinking tube during training. No evidence for extinction of current-licking behavior was observed in the group of animals that had been trained with water plus current, although they did not receive further experience with the combination of water and current during the five-day testing period.

When rats display persistent current-licking behavior, the response pattern is rather stable over time. The lick frequency during periods of uninterrupted licking is lower than in water licking (Weijnen, 1975). Only a small proportion of animals show persistent licking behavior over long periods of time.

In Figure 3, data are shown from a (male) rat that did display highly persistent current-licking behavior. The median interlick interval for each block of 1000 licks was estimated from the cumulative relative-frequency distribution, and a Spearman rank-correlation test was used to evaluate a possible trend in these medians. A positive trend was found ($p < 0.01$): the median interlick interval increased slightly with successive blocks of 1000 licks, but the actual size of the effect was small. A significant trend in the number of minutes to complete each block of lick responses was not obtained. If we consider that 10,000 lick responses were emitted in approximately 75 min, we can conclude that there is little evidence in these results that points at an influence of either fatigue or satiation. The overall rate of this rat's licking was rather slow: 4.35 licks/sec (computed from the median interlick interval). Substantial and consistent differences in lick rate can be found, probably depending on many factors.

---

[1] In this chapter data are presented in terms of the mean and the standard error of the mean (M ± SE). For statistical evaluation of the results, two-tailed nonparametric tests were used. Choice tests were evaluated with the Wilcoxon matched-pairs signed-ranks test. The Mann–Whitney U test was used for independent samples (Siegel, 1956). An effect was considered to be significant if $p \leq 0.05$ for $n \leq 14$.

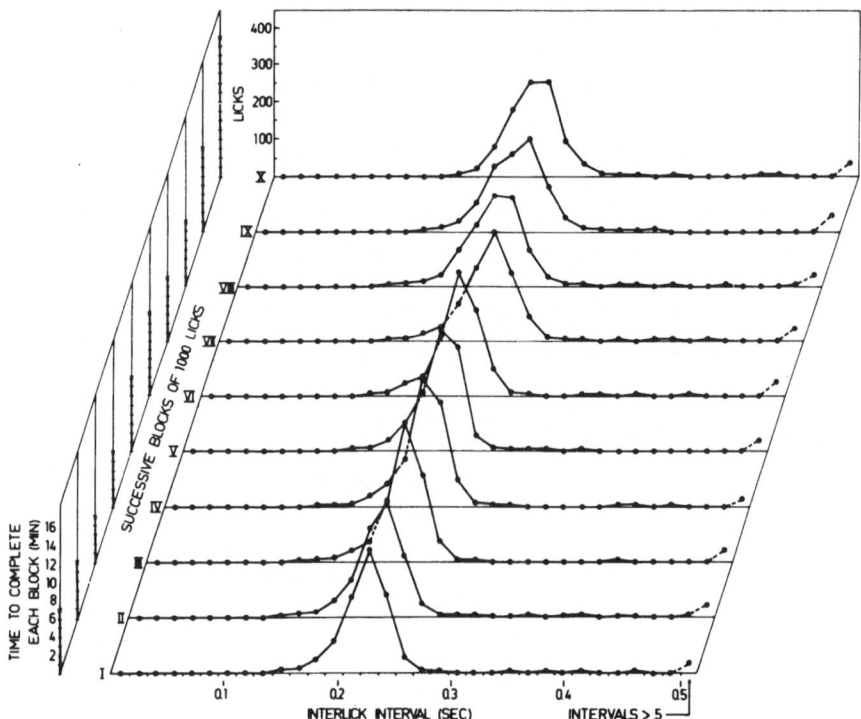

Figure 3. On-line cumputer analysis of current-licking behavior. The interlick-interval distributions are shown of 10 successive blocks of 1000 licks that were recorded in a single session that lasted approximately 75 min. Current-licking behavior was maintained by stimulation with a 50-μA current (Type II sensor).

1.3.2. Two-Choice Condition

In the single-stimulus condition, the time spent licking at the electrified rod had to be compared with the performance of other rats that had only a nonelectrified rod available in the test chamber. In this way, we could determine the contribution of other kinds of reinforcing stimulation, such as produced through orolingual contact with (relatively cool) metal objects. This is not just an academic point, as several investigators have noticed thirsty rats that lick at metal parts of their home cage.

When two licking rods are available in the test chamber and only one of them is electrified, then each subject serves as its own control, but a disadvantage of this two-choice test is the influence of position preferences that develop easily. A second test, after reversal of the positions of the electrified and the nonelectrified rods, is therefore necessary. Experience has shown that two-choice tests, each lasting 10 min, constitute an efficient way to study current-licking behavior. Training animals with water plus current reliably produces current-licking behavior during such test sessions, as is repeatedly shown in this chapter. Without reexposure to water plus current, the lick performance has been observed to drop somewhat over a period of a few weeks with daily 10-min sessions (unpublished observations), but no systematic study has been made of this partial extinction. Regular pairing of water licking with electrical stimulation maintains current licking in test sessions spread over periods of at least one year. Under these conditions no extinction of current licking has been observed.

What happens if rats are tested under two-choice conditions without prior exposure to water plus current? The results of such an experiment are given in Table 2. A significant difference was obtained in the first session only ($p < 0.01$); analysis of the Session II data that were obtained 24 hr later did not yield significant results ($p > 0.05$).

Interesting results have been obtained in recent experiments with animals that were selectively bred for good current-licking performance without prior exposure to water plus current. With these animals, current-licking behavior did not extinguish in one or two 10-min sessions, as has been found repeatedly in randomly bred animals. The average performance level, however, is still much lower than in test sessions that follow exposure to water plus current. Preliminary results also suggest that in these rats, the reinforcing properties of electrical stimulation of the tongue should be studied at lower current intensities than the routinely used 50 μA.

Table 2. *Current Licking without Prior Water-Current Pairing (Rats Reared with a Glass Waterspout)*[a]

	Time spent licking (sec)	
	Nonelectrified rod	Electrified rod
Session I	8.96 ± 2.37[b]	35.80 ± 12.26
Session II	3.27 ± 0.87	14.14 ± 5.71

[a] Current-licking performance of 12 rats that were submitted to 1 week of standard training with water alone. Glass drinking spouts were used (the significance of using glass spouts is discussed in Section 2.1.1.2). Current licking was measured for 10 min in the test chamber, equipped with 2 licking rods, on 2 successive days. The design of the experiment was balanced for position effects. At the end of the day, the rats had access to water according to the water-deprivation schedule, but the glass drinking spouts were not electrified. The same intensity of stimulation as in the single-stimulus experiment was used (see legend to Figure 2), and licking times were again measured with the manually operated keyboard.
[b] M ± SE.

1.4. Dependence of Current-Licking Behavior on Dipsogenic Conditions

Both airlicking and current-licking behavior share the requisite that the subjects should be thirsty, otherwise these behaviors can hardly be evoked. The results of current-licking studies, which were reported in the preceding part of this chapter, were obtained with rats that were kept on an arbitrarily chosen water-deprivation schedule. It was not pretended that this schedule would ensure a comparable deficit in the water balance or an equal motivational level over different experiments.

1.4.1. Influence of the Duration of Water Deprivation

In a simple experiment lacking sophistication of design, it was found that current-licking behavior is more pronounced the longer the water-deprivation period ($p < 0.001$; Friedman two-way analysis of variance). The animals were tested daily, on four successive days, without any access to water (Figure 4). These results also demonstrated again that daily exposure to water plus current is not necessary to maintain current-licking behavior over successive tests. The phenomenon is highly resistant to extinction. In

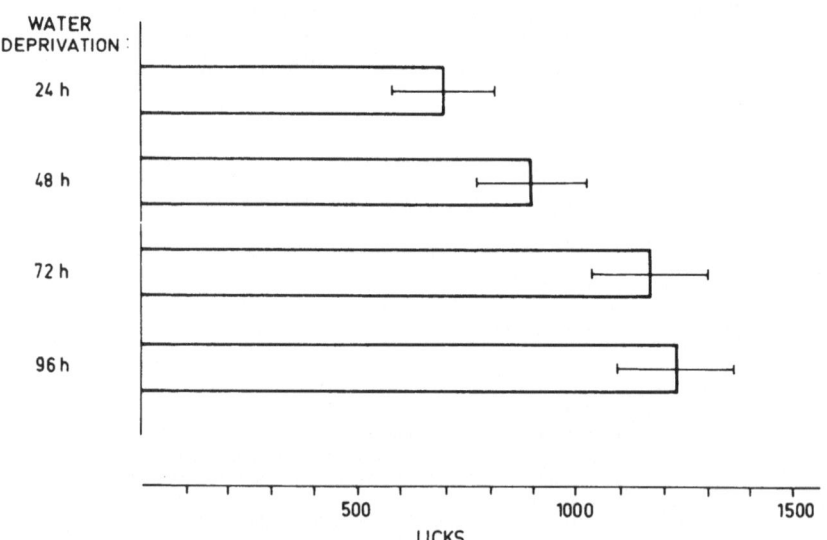

Figure 4.   Current-licking behavior of 14 "experienced" rats during 10-min sessions on four successive days of water deprivation (M ± SE). Food was continuously present in the home cage; no water was supplied between the test sessions. Licking behavior was recorded with a Type II (50 µA) sensor.

another experiment, current-licking behavior could be well demonstrated after a 72-hr period of water deprivation, as was expected. However, when *ad libitum* water was given without access to food for 1 hr preceding the test session, then only very little current licking could be observed (Weijnen, 1972, pp. 76–78).

1.4.2. Effects of Other Dipsogenic Conditions

Water deprivation constituted a simple procedure to induce current-licking behavior. It was of interest to investigate whether any dipsogenic condition would elicit current-licking behavior. Weijnen (1972, pp. 79–90) has shown that, indeed, electrical stimulation of the tongue has reinforcing properties in rats under a variety of dipsogenic conditions:

1. Treatment with hypertonic saline.
2. Eating of dry food after a period of food deprivation.
3. Administration of polyethylene glycol, a hyperoncotic colloidal solution.
4. Treatment with furosemide, a potent diuretic.
5. Central cholinergic stimulation with carbachol.

The relative potency of each of these procedures in inducing current-licking behavior is difficult to assess. An indication might have been obtained if the parameters of each procedure had been carefully chosen. If each method had resulted in ingestion of the same quantity of water per kilogram of body weight in a given period of time, then some suggestive evidence would have been procured. This has not been tried. It remains, therefore, to be shown whether, for example, depletion of intracellular and extracellular fluids have a differential influence on the level of subsequent current-licking behavior. A difference in drinking behavior engendered by these two methods of water depletion would be of interest. It could guide research that is aimed at the elucidation of the mechanism(s) involved in the reinforcing consequences of electrical stimulation of the tongue.

There exist, of course, many more ways of inducing drinking behavior in animals. However, the methods that have been used so far suggest that any dipsogenic state might be a sufficient condition for current licking to occur. Current licking engendered by a schedule of reinforcement has also been demonstrated (Figure 5). The phenomenon was less developed than schedule-induced polydipsia in the same animals under otherwise comparable conditions; not every animal displayed the effect. More research is still required to investigate schedule-induced current-licking behavior.

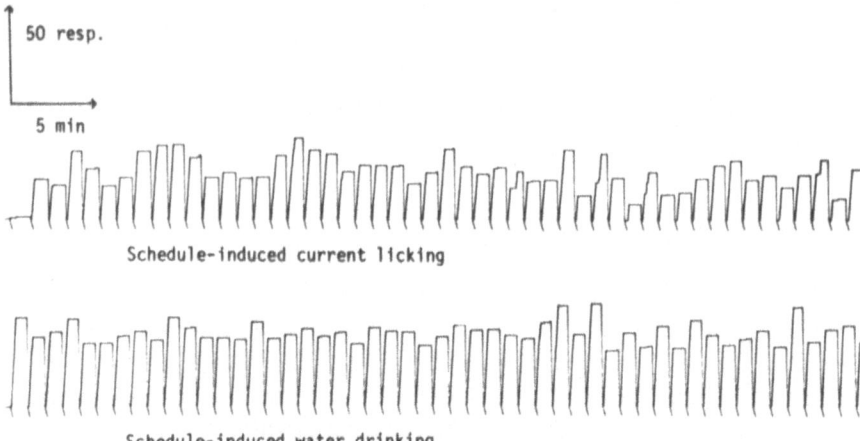

Figure 5.   Cumulative records of schedule-induced current licking and water drinking of a rat that performed well in this test situation. The body weight of the (male) rat was reduced to 85% by food deprivation. Food pellets (45 mg) were supplied on a FI-1 schedule. Either electrical stimulation or water was continuously available in two different sessions. Licking was recorded with a Type II (100 μA) sensor. Food delivery is indicated by a hatchmark on the baseline, directly after reset of the pen. The cumulative records show that immediately after pellet delivery, the animal displayed current-licking behavior or drank water.

1.5.   Current-Licking Behavior after Food Deprivation?

When water-deprived rats, which have been pretreated with water (see above), were put back in their home cage, they started to eat immediately. Their current-licking performance under this condition had been minimal. This suggested that for current-licking behavior, food deprivation might be less relevant than water deprivation.

The effect of food deprivation *per se* was tested. Animals that had displayed current-licking behavior after water deprivation were chosen as subjects. Immediately before the test sessions, each rat was given access to water, without current, for 10 min to make sure that an inadvertent dipsogenic condition would not confound the results. (Nocturnal animals eat and drink mainly in the dark, and very little, if any, water is drunk by rats under lights-on conditions. Thus, the time since light-onset in the animal quarters may function as a period of "auto deprivation.")

It appeared that testing the rats neither in undeprived conditions nor after 2 and 4 days of food deprivation resulted in an appreciable current-licking performance. Lick-contingent electrical stimulation of the tongue with 50 μA failed to maintain current-licking behavior after food deprivation of up to 4 days (Table 3).

An important aspect of the last experiment was, of course, that these rats had never experienced intake of food concomitant with electrical stimulation of the tongue. It was discussed earlier that without previous exposure to water plus current, current-licking behavior cannot be well demonstrated in water-deprived animals either.

Electrical stimulation of the tongue during intake of dry food is difficult. Presentation of a liquid diet to the rats through a pipette solved the problem. This method required a stabilized diet of suitable viscosity. To avoid neophobia effects, or any other influence of exposure to a completely new diet, the liquid diet was prepared of the usual laboratory chow. A recipe was found by trial and error: soak 1 part of dry food in 1 part of water (w/w) for approximately 12 hr; mix; add 4.5 parts of boiling water; mix thoroughly; cool to room temperature; mix before putting into the pipette (diameter of the orifice, 3 mm).

During training sessions, the animals had access to both liquid food and water. Half of the animals received liquid food with current (stainless-steel wire suspended in the liquid). Water was given without current. Only sampling of the contents of the watering tube was observed, but no bouts of drinking. It was concluded that the diet was liquid enough to prevent thirst. Table 4 shows that current-licking behavior was only weakly demonstrated during the test sessions with food- but not water-deprived rats ($p > 0.05$ for animals trained with food without current, and $p = 0.05$ for rats with experience of ingestion of liquid food plus current). These data agreed with the results of a pilot experiment. Dry-food consumption acts as a potent dipsogen. Table 4 shows that the eating of dry food for 1 hr, immediately after the first part of the experiment, resulted in current licking by the two groups (both $p < 0.05$). None of these animals had ever been exposed to water plus current after water deprivation. It was concluded that, apparently, current-licking behavior cannot be reliably obtained in animals that are only food-deprived. The phenomenon is specific for dipsogenic conditions.

Table 3. The Effect of Food Deprivation on Current Licking

	Water intake (ml) (Pretest)	Current licking (licks)	
		$0.4\,\mu A$	$50\,\mu A$
Undeprived animals	1.4 ± 0.3^a	7.8 ± 2.5	17.8 ± 5.7^b
2 days food deprivation	1.6 ± 0.3	4.9 ± 1.9	49.6 ± 25.3
4 days food deprivation	1.6 ± 0.2	7.4 ± 3.2	23.0 ± 11.1

[a] M ± SE
[b] Significantly different from 0.4-μA licks at $p = 0.05$ (for statistical details see footnote 1, Section 1.3.1).

Table 4.   Current Licking in Food-Deprived Water-Satiated Rats (Rats Reared with a Glass Waterspout)[a]

Training condition	Number of Licks			
	Before eating		After eating	
	0.4 μA	50 μA	0.4 μA	50 μA
Liquid food	41.2 $\pm$ 9.3[b]	62.7 $\pm$ 30.6	30.5 $\pm$ 8.0	192.5 $\pm$ 76.1
Liquid food + 50 μA	22.3 $\pm$ 3.8	78.0 $\pm$ 12.8	51.2 $\pm$ 8.8	488.5 $\pm$ 81.7

[a] Two groups of 6 naïve rats received 5 daily 30-min training sessions with liquid food. The animals were kept at a food-deprivation schedule that reduced their body weight to 80–85% of the *ad libitum* value. Access to water was not restricted. One group received electrical stimulation of the tongue contingent upon licking the pipette filled with the liquid diet. Current licking was recorded with Type II sensors in two 10-min test sessions immediately before and after a 1-hr period of access to dry food without water. Electrical stimulation of the tongue with 0.4 μA was used as a control in the two-choice test. It is shown later in this chapter (see Section 2.3.1) that this intensity is too low to possess reinforcing properties. The significance of rearing the rats with a glass watering spout is also discussed in a later part of this chapter (see Section 2.1.1.2).
[b] M $\pm$ SE

The performance of the rats trained with liquid food plus current suggests that a dipsogenic condition during training might not be essential for subsequent current-licking behavior. This interesting hypothesis deserves to be tested.

1.6. Lick-Contingent Electrical Stimulation of Only the Tongue

No precise knowledge is available about the current pathway through the rat in experiments involving the usual monopolar stimulation of the tongue, with the grounded floor of the test chamber as the common-reference electrode. To verify the importance of tongue stimulation *per se,* a bipolar licking rod was used (Figure 6). When the rat touched both poles of this rod with its tongue, a Type II (50 μA) lick sensor was activated (see Chapter 2). As the test cage was equipped with a Plexiglas floor, the current pathway through the rat was restricted to the tongue area between the two poles. Licking behavior at a "conventional" monopolar rod was sensed with a Type II (0.4 μA) instrument. The lick performance with this current intensity can be used as a control value; it is shown later in this chapter that reinforcing properties of a 0.4-μA current cannot be demonstrated at such low intensities (see Section 2.3.1). Figure 7 shows that the rats made significantly more licks at the bipolar rod. Lick-contingent electrical stimulation of the tongue alone is apparently sufficient to maintain current-licking behavior.

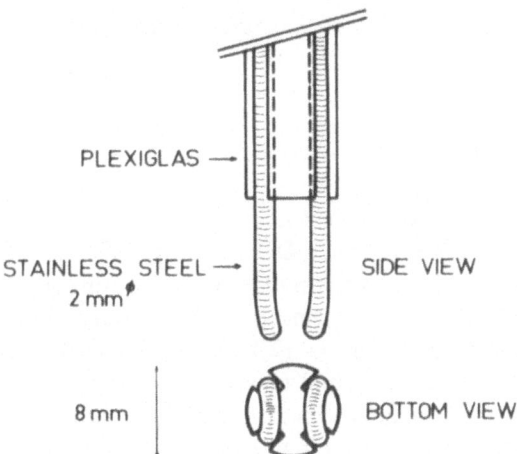

Figure 6. The bipolar licking rod.

When in other experiments rats were given the choice between stimulation with a 50-μA current via either a monopolar rod or a bipolar one, the animals preferred the former condition. It is possible that the shape of the bipolar licking rod (two small, curved wires) influenced these results. Another point is that penetration of the current into deeper structures of the tongue might be essential and that this condition is better fulfilled during monopolar stimulation.

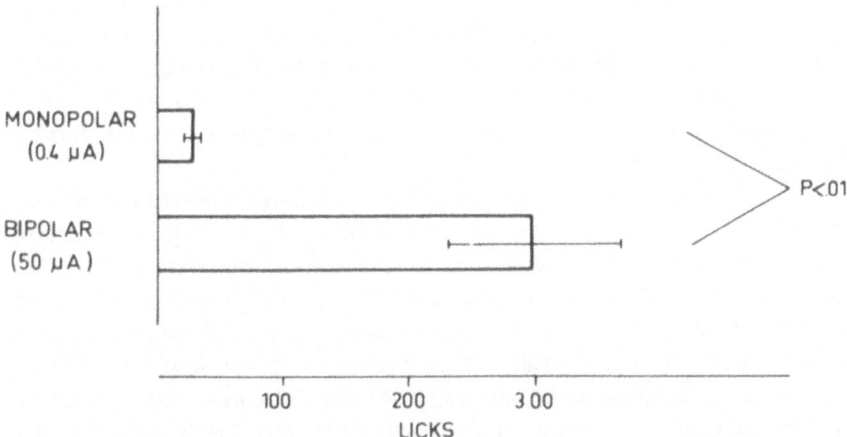

Figure 7. The reinforcing effect of bipolar electrical stimulation of the tongue *per se*. Twelve rats with current-licking experience were used in a 10-min session after 48 hr of water deprivation. A stainless-steel plate covered the Plexiglas floor under the monopolar rod only (ground contact to the Type II [0.4 μA] lick sensor).

2. Influence of Stimulus Parameters on Current-Licking Performance

2.1. The Type of Current

In the first study of current-licking behavior (Slangen and Weijnen, 1972), in which a BRS-Foringer DO-101 was employed, licking behavior was reinforced with cathodal stimulation of the tongue. This was not an explicit choice but a consequence of the power requirements of the behavioral research equipment (-12-V dc). In most of the further studies, equipment was used that required power supplies of $+24$ V, 28 V, or 30 V dc; consequently, the tongue received anodal stimulation, as the floor of the test chamber was grounded. A few experiments have been performed to investigate the influence of the type of current, during training and/or testing, on current licking.

2.1.1. Preference Studies Employing Anodal or Cathodal Stimulation of the Tongue

2.1.1.1. *Effect of Training Conditions.* Rats that had received standard training with *either* anodal *or* cathodal electrical stimulation were subsequently given a two-choice test. In this test *both* anodal and cathodal stimulation were available at an intensity of 50 μA. The total amount of time that the animal spent licking, independent of the type of current, did not reveal a differential effect of the training conditions. However, further analysis revealed that the duration of the rats' licks was longer on the rod that supplied them with electrical stimulation of the tongue that was similar to the one that had been used for the training (Figure 8). This result stressed the importance of acquired aspects in current-licking behavior. For further discussion see Section 3.6.2.

2.1.1.2. *Preference for Anodal versus Cathodal Stimulation in Untrained Rats.* The results mentioned in Section 2.1.1.1 made it interesting to investigate whether there would be a spontaneous preference for either anodal or cathodal stimulation. In a group of animals that had received standard training *without* current, a preference for cathodal stimulation was found (Figure 9). The overall licking performance was low, as could be expected after training with only water. At first, attention was focused on the physiological implications of this result. However, it soon appeared that prior, unscheduled experience with cathodal stimulation of the tongue could not be ruled out.

From weaning until the start of the experiments, the animals used for the studies of Slangen and Weijnen (1972) and Weijnen (1972) had been

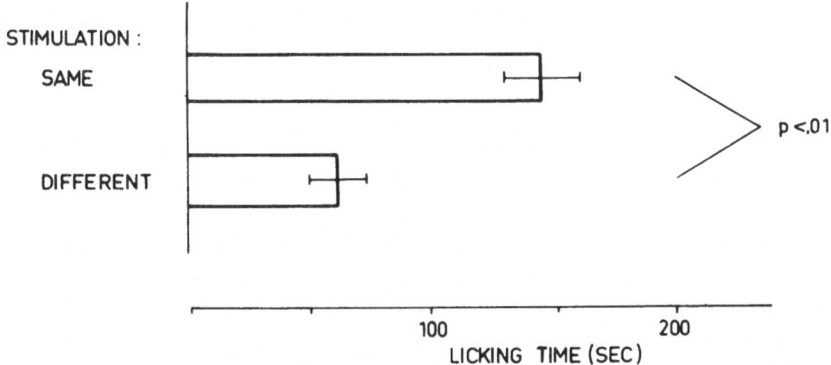

Figure 8. Influence of training conditions on stimulation preference of current-licking rats. Six rats were exposed to water plus anodal tongue stimulation (40-50 μA). A second group of 6 animals received the same standard training, except that now cathodal stimulation was supplied. During a 10-min test session, after 48 hr of water deprivation, the animals could choose between anodal and cathodal stimulation (generated by 150-V supplies with a 3-MΩ series resistor). In this figure the data are grouped according to the correspondence of the stimulation during testing to the training condition. Licking times were manually recorded.

kept in suspended galvanized (zinc) wire-mesh cages. The chrome-steel drinking spouts of the water bottles were in direct contact with these cages. An electromotive force is generated when tissue is in contact with metal. The *absolute* magnitude of the potential that is developed is not equally high for different metals and is difficult to measure. When a piece of the wire-mesh floor and a drinking spout of a water bottle were partly immersed in physiological saline, a potential *difference* of 0.9 V between the two metals was observed (oscilloscope with a 1-MΩ input impedance). In view of this observation, it is possible that some current would flow in the closed circuit

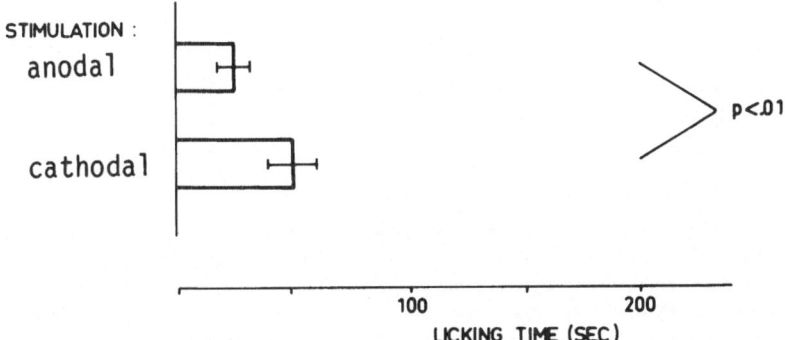

Figure 9. Stimulation preference of a group of 12 rats that had received standard training *without* current. For testing conditions, see the legend to Figure 8.

that consisted of a rat that contacted both the floor of the living cage and the attached drinking spout.

It became of interest to measure the actual current passing through an animal under optimal conditions (Figure 10). Compared to the floor of the cage, the licking spout became positive whenever the rat closed the circuit with its tongue. The direction of the current flow in this circuit was similar to the direction of the flow during cathodal stimulation of the tongue. (In following the argument about the direction of the current in the circuit, it should be realized that in the former condition, the voltage is generated by the rat; in the latter condition—during stimulation—the voltage source resides in the circuit outside the rat.)

The observed preference for cathodal stimulation in the "untrained" rats (Figure 9) confirmed the data that were described in Section 2.1.1.1. Oscilloscope inspection revealed that during licking, a potential difference of 80–100 mV was generated over the series resistor (Figure 10). Accordingly, an 8-μA to 10-μA current passed through the rat under the conditions

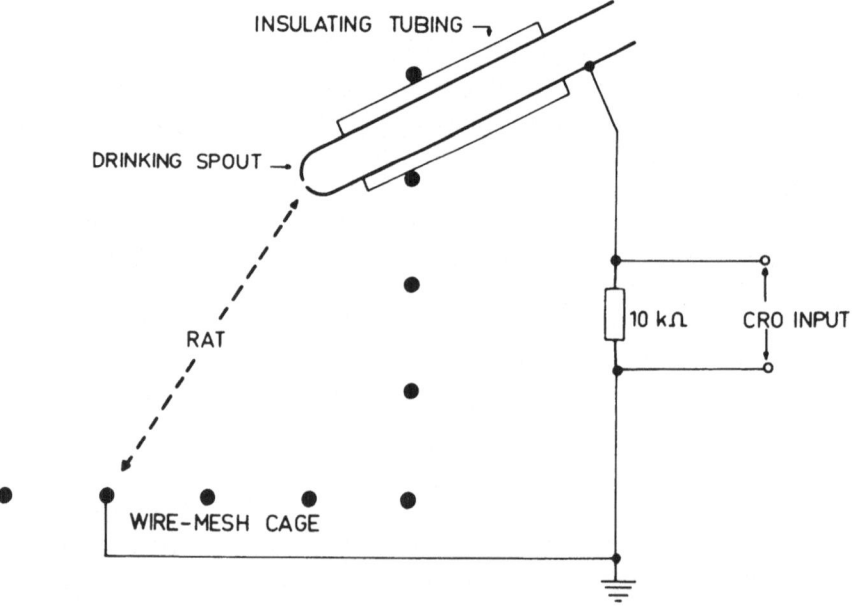

Figure 10. Measurement of the current flow through a rat that completes a circuit upon making contact with two different sorts of metal: a chrome-steel drinking spout and the galvanized wire-mesh floor (zinc). For this measurement, the direct connection between the tube and the cage was broken by insulating tubing and replaced by a 10-kΩ resistor. The voltage, developed across the resistor by the licking rat, was measured with a Tektronix-502 oscilloscope (1-MΩ input impedance). The feet of the rats ($N = 4$) and the floor of the cage were cleaned and wetted with physiological saline to reduce the contact resistance.

Table 5. Current Licking without Prior Water-Current Pairing (Rats Reared with a Glass Waterspout)[a]

	Time spent licking (sec)	
	Anodal stimulation	Cathodal stimulation
Session I	26.26 ± 5.17^b	30.10 ± 7.59
Session II	8.81 ± 2.96	17.15 ± 3.72

[a] Two-choice tests (anodal versus cathodal stimulation) involving a group of 12 rats in two 10-min sessions. Stimulation was supplied by 150-V supplies with a 3-MΩ series resistor. The regular water-deprivation schedule was used and the interval between the two tests was 24 hr. To avoid inadvertent exposure to electrical stimulation of the tongue during water intake in the living cages, the animals were reared with water bottles equipped with glass spouts. Licking times were recorded manually.
[b] M $\pm$ SE

of the experiment. Without subsequent standard training with higher current intensities, this experience—apparently—was not sufficient to establish persistent current-licking behavior.

The hypothesis was tested that rearing animals under conditions that prevent exposure to (low) levels of electrical stimulation does not result in a preference for either anodal or cathodal stimulation. To this end, the rats were reared with glass drinking tubes. Standard training (without current) was also administered with glass tubes instead of the usual ball-point tubes. When the animals were tested under the regular water-deprivation conditions, no statistically significant difference could be found in the preference for anodal or cathodal stimulation of the tongue (Table 5: Session I, $p > 0.05$; Session II, $p = 0.05$). It was concluded that there is no evidence to support the view that rats display a preference for anodal or cathodal stimulation in the *absence* of previous experience with water plus current. After exposure—experimental or inadvertent—to water plus current, rats prefer stimulation of the polarity used during training.

2.1.2. Lick-Contingent Tongue Stimulation with Alternating Current

The ample availability of lick sensors that are operated via a direct current passing through the animal has resulted in an unbalanced investigation of the reinforcing properties of dc and ac stimulation of the tongue. From an electrophysiological point of view, stimulation with alternating current has certain advantages: polarization phenomena are minimized, and metal deposit in tissue and electrolytic damage to the organism are avoided. The reinforcing properties of stimulation with alternating current have been investigated in rats that had current-licking experience with dc stimulation. Preference behavior was studied in a test chamber equipped with a pivoted

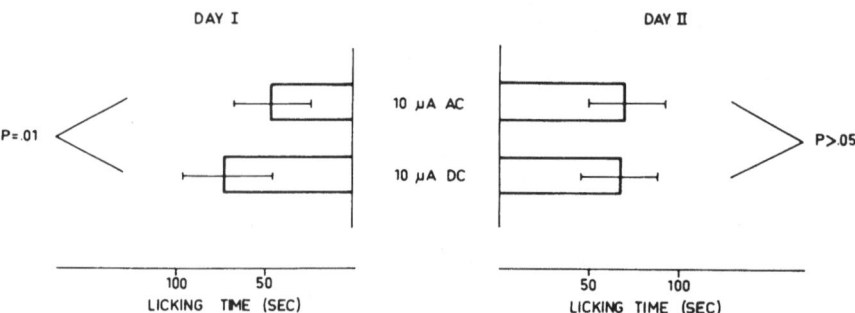

Figure 11. Two-choice tests with 10 rats with current-licking experience (dc) kept on the usual water-deprivation schedule, except that the animals received water without current for 10 min after the first session. The positions of the licking rods supplying ac and dc stimulation were reversed on Day 2.

grid floor. Movement of the rat to one end of the chamber connected the licking rod at that side, and also the grid floor, to an isolated ac power supply (150 V rms, with a 15-MΩ series resistor). Movement to the other side made the necessary dc connections (150 V, 15 MΩ). The lack of ac lick sensors made recording with a manually operated keyboard necessary.

The times spent licking in two test sessions are given in Figure 11. On the first test day, a preference for 10 μA dc was obtained ($p = 0.01$); a statistically significant difference was not found on the second day ($p > 0.05$). The Day 1 results might reflect the prior experience with dc stimulation. The importance of this difference should not be overestimated; the main objective of the experiment was to investigate whether ac stimulation would have reinforcing properties. This was demonstrated and has been replicated with other current intensities. No extensive study has been made of this particular subject.

2.2. Current Intensity during Training and Subsequent Current-Licking Behavior

Under standard training conditions, the animals are exposed to a current intensity of approximately 40–50 μA during water licking (see Section 1.2.4). In two experiments, the effects of using other intensities were studied.

2.2.1. Training with Water plus 12.5 μA

When animals were trained with water plus either anodal or cathodal stimulation of the tongue, then in subsequent test sessions, the rats showed a preference for the polarity that was the same as that which was present during training (Section 2.1.1.1). Standard training, with water plus approximately 40–50 µA, resulted in a preference for the higher of the two intensities taken from the range 1–50 µA (Section 2.3.2.1). The hypothesis has been tested that training with water plus a low current intensity (12.5 µA) would result in a preference for this intensity, if the rats could choose between electrical stimulation of the tongue with 12.5 µA or the frequently employed 50 µA.

It appeared (Figure 12), however, that 50 µA was preferred and not the intensity that had been used in training. These results point to the existence of an optimal current intensity (50 µA or perhaps higher) and not to a training-dependent value.

2.2.2. Training with Water plus 12.5 µA, 50 µA, or 200 µA

The effect of the current intensity during training on subsequent current-licking performance was further investigated. The regular standard training procedure was followed, except that the (male) rats were exposed to a current intensity of 12.5 µA, 50 µA, or 200 µA. Two-choice tests revealed that all animals showed current-licking behavior (Table 6). In the tests, 50 µA was highly preferred over 0.5 µA by each rat from all three groups (for the use of 0.5 µA as a control level of stimulation, see Section 2.3.1). However, among the groups, no difference in the level of current-licking behavior was found that could be attributed to the current intensity that was employed during training (Kruskal–Wallis one-way analysis of variance over the 50-µA licks, $p \gg 0.10$).

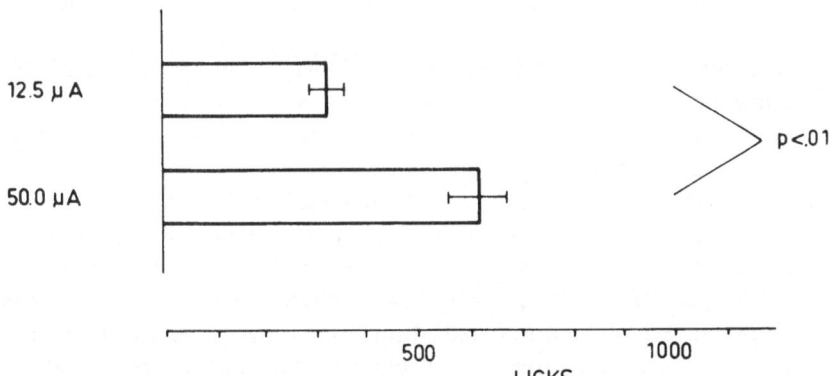

Figure 12. A 10-min two-choice test with 12 rats that had received standard training with water plus 12.5 µA. Current-licking behavior was recorded with Type II sensors (12.5 and 50 µA, respectively).

Table 6.   *Effect of the Current Intensity during Training*[a]

| Training condition | Number of licks per 10-min session | | | |
| | Day I | | Day II | |
	50 μA	0.5 μA	50 μA	0.5 μA
Water + 12.5 μA	841.5 ± 177.0[b]	89.5 ± 12.8	1078.7 ± 190.6	61.7 ± 6.1
Water + 50 μA	963.5 ± 118.1	90.7 ± 15.1	1132.5 ± 227.6	69.8 ± 20.8
Water + 200 μA	842.0 ± 123.0	74.7 ± 22.9	956.2 ± 68.4	52.2 ± 8.3

[a] Current-licking behavior after standard training with 12.5 μA, 50 μA, or 200 μA in three groups of 6 male rats that were kept on the water-deprivation schedule. Licking was recorded with Type II sensors during 10-min two-choice sessions.
[b] M ± SE

The last two experiments strongly suggest that the current-licking performance does not depend on the intensity that is used during training.

2.3. Intensity of Stimulation and Current-Licking Performance

2.3.1. The Threshold for the Reinforcing Effect of Lick-Contingent Electrical Stimulation of the Tongue

In the study of Slangen and Weijnen (1972), current-licking behavior was still obtained at intensity levels as low as approximately 1 μA. When the intensity was lowered to 0.5 μA, the animals ceased licking. Subsequent studies of the threshold for the reinforcing effect, with the two-choice procedure confirmed these results (Weijnen, 1972). A summary is given in Table 7.

In another experiment (Weijnen, 1972), when the rats could choose between anodal or cathodal stimulation with an intensity of 0.5 μA, no preference for either type of stimulation was observed. The time spent licking was low and comparable to the data for the same intensity in Table 7. With some rats from a group of animals that were chosen for their superior current-licking performance, we have recently obtained results that indicate reinforcing properties of even a 0.5-μA (anodal) current in selected subjects.

A threshold of 0.5–1 μA is very low compared to thresholds for brain stimulation in self-stimulation and stimulation-induced behavior studies. Moreover, the relatively large area of contact between the tongue and the licking rod (diameter, 8 mm) results in a very low current density. One should, of course, consider that the current density is not constant but is the highest upon making (or breaking) contact when the contact area is minimal. A short pulse of relatively high current density might in itself already provide sufficient reinforcing stimulation (see Section 2.4). It is also possible that physiological summation counterbalances the effect of stimulating a large area with a low-density current.

Based on the presented evidence, it was concluded that the use of lick sensors that are operated when a current smaller than or equal to 0.5 μA flows through the animal is justified for the recording of nonreinforced licking behavior. Any positively reinforcing effect associated with the use of these sensors as control devices in two-choice tests would, at worst, make the results obtained with the stimulation intensity under investigation more conservative.

2.3.2. The Range of Current Intensities That Can Have Reinforcing Properties

2.3.2.1. Test Chambers with a Plate Floor. Good contact with the floor of the test cage is ensured when the floor consists of a (stainless-steel) metal plate. At the same time, the relatively large contact surface reduces the density of the current passing through this area. By these means, the behavioral consequences of stimulating the feet are attenuated.

Weijnen (1972, p. 57) showed that when a rat that has been exposed earlier to water plus current can choose between two current intensities

Table 7. Current Licking in a Two-Choice Situation at Low Stimulation Levels[a]

	Day I		Day II		
1 μA	91.3 ± 35.5[b]		82.7 ± 29.9		$N = 10$
		(p 0.01)		($p < 0.01$)	
No current	22.5 ± 7.0		25.3 ± 10.7		
0.5 μA	34.6 ± 11.3		29.1 ± 8.0		$N = 10$
		($p = 0.05$)		($p > 0.05$)	(same rats as
No current	12.9 ± 2.2		19.1 ± 7.7		above)
0.5 μA	14.4 ± 4.0				$N = 12$
		($p > 0.05$)	(no second session)		
No current	11.2 ± 1.8				

[a] Time spent licking during 10-min sessions, measured with a manually operated keyboard. Rats with current-licking experience were used for these experiments. Anodal tongue stimulation was available (30-V dc supply with either 30 MΩ or 60 MΩ in series, resulting in, respectively, 1 and 0.5 μA). The rats were kept on the usual water-deprivation schedule. There was a 24-hr interval between the two sessions. Effects of position preferences were minimized as the electrical tongue stimulation during the second session was made available from the rod that was unelectrified during Session I (for statistical details see footnote 1, Section 1.3.1).
[b] M ± SE

taken from the range 1–50 μA, then the higher of the two intensities is preferred. Two-choice tests, employing combinations of 50 μA, 100 μA, and 200 μA, did not result in indications of differences in reinforcing properties among these intensities.

Intensities higher than approximately 300 μA evoke aversive reactions of the animal and suppress licking behavior. The transition zone between the higher stimulus intensities that were used in the two-choice tests and the clearly aversive current values is marked by a negative correlation between stimulus intensity and tongue-contact duration (Weijnen, 1972, p. 60). A lasting avoidance of the licking rod after stimulation with these high current intensities has not been observed. With lower intensities available, licking behavior was resumed without observable aftereffects of the previous, aversive experience. Similar observations were made after suppression of water licking at high current intensities (up to 500 μA). Fast recovery of a comparable type of passive-avoidance behavior has been reported by Braud and Prytula (1969). They observed that suppression of licking a sugar solution by contingent tongue shock extinguished almost immediately, while contingent paw shock resulted in long-lasting suppression.

2.3.2.2. Test Chambers with a Grid Floor. Aversive reactions to an electrical current passing through the animal might occur at lower current levels if the test chamber were equipped with a grid floor. The smaller area of contact between the feet and the floor in this situation increases the current density. It is generally assumed that the density of the current is a critical factor in these studies.

Two-choice experiments with current intensities between 50 and 200 μA have not been performed. However, the results of single-stimulus tests clearly indicate that an optimal level of stimulation is reached sooner when a grid floor is used (Figure 13). For this study 2.4-mm grids were used, spaced 12.7 mm apart (center to center). Current-licking behavior could still be obtained with 100-μA and 200-μA stimulation, but at a much lower level than with 50 μA. This marked difference contrasts with the results of experiments employing a test cage with a plate floor (see above).

Further evidence of the emergence of aversive components of electrical stimulation at 100 μA, in experiments with test cages equipped with a grid floor, was obtained with rats that could choose between water and water plus current (Weijnen, 1972, p. 68). No significant difference was found when the animals could choose between water with and without 50-μA stimulation. But the rats preferred drinking water from the nonelectrified tube if licking the other one resulted in water intake paired with 100-μA stimulation. On the day after this experiment, the animals displayed current-licking behavior in a two-choice test involving 0.4 and 100 μA. Apparently, tongue stimulation with this current intensity was reinforcing, in absence of water, in these animals, too.

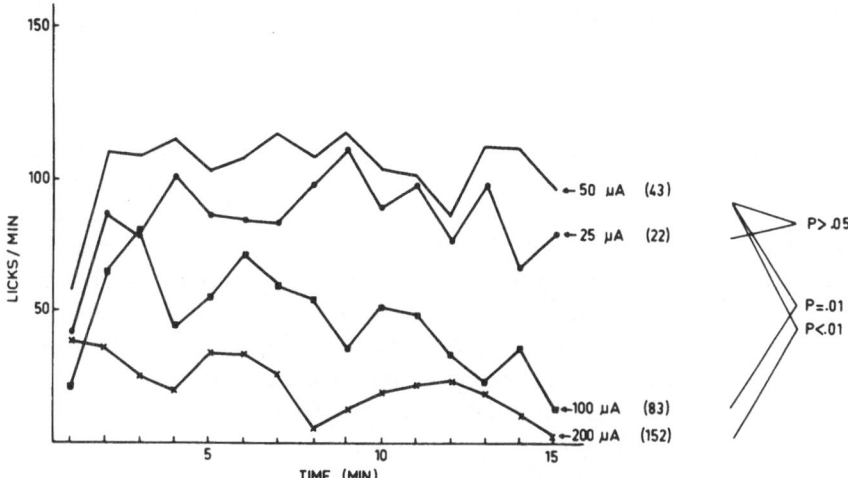

Figure 13. Mean number of licks per minute for a group of animals (N = 14) tested in a chamber equipped with a grid floor, for four nominal current intensities obtained in 15-min sessions. The median intensity of the current actually passing through the animal, estimated from oscilloscope observations, is shown in parentheses. The animals had current-licking experience in a test chamber equipped with a plate floor and had also participated in the experiment reported in Section 2.3.2.3 (see Figure 14). The licking rod was put in the side position (see Figure 1) and was connected to a Type II sensor set at 25 μA, 50 μA, 100 μA, or 200 μA. Half of the rats received the intensities in an ascending order; the other animals were tested with the highest intensity first. The total number of licks per animal over the 15-min session was used for statistical evaluation of the results (see footnote 1, Section 1.3.1).

Aversive properties of high current intensities can also be demonstrated by the recording of water or current licking with a tube or rod that is not protected by a glass sleeve (see Chapter 2, Figure 4). Under these conditions, the animals can easily receive electrical stimulation of the front feet, the nose, or the lips. Aversive responses to this kind of stimulation were observed with current intensities of 100 μA and higher. In two-choice situations, a protected tube or rod is preferred over a "naked" one at these current intensities.

2.3.2.3. Negatively Reinforcing Properties of Foot Shock with a Pulsating Direct Current. When foot shock is applied via a grid floor, the aversive threshold is reported to be as low as 15 μA (dc) in nondeprived rats (Campbell and Masterson, 1969). This value contrasts sharply with the high current intensities that still have positively reinforcing properties in the study of current-licking behavior.

Campbell and Masterson used a spatial preference technique in which the animal was allowed to choose between the absence and the presence of electrical stimulation of the feet. Movement of the rat from one side of the cage to the other end operated a microswitch that controlled the presence or absence of the foot shock and also activated timing circuits. The last 6 min of 15-min sessions were used for threshold calculation; asymptotic preference had been reached by that time. The authors' definition of the concept "aversion threshold" was: "The behaviorally halfway point between chance performance and 100% performance" (p. 13). Therefore, the threshold was attained when the animal avoided foot shock 75% of the time.

The preference/aversion for foot stimulation has been investigated in rats that had current-licking experience. In order to mimic resemblance to the current-licking situation, the animals were kept under the usual water-deprivation conditions, and the stimulation was presented in pulses of 80 msec at a rate of 5 pulses/sec (Grass, model S4 square-wave stimulator). Figure 14 shows the results. The avoidance threshold was not calculated, as the behavior underlying these data was quite heterogeneous. Most of the rats started to lick the grids at least during a part of the session. Therefore, the effects of pulsating stimulation of the feet *per se* could not be studied. The amount of tongue stimulation received depended, of course, on the synchrony between the pulsating current and the lick response. For the duration of the contact between the tongue and the grid floor, the intensity of the current passing through the animals increased because of the superior quality of the contact of the moist tongue with the grid floor (oscilloscope observation). Some rats avoided the electrified side of the chamber without licking, especially at the higher current intensities. With the 50-μA stimulation, the mean percentage of time spent on the shock side never fell below chance level, but good average preference levels were not obtained. It is probable that without licking behavior, better avoidance would have been observed. There is little doubt about the aversive properties of the 200-μA current in this experiment.

An interesting observation of the difference in the reinforcing value of electrical stimulation of the tongue and the feet was made during another experiment. The grid floor was continuously electrified, and a high intensity of stimulation was used. It was observed that a rat lifted one of its front feet each time it contacted, with its tongue, a rod of the floor of opposite polarity. This behavior stopped when the rat put its foot on the rod that was being licked.

Taken together, the results suggest that pulsating electrical stimulation of the feet alone possesses no positively reinforcing properties in water-deprived rats with current-licking experience.

2.4. Varying the Duration of Stimulation per Lick

There is evidence that in gustation, the initial—phasic—component of the neural response to sensory stimulation has a relatively high information

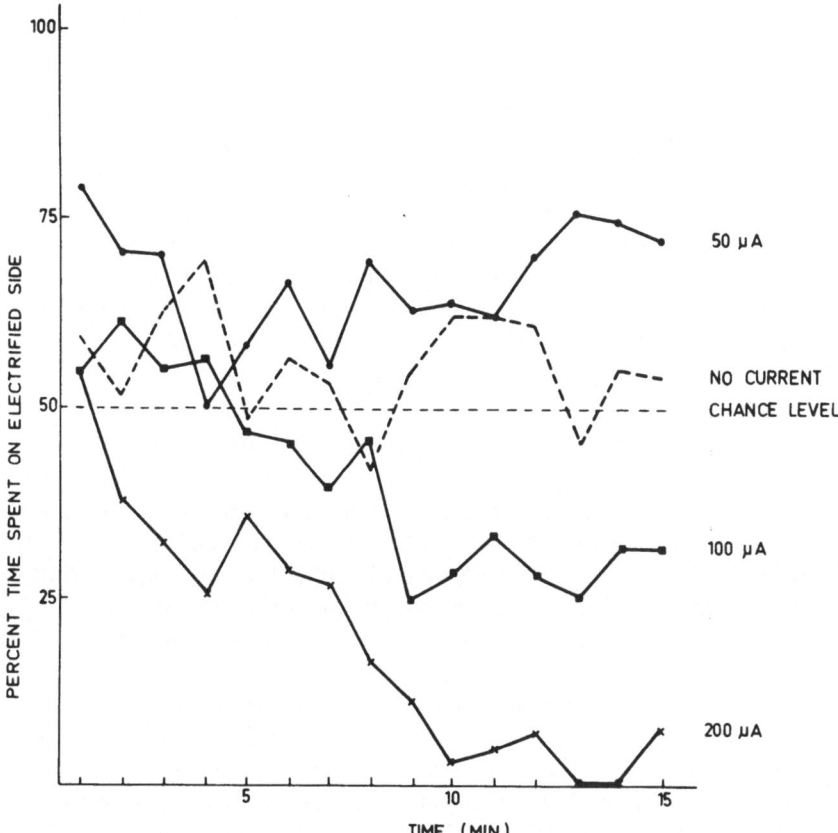

Figure 14. The effect of presenting foot shock with a pulsating current (at lick frequency) to water-deprived rats in a spatial-preference situation that was built according to the description of Campbell and Masterson (1969). The animals could terminate the foot shock by moving to the safe side of the test chamber; returning to the former side automatically switched on the foot shock. Mean values obtained with 14 rats are shown. Based upon preliminary experiments, the following shock intensities were chosen: 0, 50, 100, and 200 μA (short-circuit values). A 150-V dc source was used with appropriate series resistors. The polarity of the grids alternated from rod to rod. Each animal was tested at every intensity level; half of the group of rats started with 200 μA and the other 7 animals were tested in an ascending order of stimulation. The position of the shock side was balanced over the four sessions. Only one intensity was used per 15-min session. The experiment was started as soon as the animal had moved to the shock side; otherwise a subject that rested immobile would display spuriously high "avoidance" behavior.

content. The subsequent, more stable—tonic—component is considered to be of less importance (Faull and Halpern, 1972). The neural responses to electrical and gustatory stimulation are quite similar (Smith and Bealer, 1975). At low-to-moderate current intensities, the average duration of the tongue contact with the licking rod is roughly 80 msec. It appeared that after the stimulation time was reduced to 10% of the tongue-contact duration (8 msec), current-licking could still be obtained (Figure 15). When the animals could choose between stimulation for 8 msec per lick and stimulation for the full duration of the contact time, then the latter condition was preferred (Weijnen, 1972, p. 72). Type I (40–50 μA) sensors were used in this study. The fact that short-duration stimulation is a sufficient condition to obtain current-licking behavior stresses the importance of good control of the stimulus intensity during the initial phase of the contact of the tongue with the licking rod. It is in this very phase of the contact that the capacitance of the input circuit of the lick sensor can affect the current intensity (see Chapter 2, Section 2.2).

2.5. Ambient Light Level and Current-Licking Behavior

Rats are nocturnal animals and display relatively little spontaneous activity when the lights are on. The dependence of their behavior on the light/dark cycle is also seen in chronic current-licking experiments (lasting 24 hr or longer). Unless the dipsogenic stimulation is very strong, rats display little current-licking behavior in these chronic experiments so long as the lights are on.

Striking examples of this phenomenon were obtained in the study of the effects of various types of dipsogenic stimulation on current licking (Weijnen, 1972, pp. 86–88). It appeared that sudden changes in the light/dark cycle also changed the period in which the rats show current licking. Treatment with a potent diuretic (furosemide, 5 mg/kg sc) at noon induced current licking during the dark period only (19.00–05.00 hr). Repeating the experiment under abruptly altered light/dark conditions (lights off from 22.00 to 05.00 hr) delayed the onset of current licking to 22.00 hr. But, again, this behavior stopped as soon as the lights were put on (Fig. 16). This

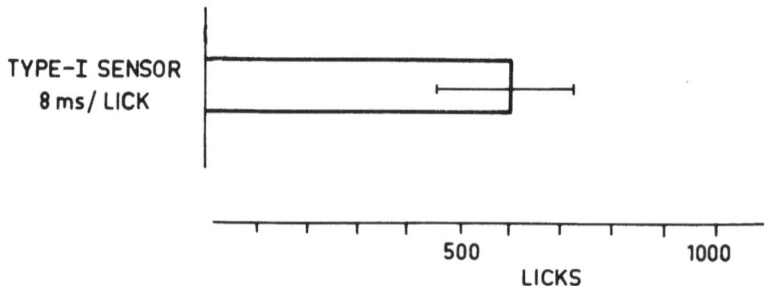

Figure 15.   Current licking under reduced stimulation-duration conditions. Fourteen rats with current-licking experience were used in a 10-min session after 48 hr of water deprivation. For technical details on the reduction of stimulation-duration, see Weijnen (1972, p. 71).

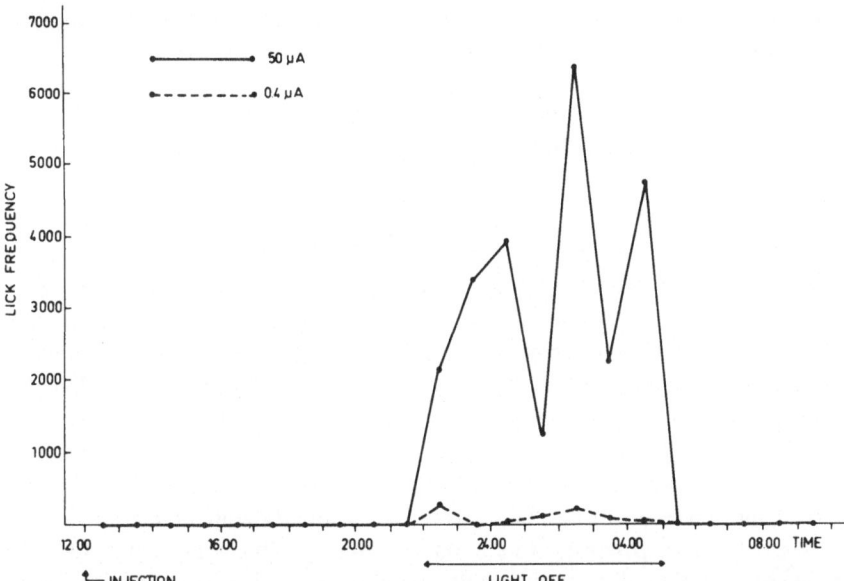

Figure 16. Current-licking performance of a group of 4 rats after treatment with furosemide (5 mg/kg of a 2.5 mg/ml solution, sc). A 75-W lamp suspended 75 cm above the test cage provided the illumination. The onset of the dark period was changed from 19.00 to 22.00 hr for this experiment. A modified living cage, equipped with a wire-mesh floor served as the test situation.

influence of the light/dark cycle does not depend on the particular dipsogenic treatment that is used; similar effects were obtained with a moderate level of water deprivation. Strong dipsogenic stimulation reduces the inhibiting effect of light on current licking during long sessions (unpublished observations). A possible effect of the illumination conditions on current licking during short test sessions remains to be investigated. Most of the experimental results that are reported in this chapter were obtained during the day, in a dimly lit test chamber.

3. Nature of the Reinforcing Effects of Electrical Tongue Stimulation

3.1. Electrical Stimulation of the Anterior Tongue

During current-licking behavior, only the anterior part of the dorsal surface of the tongue contacts the licking rod. Afferent innervation of this part of the tongue is supplied by the lingual branch of the trigeminal nerve and the chorda tympani (see Chapter 1). Nerve fibers may be directly stimulated by the current flow, although it is also possible—at least when taste is involved—that receptors are stimulated electrolytically. This holds particularly for low anodal current intensities (Beidler, 1970; Bujas, Rohaček, and Kovačić, 1975).

3.1.1. Sensory Effects

The chorda tympani nerve innervates chiefly gustatory receptors in the anterior two-thirds of the tongue and also tactile, temperature, and pain receptors. The trigeminal portion of the lingual nerve carries mainly mechanosensitive, painsensitive, and thermosensitive fibers. Kawamura, Okamoto, and Funakoshi (1968) have presented evidence showing that this latter nerve also responds to strong chemical stimulation (pain?).

The sensory effect of electrical tongue stimulation, as evidenced in humans, has a strong taste component. The review of Bujas (1971) reveals that at the make of an anodal current, a "distinctly sour or metallic-sour" taste is reported. Breaking this anodal current produces no new taste. The make of a cathodal current results in a taste quality that is difficult to characterize: "alkaline-bitter, alkaline bitter-sweet," more rarely "metallic-sour-sweet, bitter-salty or salty." Breaking the cathodal current provokes a "clear sour taste, lasting briefly and coupled with a sweet taste, especially when the posterior part of the tongue is stimulated" (p. 181). However, the kind of taste sensation is also strongly dependent on a variety of other stimulus parameters, such as pulse length and frequency. Research by Von Békésy (1964) and Plattig (1969, 1971) led to the conclusion that with properly chosen parameters, any of the four basic taste qualities can be provoked. These results were obtained with highly trained subjects who were probably experienced in discriminating taste sensation from other sensory consequences of the electrical stimulation. Large individual differences existed among the subjects.

The mean stimulation threshold in man ranges from 2 to 7 μA for anodal current, from 60 to 150 μA for cathodal current, and from 5 to 15 μA for breaking the cathodal current. The current density of the anodal threshold current is relatively unimportant, if changes in the contact area are moderate. This effect could be explained by spatial summation (Bujas, 1971; Bujas *et al.*, 1975). Apart from evoking a certain taste, electrical tongue stimulation might also be able to change taste sensation when applied during chemical stimulation of the tongue (Martonyi and Valenstein, 1971). More research is required to clarify this phenomenon.

3.1.2. Motor Effects

There is good evidence that afferent stimulation of the tongue participates in the control of tongue movements. Effects of mechanical, thermal, and taste stimuli on the linguo-hypoglossal reflex were investigated in the rat by Yamamoto (1975). Cooling the tongue enhanced, and warming depressed, the reflex that was produced by mechanical stimulation of the tongue. Modifying effects of taste stimulation were also found. Electrical stimulation of lingual- and glossopharyngeal-nerve afferents can produce reflex discharges of hypoglossal-nerve fibers (Blom and Skoglund, 1959; Blom, 1960). The implications of these results for the study of current-licking behavior deserve attention but are as yet not clear.

3.1.3. Secretory Effects

Mechanical and gustatory stimulation of the tongue can elicit salivation. It is generally accepted that secretion of saliva by the submandibular and the sublingual glands is controlled by the chorda tympani. However, there is also evidence for some contribution of the trigeminal part of the lingual nerve (Hellekant and Kasahara, 1973b). In a different study, Hellekant and Kasahara (1973a) investigated the lingual–chorda tympani reflex. Mechanical stimulation of the tongue by a stimulator driven brush increased the efferent activity of the chorda tympani. A direct relationship between efferent chorda tympani activity and secretion of saliva was established (Hellekant and Hagstrom, 1974).

The possibility might exist that electrical stimulation of the tongue also results in secretion of saliva and that this salivary secretion has reinforcing properties in water-deprived animals. Current-licking behavior was therefore also studied in desalivated animals (see Section 3.3).

3.2. Effects of Nerve Lesions on Current-Licking Behavior

3.2.1. Chorda Tympani Nerve Lesions

Section of the chorda tympani is possible at two different sites: the portion of the nerve between the intersection with the trigeminal part of the lingual nerve and the tympanic bulla, and the portion inside the middle ear. Pfaffmann (1952) has shown that after a period of approximately two months, extensive nerve regeneration can be demonstrated. It is therefore important to study the effects of the lesion within one month after the operation.

3.2.1.1. Alcohol Lesions. Before more sophisticated surgical techniques were available for this study, bilateral lesions of the chorda tympani were made with alcohol. Under light ether anesthesia, the tympanic membrane was ruptured with a blunt needle that was put on a syringe. Enough alcohol (96%) was applied through the needle to fill the inner ear and part of the ear canal. The efficacy of the lesion was tested in pilot experiments. Unilateral lesions resulted in atrophy of the submandibular–sublingual glands at the ipsilateral side. Rats with bilateral lesions showed prandial drinking. This eating–drinking pattern is characteristic for desalivated rats (Kissileff, 1969). Much food was spilled during eating. When food and water were not simultaneously available, then the animals did not manage to eat enough food to maintain body weight. The drinking bottles became soiled by accumulation of food particles. Impairment of preference for hypotonic saline solutions was also observed. These signs of functional damage of efferent fibers in the chorda tympani were taken as suggestive evidence that the afferent fibers had also been damaged. The clear desalivation symptoms suggested that the alcohol injection in the middle ear had also destroyed efferent innervation of the parotid gland (the tympanic branch of the glossopharyngeal nerve).

Current-licking behavior in alcohol-treated rats has been investigated in several studies.

In one experiment, a group of rats with ample current-licking experience showed a dramatic decrease in current licking after bilateral alcohol application to the middle ear, while a control group that was treated with 0.9% saline performed normally (Figure 17). These results were obtained 2 weeks after the alcohol injections, during an all-night session that started after 35 hr of water *and* food deprivation.

In a different study, two groups of six naïve rats were used. Following alcohol or control treatment (0.9% saline) and a 2½-week period of rest, the animals received standard training with water plus current. Then, 4 weeks after the injections the animals were tested in 10-min two-choice tests on two successive days. They had been kept on the regular water-deprivation schedule since standard training. Normal current-licking behavior was observed in the saline-treated rats; however, neither on the first nor on the second day was any evidence obtained that the experimental animals preferred to lick at the rod supplying electrical stimulation with a Type I sensor (Weijnen, 1972, pp. 94–95). During standard training, no differences in water-licking behavior had been noticed between the two groups. From this observation it was concluded that licking behavior *per se* was not affected by alcohol lesions in the middle ear.

A major problem in experiments with rats lacking saliva is that they do not eat dry food in the absence of water. Consequently, water deprivation functionally means water *and* food deprivation in these animals (Mendelson, Zec, and Chillag, 1972). Eating of dry food is a potent dipsogen. From this, we should conclude that the dipsogenic condition of the experimental and the control group of the second experiment was not comparable, which complicates the interpretation. In the first experiment the rats had been tested after deprivation of water *and* food.

The results of these experiments did not contradict the hypothesis that

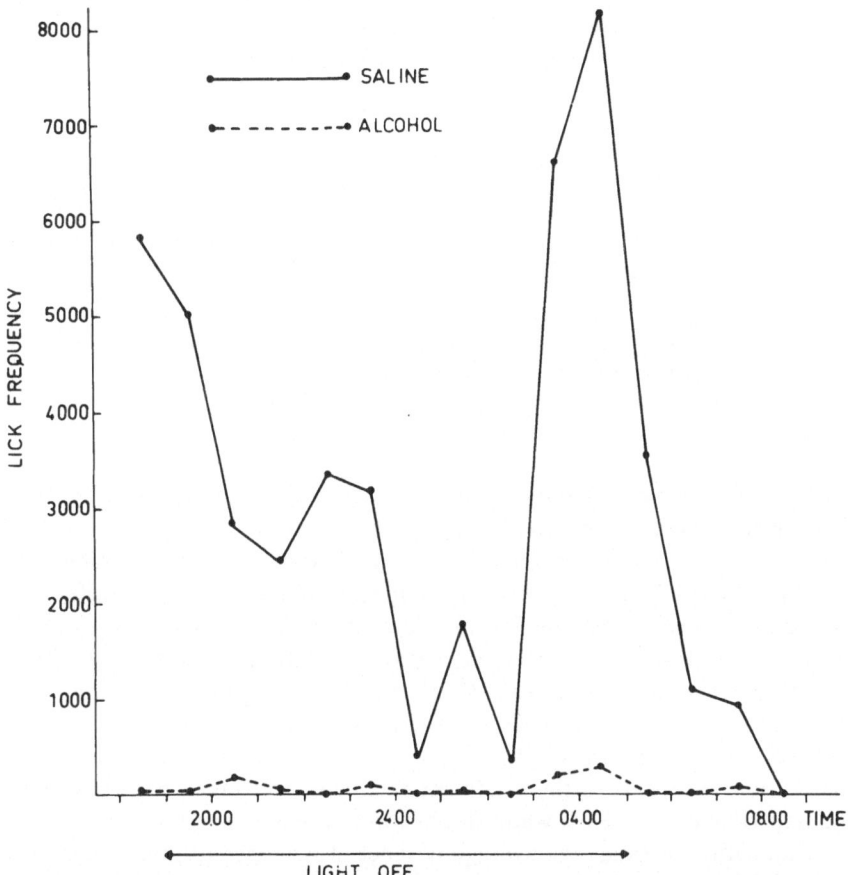

Figure 17. The effect of bilateral alcohol injections in the middle ear on subsequent current-licking behavior compared with the effect of saline control injections during an all-night group test. Two groups of 5 rats were used. Each group of rats had one licking rod available connected with a Type I sensor. Food and water had been removed 35 hr before the start of the session.

an intact chorda tympani nerve is a necessary condition for current-licking behavior. But further investigation of the effects of desalivation *per se* is required. Studies performed with desalivated rats support the hypothesis that intact afferent fibers of the chorda tympani nerve are essential for current-licking behavior (see Section 3.3).

3.2.1.2. Transection of the Chorda Tympani Nerve. When the chorda tympani is transected bilaterally between the intersection of this nerve and the trigeminal portion of the lingual nerve and the bulla, then the innervation of the parotid glands is spared and only the submandibular and sublin-

*Table 8.   Current-Licking Behavior in Rats after Bilateral
Transection of the Chorda Tympani Nerve*

| | Number of licks per 10-min session (Type II sensors) | | |
	50 μA	0.4 μA	
Session I	83.3 ± 17.9[a]	58.7 ± 21.1	($p > 0.05$)
Session II	77.2 ± 25.3	47.8 ± 15.4	($p > 0.05$)
	(airlicking)		
Session III	65.8 ± 12.0	69.7 ± 14.0	($p > 0.05$)

[a] M ± SE (for statistical details see footnote 1, Section 1.3.1).

gual glands become nonfunctional. Details of the operation, which was developed by Meiselman, have been given by Halpern (1973).

Six naïve rats were operated and received standard training (water plus current) two weeks later. Two-choice current-licking tests were performed in the third week under the regular water-deprivation conditions. No evidence for reinforcing effects of electrical stimulation of the tongue with a 50-μA current was obtained (Table 8). With appropriate reinforcing stimulation of the tongue, however, these rats could lick as well as normal subjects, as was shown by stimulation of the tongue with air contingent upon a lick. Previous to this study the rats had never taken part in airlicking experiments. An 85% water-deprivation schedule was used. After two familiarization sessions, the average number of licks per 10-min session was 2349 and 2109 on two subsequent days. The licking rate was 6.5–7.0 licks/sec. On the following day the tube supplying air was removed and two licking rods were reinstalled. Again, current-licking behavior could not be obtained (Table 8). It was concluded that bilateral transection of the chorda tympani nerve abolishes the reinforcing effect of electrical tongue stimulation but that licking behavior as such is not affected. Rats that have undergone this operation are only partially desalivated; the efferent innervation of the parotid glands is still intact. Prandial drinking patterns were not observed.

3.2.2. Transection of the Trigeminal Portion of the Lingual Nerve

When the chorda tympani nerve had been cut bilaterally, reinforcing properties of lick-contingent electrical tongue stimulation could not be demonstrated. Sensory consequences of stimulation of the anterior part of the tongue, traveling to the brain via the remaining trigeminal portion of the lingual nerve, apparently lacked reinforcing properties. Current licking has been studied after bilateral transection of this nerve (Weijnen, unpublished observations). The results of these experiments showed that, indeed, current-licking behavior could still be obtained. However, some animals did not perform well. This effect could be due to the fact that after the operation, all animals showed damage of the tongue caused by biting. The consequences of this damage are difficult to investigate. Normal afferent functioning of this part of the tongue via the chorda tympani was no longer certain.

3.3. Current-Licking Behavior in Desalivated Rats

Experiments with desalivated rats were necessary to investigate the possibility that reflex secretion of saliva in response to electrical tongue stimulation might constitute the reinforcing effect (see the discussion in Sections 3.1.3 and 3.2.1). The submandibular–sublingual glands can be easily extirpated. Removal of the parotid glands is very difficult, but the ducts can be ligated. After the operation, the animals develop all the "classical" symptoms of desalivation (see Section 3.2.1.1). Not all secretion of fluid is arrested after the treatment, but the contribution of the remaining sources is negligible (Schneyer and Schneyer, 1959).

A group of six naïve rats was desalivated and subjected to standard training one month later. This relatively long delay was necessary, as it took the animals a few weeks to stop losing body weight. Rats that had received alcohol injections in the middle ear recovered faster. The cause of the difference is not clear. This observation raised the question whether the alcohol injections had completely stopped secretion of saliva from the submandibular–sublingual and parotid glands. Behavioral signs of desalivation do not necessarily guarantee that secretion is completely blocked. The contribution of sympathetic nervous innervation of the glands in the secretion of saliva might have been underestimated. It is also possible that ligation of the parotid ducts induced painful swelling of the glands, which might have affected eating behavior.

During standard training, the animals were not given access to the usual dry lab chow, but to a moist mash. This mash was made drier every day (Mendelson, Zec, and Chillag, 1972). By this procedure, the animals could eat in the absence of water. The relatively low ultimate water:food ratio of the mash (1:2) made it possible that eating still contributed to the dipsogenic state of the animals. Current licking was investigated while the rats were maintained on the mash diet. The results are presented in Table 9. All animals preferred tongue stimulation with 50 μA on both test days. The

Table 9.   Current-Licking Behavior in Desalivated Rats

| | Number of licks per 10-min session (Type II sensors) | | |
	50 μA	0.4 μA	
Session I	216.5 $\pm$ 73.9[a]	64.5 $\pm$ 14.9	(p = 0.05)
Session II	197.5 $\pm$ 65.3	28.2 $\pm$ 13.3	(p = 0.05)

[a] M $\pm$ SE (for statistical details see footnote 1, Section 1.3.1).

total number of licks was relatively low, as was the quantity of water drunk after the tests in a 10-min period. It is likely that a more pronounced dipsogenic state of the animals would have resulted in better-developed current-licking behavior.

The same animals were used for another experiment eight months later. Prandial drinking behavior could still be observed at that time. All the rats displayed current-licking behavior after injection with hypertonic saline (750 mg/kg of a 15% solution sc). With another group of desalivated rats, current licking was obtained in all-night sessions after treatment with a diuretic (furosemide, 5 mg/kg).

In conclusion, desalivated rats can display current-licking behavior. This phenomenon, therefore, cannot be based on reflex secretion of saliva.

3.4. Relative-Reinforcing Attribute of Electrical Tongue Stimulation

Under dipsogenic conditions, rats can be reinforced with water, oral cooling, and electrical stimulation of the tongue. Some research has been carried out to compare the reinforcing properties of these different types of stimulation.

3.4.1. Water Compared with Electrical Tongue Stimulation

When water-deprived rats can choose between drinking tap water at room temperature and current licking, then the results are unambiguous: water is preferred (Figure 18). A parametric study of this subject has not been performed, but the results obtained with a current value taken from the range of optimal reinforcing intensities might be generalized without great risks. Additional evidence for the difference in the reinforcing effect of water and electrical stimulation is supplied by the observation that if two drinking tubes are available in the test chamber, the rate of alternation between the tubes is rather low. However, relatively high alternation rates are observed when two licking rods are present, the rate of alternation depending on the intensity of and the difference between the two currents.

3.4.2. Water Compared with Water plus Current

In choice tests, rats prefer or avoid fluids with a certain taste. Effects of a temperature difference can also be observed. Specific information on these topics is presented in other chapters of this book. The effects of the extra stimulation supplied by the taste or the temperature apparently add to the compound stimulus pattern that is generated when water is drunk. The possibility of a difference in the reinforcing properties of water and of water plus current has been investigated repeatedly by the author over a range of current intensities. A consistent preference for water plus current has never been obtained. Once water-deprived rats start licking water from a drinking tube in a two-choice situation, response momentum keeps the animal licking (Pfaffmann, Fisher, and Frank, 1967). A position preference is readily established. Both factors result in large variations in the experimental data and might therefore easily conceal a subtle preference, if at all present.

Earlier in this chapter (Section 2.3.2.2), it was reported that water plus 100 μA is even avoided, although this intensity maintained current licking in the same situation (test chamber equipped with a grid floor). Chronic tests with water and water plus current available in the living cage have also failed to demonstrate any preference for water plus current. In all these experiments, tap water was used at room temperature.

3.4.3. Airlicking Compared with Current Licking

Airlicking behavior can be obtained in rats without great difficulties. It requires no special training procedure. Persistent licking behavior will

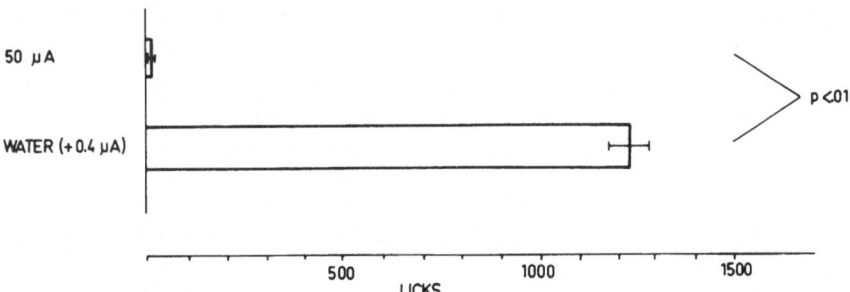

Figure 18. Preference behavior of 20 rats with current-licking experience after 48 hr of water deprivation. Water and lick-contingent electrical tongue stimulation, supplied by a Type II (50 μA) sensor, were available in a 10-min test. Water licking was recorded with a Type II (0.4 μA) sensor.

develop soon in most water-deprived animals. Current licking, on the other hand, extinguishes rapidly if the rats are not trained with water *plus* current. Only very few animals lick as persistently as airlicking rats. Current-licking animals will also more readily stop licking when disturbed.

Parametric studies aimed at comparing the reinforcing properties of the stimulation supplied by an air stream and electrical tongue-stimulation have not been performed. The only available evidence (Weijnen and Mendelson, unpublished results) was obtained with a group of 18 male and female rats of two different strains (Long Evans and Sprague Dawley). The animals had a history of both airlicking and current licking and were kept on an 85% water-deprivation schedule. Two-choice tests lasting 10 min were performed on two subsequent days. Airlicking was sensed with a Type II (0.4 μA) sensor; the air stream was made available contingent upon the lick response and had a pressure of 4-cm water. Current licking was also recorded with a Type II sensor; the intensity of the current was set at 50 μA. The positions of the licking rod and the tube supplying air were interchanged after the first session to minimize the influence of position effects. The operate/reset value of the timing circuits was 200 msec. For more information on the technique of measuring time spent licking in a lick-contingent situation, see Chapter 2, Section 3.2. The pooled data of the two sessions showed that on the average, the rats spent 65% of the time airlicking and only 1.7% current licking. Under the conditions of the experiment, the preference for airlicking seemed to be very clear.

3.4.4. Airlicking Compared with Licking Air plus Current

The situation is less clear when rats can choose between airlicking and licking air concomitant with electrical tongue stimulation. Weijnen and Mendelson (unpublished results) investigated the possibility that the reinforcing stimulation supplied by an air stream would be additive to the effect of lick-contingent electrical stimulation of the tongue. Thirty rats participated in this experiment. The experiment history of the rats and the testing situation were similar to the conditions of the experiment described above. Preference for air *plus* current was displayed only during the first session ($p < 0.001$, Wilcoxon matched-pairs signed-ranks test).

3.5. Reinforcing New Behavior with Electrical Stimulation of the Tongue

Reinforcers have the power to produce the learning of new responses. Would electrical tongue stimulation strengthen new behavior? A 10-μA current, supplied by a Type II sensor, was made available from a licking rod

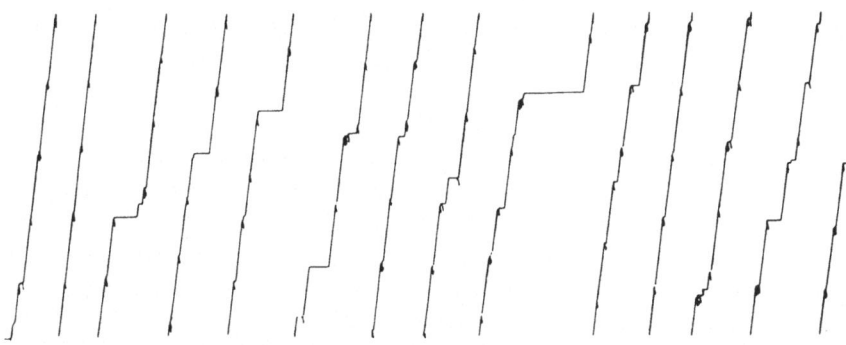

50 LICKS

1 MIN

Figure 19. Lever pressing for lick-contingent electrical tongue stimulation. Cumulative record of the licking performance in a 30-min session. A 10-μA current was available for 15 sec upon a press of the lever. Lever pressing was registered on the event trace. The hatchmarks on the cumulative record indicate 0.4-μA licks. For further details see Weijnen (1972, p. 44).

for 15 sec upon the pressing of a lever. Between these periods, the current intensity was lowered to 0.4 μA. A cumulative record of a water-deprived male rat that performed well is given in Figure 19. Licking stopped abruptly when the current intensity dropped from 10 to 0.4 μA after each 15-sec period; a new lever-press made the reinforcing stimulation available again. Other current intensities have also been employed (Weijnen, 1972, p. 57). New learning was produced and could be maintained by lick-contingent electrical stimulation of the tongue.

3.6. Current Licking Maintained by Primary or Conditioned Reinforcement?

3.6.1. Sensory Reinforcement

Current-licking behavior as displayed by rats without previous experience of water *plus* current might be accounted for by "sensory reinforcement." That sensory stimulation or, better defined, brief changes in sensory stimulation can have reinforcing properties is well known and has been demonstrated for different sense modalities (cf. Tapp and Long, 1968; Glow and Russell, 1975). Not only adequate but also inadequate stimulation of peripheral receptors can produce these effects (see Section 1.1). A specific deprivation state of the animal is not required, but hunger and thirst may increase the frequency of a response that is reinforced by a change in sensory stimulation (Tapp, Mathewson, and Simpson, 1968). Rates of responding are relatively low when sensory change serves as a reinforcer. Satiation effects are clearly obtained within sessions (Campbell, 1972).

Some cooling of the tongue may result from licking at a metal rod. This was evidenced by temperature measurements with a thermistor placed in the tip of a licking rod. Temperatures exceeding 30°C were observed at a room temperature of 26°C. The cooling of the tongue cannot in itself have constituted the reinforcing stimulation in the current-licking studies. The two-choice procedure eliminated this possibility. Moreover, no preference for either rod was observed in a two-choice test in which rats could choose between lick-contingent electrical tongue stimulation, supplied by Type I sensors, by licking a rod kept at 37.5°C or by licking a nonwarmed rod (Weijnen, unpublished observations).

We cannot exclude the possibility that nerve fibers that respond to cooling of the tongue are also activated by electrical stimulation. But this hypothesis does not explain the great increases in the reinforcing consequences of electrical tongue stimulation after training with water *plus* current. It is as yet not evident whether or not oral cooling in water-deprived airlicking or cold-licking rats qualifies as a sensory reinforcer. The physiological consequences of oral cooling are poorly understood.

3.6.2. Evidence in Favor of a Conditioned-Reinforcer Interpretation of Current Licking

Pairing water intake with a new stimulus—electrical stimulation—strongly increased the reinforcing properties of lick-contingent electrical tongue stimulation (see Section 1.3). Standard training with either anodal or cathodal stimulation resulted in a preference for electrical tongue stimulation with a current that was similar to the training situation (see Section 2.1.1).

Generalization across deprivation states of the animal did not occur after standard training of water-deprived rats with water *plus* current. Neither nondeprived rats (see Section 1.4) nor food-deprived animals (see Section 1.5) displayed current licking.

3.6.3. Challenge to the Conditioned-Reinforcement Interpretation

A conditioned-reinforcer hypothesis cannot fully explain the current-licking phenomenon. Without previous experience with water *plus* current, some reinforcing properties could already be obtained (see Section 1.3). It is not clear how electrical stimulation that was paired with water intake could have become an effective conditioned reinforcer that is very resistant to extinction. Efficient procedures for establishing a conditioned reinforcer usually require that the stimulus shortly precede the primary reinforcer. During standard training, the electrical stimulation coincides with the stimulation supplied by water (tap water at room temperature). It is further difficult to understand how, in the standard-training procedure, electrical tongue stimulation could acquire informative, uncertainty-reducing, value. Rats have no difficulties in detecting tap water at room temperature. Extra stimulation supplied by the current seems rather redundant. Moreover, stimuli that are presented concurrently with primary reinforcement usually do not become secondary reinforcers. In general, conditioned reinforcers are relatively weak reinforcers: the effect extinguishes rapidly. This is not the case with electrical stimulation of the tongue. For a general discussion of the establishment of secondary reinforcers, see Kelleher and Gollub (1962) and Bolles (1975, Chapter 13).

One would expect optimal current-licking behavior to be obtained when conditioning and testing conditions are similar. The experiments with anodal and cathodal stimulation are in line with this prediction. However, this outcome was not achieved when the intensity of the current was varied (see Sections 2.2.1 and 2.2.2). There is a range of optimal intensities independent of the current intensity that is used during standard training (see Sections 2.3.2.1 and 2.3.2.2).

Increasing the current intensities to values that resulted in clearly aversive reactions could not inhibit current-licking behavior at subsequent lower levels of stimulation (Section 2.3.2.1). Pairing of liquid-food consumption with electrical tongue stimulation in food-deprived animals resulted in very little current-licking behavior by hungry rats in subsequent tests. Dipsogenic stimulation, however, was an effective condition for this behavior, although no pairing of water with current had taken place in these animals (Section 1.5).

Additional research was performed to further investigate conditioned aspects of current-licking behavior.

3.6.3.1. Establishing Lick-Contingent Electrical Stimulation as a Conditioned Stimulus in an Avoidance Situation. An interpretation of current-licking behavior in terms of a conditioned-reinforcement hypothesis implies that the electrical stimulation of the tongue constitutes a stimulus that has acquired *positively* reinforcing properties by repeated pairing with the unconditioned reinforcer: water. This hypothesis predicts: lick-contingent electrical stimulation of the tongue that is established as a *negatively* reinforcing stimulus is ineffective in maintaining current-licking behavior. An experiment was designed to test the hypothesis.

If licking at one of two drinking tubes is punished by foot shock, water-deprived rats soon learn to avoid this foot shock if it is possible to discriminate between the tubes. Lick-contingent electrical stimulation was used as a conditioned stimulus that signaled foot shock in this passive-avoidance, or punishment, situation. To adapt the animals to the water-deprivation schedule, 12 naïve rats first received one week of standard training *without* current. The following week, passive-avoidance training began. The animals could obtain water from either of the two drinking tubes. Every 25th lick at one of these tubes that was recorded with a Type I sensor was automatically followed by unscrambled foot shock lasting 0.4 sec. The intensity of the sensing current was approximately 30–40 μA (see Chapter 2, Figure 9). The intensity of the foot shock was set at 150 μA ac (short-circuit value). Of the animals 6 received passive-avoidance training: the other rats were subjected to the same procedure except that no shock was given. They served as controls.

   Fast acquisition of passive-avoidance behavior was obtained in the experimental group (Figure 20). The total water intake did not differ from the intake of the control group. Calculated over the whole training period, the experimental group received an average of 1.1 foot shocks/animal/session, compared to 27 "sham shocks" for the control animals. This does

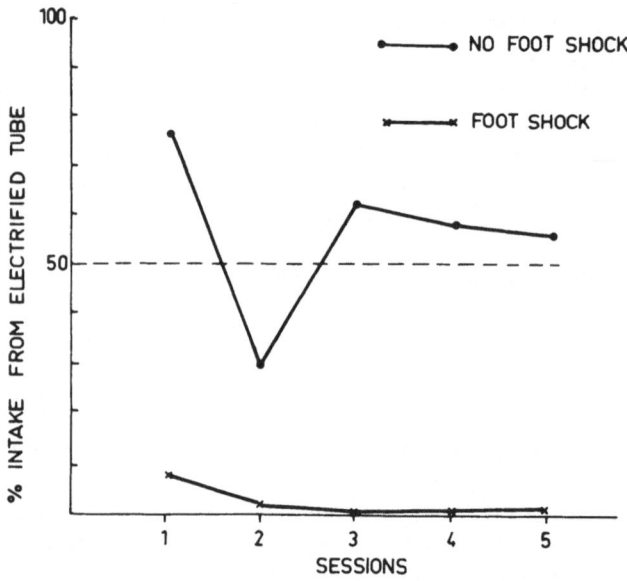

Figure 20.   Percentage of water intake from the bottle connected with the Type I sensor for two groups of rats. One group received foot shock after every 25th lick. Drinking from the other bottle had no scheduled consequences. There were five daily 10-min sessions. A test chamber equipped with a grid floor was used. The position of the tube connected with the sensor was changed every day. Foot shock was supplied by a 300-V ac supply with a large resistor in series with the rat. During foot shock, the sensor circuit was disconnected; between shocks the grid floor was grounded.

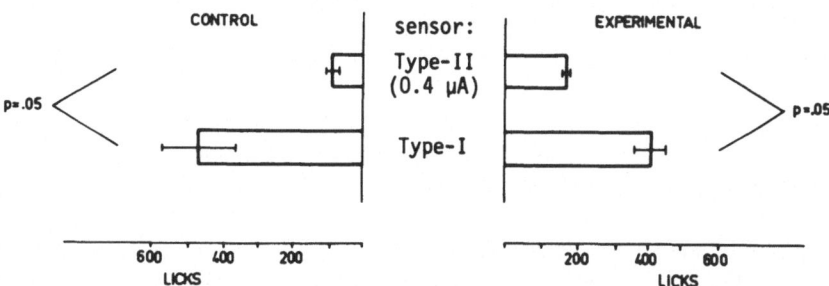

Figure 21. Current-licking behavior of rats after training that established electrical tongue stimulation as a conditioned stimulus in an avoidance situation (see also legend to Figure 20).

not mean that the rats stopped licking the water bottle that was connected with the Type I sensor as soon as foot shock was received. In 40% of the sessions, licking stopped before the 25th lick, so that no foot shock had to be applied. Inspection of the cumulative records showed that when the experimental animals licked the tube that was connected to the sensor, licking was frequently interrupted by pauses. The control group displayed very regular licking behavior. It was concluded that lick-contingent electrical stimulation of the tongue served as an effective conditioned stimulus and had acquired negatively reinforcing properties.

After the regular weekend regime (1 night *ad libitum* water and food, followed by 48 hr of water deprivation), current licking was investigated in the same test situation. In Figure 21 the results of the test session are presented. Both groups displayed significant current-licking behavior; all animals showed a preference for electrical stimulation of the tongue supplied by the Type I sensor. The modest level of significance ($p = 0.05$) may be related to the small size of the groups ($N = 6$). The current-licking behavior of the experimental animals becomes more impressive if we realize that these rats had received very little experience with the pairing of water and current. During avoidance training, a mean number of 33.3 licks/rat/session was recorded with the Type I sensor for these animals, compared with 679.7 licks for the controls. In a second session, balanced for position effects, similar results were obtained. When the training situation, but without foot shock, was reinstated for two sessions, the experimental animals still preferred water without current, although not so clearly as during training.

Similar results had been obtained in a preliminary experiment in which suppression of water licking was studied in a conditioned emotional-response paradigm.

One could argue that the sound produced by the relay of the sensor might have been established as the real discriminative stimulus, as the second drinking tube had not been connected to a sensor during training. But if this were true, then the animals should have avoided licking at *both* rods during the current-licking test, as each of the licking rods was now connected to a sensor. Recent studies have shown that rats can easily discriminate water from water plus current at comparable levels of current intensity without any auditory feedback.

It is difficult to see how the procedure employed in the above experiment could have been successful in establishing the sensor current as a positive conditioned reinforcer. One could postulate that the consequences of electrical stimulation of the tongue that was paired with water licking were substantially different from electrical stimulation without water, with the consequence that this stimulation did not serve as a conditioned stimulus during the test session. But then the reason for testing the conditioned-reinforcement hypothesis is no longer present.

3.6.3.2. Electrical Stimulation of the Tongue in a "Poison-Avoidance" Paradigm. There is ample evidence that a brief exposure to gustatory stimulation that is followed immediately, or after a moderately long delay, by noxious effects for the rat can result in a strong avoidance of that stimulation (see, e.g., Revusky and Garcia, 1970; Rozin and Kalat, 1971). This phenomenon is known as *poison avoidance* or *conditioned taste aversion*. In most of the poison-avoidance research with the rat as subject, gustatory stimulation has been employed as the conditioned signal. However, it has been shown that the temperature of a fluid can also serve as a cue (Nachman, 1970).

In the present context, the first thing to study was whether electrical tongue stimulation would also be an adequate stimulus in this situation: subsequent current-licking behavior tests would be of interest only if positive results were obtained.

Thirteen naïve rats participated in the experiment. On the last day of training, water licking was paired with electrical tongue stimulation. (Type II 50-μA sensor). Within a few minutes after the completion of this 10-min session, 7 rats were treated with cyclophosphamide (50 mg/kg of a freshly prepared 2% solution, i.p.). The other 6 rats, serving as controls, received physiological saline. One night of *ad libitum* access to food and water was given. Water and water *plus* current were made available in a 10-min test after 48 hr of water deprivation. In Table 10, it is shown that strong suppression of intake of water *plus* current was obtained in the experimental animals. These rats made only a few licks, in small bursts, at the drinking tube connected with the Type II (50 μA) sensor. The median length of the first burst was two licks.

Current-licking behavior was investigated in two additional sessions. The same test chamber was used; each session was preceded by 48 hr of water deprivation. Between these sessions, the water-intake preference (with

Table 10. *Preference Behavior after Poison-Avoidance Training*

	Intake per 10-min session		
	Water	Water + current	Licks
Experimental rats (N = 7)	2.29 ± 0.53[a]	< 0.05 —	8.9 ± 1.4
Control rats (N = 6)	2.73 ± 0.57	3.37 ± 0.71	648.8 ± 124.5

[a] M ± SE, water intake in ml.

and without current) was studied again, showing that all experimental animals still avoided water plus current. The control animals displayed good current-licking behavior (Table 11). Not all of the animals that had been treated with cyclophosphamide licked more at the rod supplying electrical tongue-stimulation with a 50-μA current. But all of the 4 rats that preferred 50 μA in Session I also preferred 50 μA in Session II. The mean number of licks (50 μA) increased from Session I to II.

The results strongly suggest that at least in some animals, lick-contingent electrical stimulation of the tongue had positively reinforcing effects, in spite of the poison-avoidance training procedure. It seems that both conditioned aversion *and* positively reinforcing effects of the electrical stimulation can be obtained in the same animal. A similar conceptual paradox has been described by Wise, Yokel, and De Wit (1976). These authors reported both positive reinforcement and conditioned aversion from the same drug injection.

In two different paradigms, rats learned to avoid electrical tongue stimulation that was paired with water intake and, respectively, foot shock or illness. When electrical stimulation was presented without water, however, its positively reinforcing effects could still be demonstrated.

A conditioned-reinforcer hypothesis, assuming a learned association

Table 11. *Current-Licking Behavior after Poison-Avoidance Training*

	Number of licks per 10-min session (Type II sensors)			
	Experimental rats		Control rats	
	50 μA	0.1 μA	50 μA	0.1 μA
Session I	60.7 ± 28.3[a]	29.3 ± 8.6	699.8 ± 187.9	53.8 ± 17.2
Session II	329.7 ± 111.9	111.9 ± 31.8	1005.5 ± 324.9	130.8 ± 25.8

[a] M ± SE

between water and a neutral stimulus, is too simple to account for the data reported in this section. Electrical stimulation of the tongue constitutes a nonneutral stimulus in rats under dipsogenic conditions. But the reinforcing properties of this stimulation are weak if the animals have not been previously exposed to water intake concomitant with electrical stimulation of the tongue. In the author's opinion, the pairing of electrical tongue stimulation with water intake strongly boosts the intrinsic positively reinforcing effects of lick-contingent electrical tongue stimulation under dipsogenic conditions (sensitization?).

The consequences of electrical tongue stimulation might be comparable to the effects of two groups of stimuli that are of particular relevance to rats: gustatory and olfactory stimuli. They are believed to be projected to brain structures concerned with reinforcement.

The fact that the same response (licking) is required for both water intake and lick-contingent electrical stimulation of the tongue might contribute to the reinforcing effects of this stimulation. Hearst and Jenkins (1974) wrote in the concluding remarks of their monograph on "Signtracking": "directed movements of animals are strongly controlled by signals of appetitive objects," and: "Animals ... frequently contact the source of such a positive signal. Moreover, in certain cases, they exhibit responses to the signal that are about as similar to the consummatory response pattern elicited by the goal object as the physical properties of the signal source will allow" (p. 45).

4. Summary

In this chapter, the reinforcing properties are examined of lick-contingent electrical stimulation of the tongue in rats: current licking.

Persistent current-licking behavior can be demonstrated in animals that have been exposed to electrical stimulation of the tongue concomitant with water intake. Without this previous experience, the phenomenon cannot be obtained at comparable performance levels; in most animals extinction of the reinforcing properties takes place within a few test sessions. Current-licking behavior requires the animal to be in a dipsogenic state. Testing undeprived rats or rats that are food-deprived does not result in an appreciable current-licking performance.

Most studies have employed anodal tongue stimulation; however, the reinforcing properties of cathodal and alternating-current stimulation have also been demonstrated. The threshold for the reinforcing effect of electrical tongue stimulation is extremely low: 0.5–1 μA. Intensities of 50–100 μA are optimal values for current-licking behavior. These intensities can evoke aversive responses of rats when the stimulation is restricted to the feet of the animals.

The nature of the sensory consequences of electrical tongue stimulation in rats in unknown. However, a taste component is likely. Bilateral lesions of the chorda tympani nerve abolish the reinforcing effects of the stimulation, but transection of the trigeminal portion of the lingual nerves, which also innervate the anterior part of the tongue, does not. Desalivated rats can display current-licking behavior; accordingly, the phenomenon does not depend on reflex secretion of saliva.

Choice tests in water-deprived rats have shown that water drinking and airlicking are preferred over current licking. A preference for water with concomitant electrical tongue stimulation over water alone has never been consistently demonstrated.

The investigation into the nature of the reinforcing properties of electrical tongue stimulation has evidenced conditioned aspects. But a conditioned-reinforcer hypothesis cannot fully account for the data obtained in the study of current-licking behavior. It is the author's opinion that in rats under dipsogenic conditions, the pairing of electrical tongue stimulation with water intake strongly elevates the intrinsically reinforcing effects of lick-contingent electrical tongue stimulation.

ACKNOWLEDGMENTS

The skillful assistance of research students and technicians provided many of the data on which this chapter is based. Their dedicated help is acknowledged with many thanks.

References

Beidler, L. M., 1970, Physiological properties of mammalian taste receptors, *in* "Taste and Smell in Vertebrates" (G. E. W. Wolstenholme and J. Knight, eds.), Churchill, London, pp. 51-71.

Blom, S., 1960, Afferent influences on tongue muscle activity: A morphological and physiological study in the cat, *Acta Physiol. Scand.*, Suppl. **170**:1-97.

Blom, S., and Skoglund, S., 1959, Some observations on the control of the tongue muscles, *Experientia* (Basel) **15**:12-13.

Bolles, R. C., 1975, "Theory of Motivation" (2nd ed.), Harper and Row, New York, Chapter 13.

Braud, W. G., and Prytula, R. E., 1969, Licking suppression and recovery as functions of contingency, locus, and frequency of punishment, *Psychon. Sci.* **14**:16-18.

Bujas, Z., 1971, Electrical taste, *in* "Handbook of Sensory Physiology (Vol. IV, Chemical senses, 2) (L. M. Beidler, ed.), Springer, Berlin, pp. 180-199.

Bujas, Z., Rohaček, A., and Kovačić, M., 1975, Electrode area and electrical taste thresholds, *Acta Inst. Psychol. Univ. Zagrab* **76**:25–29.

Campbell, B. A., and Masterson, F. A., 1969, Psychophysics of punishment, *in* "Punishment and Aversive Behavior" (B. A. Campbell and R. M. Church, eds.), Appleton-Century-Crofts, Meredith Corporation, New York, pp. 3–42.

Campbell, H. J., 1968, Peripheral self-stimulation as a reward, *Nature* **218**:104–105.

Campbell, H. J., 1972, Peripheral self-stimulation as a reward in fish, reptile and mammal, *Physiol. Behav.* **8**:637–640.

Faull, J. R., and Halpern, B. P., 1972, Taste stimuli: Time course of peripheral nerve response and theoretical models, *Science* **178**:73–75.

Glow, P. H., and Russell, A., 1975, The effects of a dexamphetamine-amylobarbitone sodium mixture on the reward value of different sensory changes, *Psychopharmacologia* (Berlin) **41**:181–185.

Halpern, B. P., 1973, The use of vertebrate laboratory animals in research on taste, *in* "Methods of animal experimentation (Vol. IV, Environment and the Special Senses)" (W. I. Gay, ed.), Academic Press, New York, pp. 225–362.

Harrington, G. M., and Linder, W. K., 1962, A positive reinforcing effect of electrical stimulation, *J. Comp. Physiol. Psychol.* **55**:1014–1015.

Hearst, E., and Jenkins, H. M., 1974, "Sign-Tracking: The Stimulus-Reinforcer Relation and Directed Action," Psychonomic Society, Austin, Texas.

Hellekant, G., and Hagstrom, E. C., 1974, Efferent chorda tympani activity and salivary secretion in the rat, *Acta Physiol. Scand.* **90**:533–543.

Hellekant, G., and Kasahara, Y., 1973a, The linguo-chorda tympani reflex—An electrophysiologically undescribed reflex, *Acta Physiol. Scand.* **87**:199–207.

Hellekant, G., and Kasahara, Y., 1973b, Secretory fibres in the trigeminal part of the lingual nerve to the mandibular salivary gland of the rat, *Acta Physiol. Scand.* **89**:198–207.

Hendry, D. P., and Rasche, R. H., 1961, Analysis of a new nonnutritive positive reinforcer based on thirst, *J. Comp. Physiol. Psychol.* **54**:477–483.

Kawamura, Y., Okamoto, J., and Funakoshi, M., 1968, A role of oral afferents in aversion to taste solutions, *Physiol. Behav.* **3**:537–542.

Kelleher, R. T., and Gollub, L. R., 1962, A review of positive conditioned reinforcement, *J. Exp. Anal. Behav.* **5**:543–597.

King, N. W., Justesen, D. R., and Simpson, A. D., 1970, The photo-lickerandum: A device for detecting the licking response, with capability for near-instantaneous programming of variable quantum reinforcement, *Behav. Res. Methods Instrum.* **2**:125–129.

Kissileff, H. R., 1969, Food-associated drinking in the rat, *J. Comp. Physiol. Psychol.* **67**:284–300.

Martonyi, B., and Valenstein, E. S., 1971, On drinkometers: Problems and an inexpensive photocell solution, *Physiol. Behav.* **7**:913–914.

Mendelson, J., Zec, R., and Chillag, D., 1972, Effects of desalivation on drinking and air licking induced by water deprivation and hypertonic saline injections, *J. Comp. Physiol. Psychol.* **80**:30–42.

Nachman, M., 1970, Learned taste and temperature aversions due to lithium chloride sickness after temporal delays, *J. Comp. Physiol. Psychol.* **73**:22–30.

Pfaffmann, C., 1952, Taste preference and aversion following lingual denervation, *J. Comp. Physiol. Psychol.* **45**:393–400.

Pfaffmann, C., Fisher, G. L., and Frank, M. K., 1967, The sensory and behavioral factors in taste preferences, *in* "Olfaction and Taste," II (T. Hayashi, ed.), Pergamon Press, Oxford, pp. 361–379.

Plattig, K.-H., 1969, Über den elektrischen Geschmack. *Z. Biol.* **116**:161–211.

Plattig, K.-H., 1971, Taste sensations and evoked brain potentials after electric stimulations of tongue in man, *in* "Gustation and Olfaction" (G. Ohloff and A. F. Thomas, eds.), Academic Press, London, pp. 73–86.

Revuski, S., and Garcia, J., 1970, Learned associations over long delays, *in* "The Psychology of Learning and Motivation: Advances in Research and Theory," IV (C. H. Bower and J. T. Spence, eds.), Academic Press, New York, pp. 1–84.

Rozin, P., and Kalat, J. W., 1971, Specific hungers and poison avoidance as adaptive specializations of learning, *Psychol. Rev.* **78:**459–486.

Schneyer, C. A., and Schneyer, L. H., 1959, Electrolyte levels of rat salivary secretions, *Proc. Soc. Exp. Biol. Med.* **101:**568–569.

Siegel, S., 1956, "Nonparametric Statistics for the Behavioral Sciences," McGraw-Hill, New York.

Slangen, J. L., and Weijnen, J. A. W. M., 1972, The reinforcing effect of electrical stimulation of the tongue in thirsty rats, *Physiol. Behav.* **8:**565–568.

Smith, D. V., and Bealer, S. L., 1975, Sensitivity of the rat gustatory system to the rate of stimulus onset, *Physiol. Behav.* **15:**303–314.

Tapp, J. T., and Long, C. J., 1968, A comparison of the reinforcing properties of stimulus onset for several sense modalities, *Can. J. Psychol.* **22:**449–455.

Tapp, J. T., Mathewson, D. M., and Simpson, L. L., 1968, Effects of hunger and thirst on reinforcing properties of light onset and light offset, *J. Comp. Physiol. Psychol.* **66:**784–787.

von Békésy, G., 1964, Sweetness produced electrically on the tongue and its relation to taste theories, *J. Appl. Physiol.* **19:**1105–1113.

Weijnen, J. A. W. M., 1972, Lick-contingent electrical stimulation of the tongue: Its reinforcing properties in rats under dipsogenic conditions, doctoral dissertation, University of Utrecht (University Microfilms, Ann Arbor, Mich., Order No. 76-6040).

Weijnen, J. A. W. M., 1975, Orale sensorische stimulatie bij de rat: Effecten op lik- en drinkgedrag, *Gedrag, Tijdschr. Psychol.* (Nijmegen), **3:**314–326.

Wise, R. A., Yokel, R. A., and de Wit, H., 1976, Both positive reinforcement and conditioned aversion from amphetamine and from apomorphine in rats, *Science* **191:**1273–1274.

Yamamoto, T., 1975, Linguo-hypoglossal reflex: Effects of mechanical, thermal and taste stimuli, *Brain Res.* **92,**499–504.

6

Temperature of Ingested Fluids: Preference and Satiation Effects (Pease Porridge Warm, Pease Porridge Cool)

Richard M. Gold and Robert G. Laforge

1. Introduction

In the laboratory, the air, food, and water all typically stay at one stable temperature, and when that temperature is experimentally manipulated, as by cold stress, the air, food, and water all tend to change together. As a result of this confounding, several reports have appeared to demonstrate that water intake varies with changes in the temperature of the air to which an animal is exposed.

For example, Hamilton (1963) has shown that rats housed (for 21 days) in warm (35°C) environments drink 133% more water than when kept at the standard laboratory temperature (24°C). No controls for water temperature were mentioned. Thus, it is probably safe to assume that water temperatures shared the manipulated temperature of the environment. The importance of this sharing has now become apparent.

As another example, Hainsworth, Stricker, and Epstein (1968) exposed satiated rats to a range of ambient temperatures from 28°C to 40°C for 6 hr. They found increased water intake with an increase in ambient temperature. They suggested that the increased drinking in the elevated ambient

Richard M. Gold and Robert G. Laforge • Psychology Department, University of Massachusetts, Amherst, Massachusetts. Supported by USPHS grants MH13561 and MH 26251 to R. M. Gold.

conditions occurred in response to the level of dehydration that was brought about during the heat exposure. This may be, but in addition, a possible direct contribution of water temperature to the increased water consumption was overlooked.

In a similar manner, many others have neglected to control for the influence of water temperatures (e.g., Budgell, 1970a; Fregly, 1971; Grace and Stevenson, 1971; Carlisle, 1973; Nelson, Fregly, and Tyler, 1974; Box, Montis, Yeomans, and Stevenson, 1973; Banet and Seguin, 1970; McFarland and Wright, 1969). In all of these studies, the changes in ambient temperature must have influenced more than skin temperature. Nevertheless, most authors downplay or even ignore the temperatures of the food and water supplies that inadvertently were also being manipulated. Any behavioral changes, such as in food or water intake, that resulted in these studies were implicitly interpreted as responses to the temperature changes in the nonnutritive parts of the environment.

In contrast to this typical laboratory situation, the temperatures of the food, water, and air do not fluctuate in synchrony in naturalistic settings. Independent temperature cycles occur both daily and seasonally. For example, animals that feed during the day would experience air warmer than the water they drink. For nocturnal feeders, on the other hand, the temperature of the water source may be warmer than the night air. In spring, the water temperature of snow-fed streams sharply contrasts with the warm temperature of the air. Conversely, water temperatures in late autumn tend to be warmer than the air.

Animals that live in severe cold are forced to get their water by licking ice and snow, which produces an extreme degree of orolingual cooling. Conversely, carnivores and insects that prey on mammals tend to eat or drink warm food. Those animals that burrow or cuddle together in nests during the winter, or grow a thick winter fur, create a warm environment for their skin, whereas their water, and possibly their food, remains cold. Humans in modern society go to great lengths to independently warm or cool their air, food, and beverages.

Differential air, fluid, and food temperatures are thus the rule rather than the exception, both in nature and in society. The possible effects of each of these temperatures on food and water intake would be most difficult to evaluate in natural or in societal settings. In the laboratory, however, the last few years have seen a growing concern with the respective influences of air and water temperatures on food and water intakes. These laboratory studies have already demonstrated that changes in water temperature can by themselves dramatically modify water intake. The magnitude of the intake changes produced by water-temperature manipulations appears in many cases to be sufficient to account for intake changes previously demonstrated via air-temperature shifts. In the few studies in which the effects of air- and water-temperature manipulations on water intake have been compared, the water-temperature manipulation produced all of the intake changes (Nelson, Fregly, and Tyler, 1975; Carlisle and Lavdenslager, 1976).

2. Origins

Much of the early work on oropharyngeal factors in drinking is reviewed by Epstein (1967) and Nicolaidis (1969). A major role for oropharyngeal cues in the regulation of water intake was inferred from the relative failure of intravenous or even intragastric infusions of water to suppress oral water intake. Only when ambient-temperature water was delivered to the stomach via a nasopharyngeal tube, thereby imparting a local cooling sensation as it passed through the nose and the pharynx, could oral water intake be totally suppressed (Miller, Sampliner, and Woodrow, 1957; Kissileff, 1960).

Recent attention has turned in a more naturalistic direction with direct manipulation of the temperature of the water offered for ingestion. These recent studies owe their conception to the discovery of airlicking by Hendry and Rasche in 1961. They found that thirsty rats will lick at a stream of air. The airstream appears to partially satiate thirst, even though its only effect on water balance is to evaporate saliva.

Hendry and Rasche's original discovery of airlicking has been followed up extensively; however, most that work is not the subject of this chapter. We follow here only the line of research that leads to the influence of water temperature on water intake. That line goes from airlicking to Mendelson and Chillag's 1970 report of cold licking.

Mendelson and Chillag offered thirsty rodents cold metal to lick. The thirsty rodents licked the cold metal just as airlickers lick an airstream, thus showing that evaporative cooling is probably the effective parameter of the airlicking. Subsequent research (see Chapter 4) confirms this interpretation (Mendelson and Zec, 1972; Mendelson, Chillag, and Paramesvaran, 1973).

Upon reading Mendelson and Chillag's report, we began a series of studies looking into the effect of orolingual cooling by the more natural medium of water presented at different temperatures. The initial report of this work (Kapatos and Gold, 1972) showed that up to body temperature (37°C), water intake increases directly with water temperature. That is, thirsty animals drink more warm water and less cool water, see Figures 1 and 2. At still higher temperatures, water intakes decline. (Since the closed mouth is normally at 37°C, we consider intakes of 37°C water as the

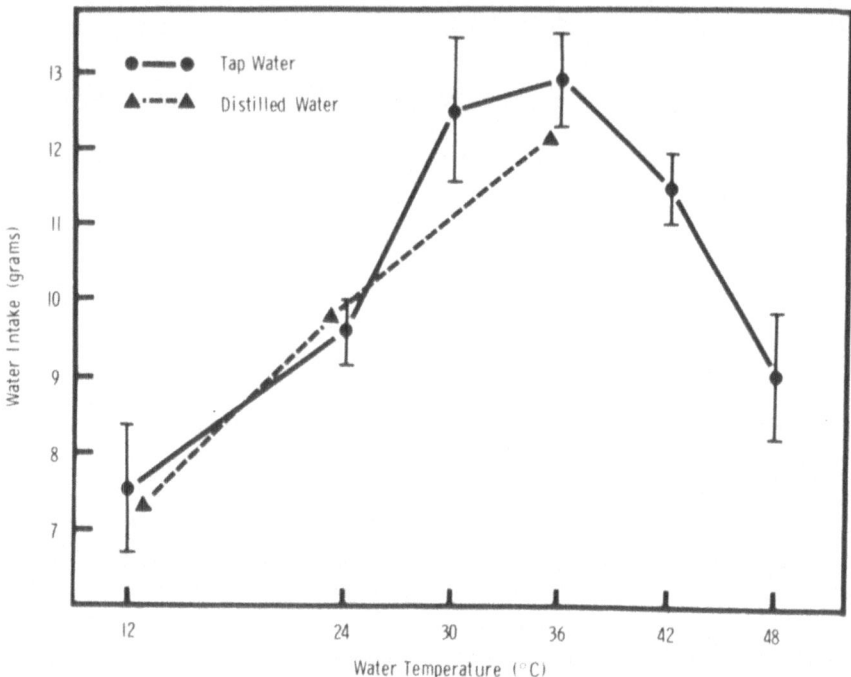

Figure 1. Tap-water and distilled-water intakes as a function of water temperature. Female rats obtained all of their water in a daily 30-min session. Only one water temperature was available per day to any one rat (no-choice). Food was available *ad libitum* except during the drinking sessions. (From Kapatos and Gold, 1972; copyright 1972 by the American Association for the Advancement of Science. Reprinted by permission.)

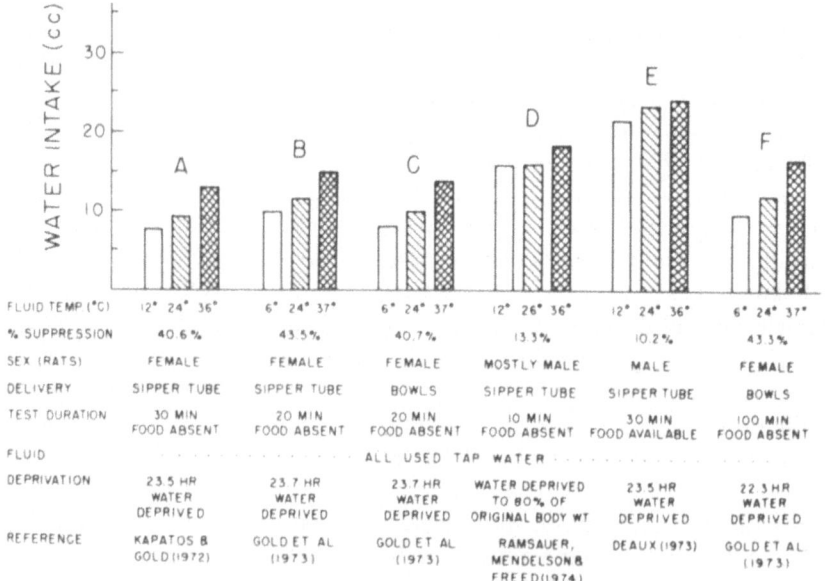

	A	B	C	D	E	F
FLUID TEMP (°C)	12° 24° 36°	6° 24° 37°	6° 24° 37°	12° 26° 36°	12° 24° 36°	6° 24° 37°
% SUPPRESSION	40.6%	43.5%	40.7%	13.3%	10.2%	43.3%
SEX (RATS)	FEMALE	FEMALE	FEMALE	MOSTLY MALE	MALE	FEMALE
DELIVERY	SIPPER TUBE	SIPPER TUBE	BOWLS	SIPPER TUBE	SIPPER TUBE	BOWLS
TEST DURATION	30 MIN FOOD ABSENT	20 MIN FOOD ABSENT	20 MIN FOOD ABSENT	10 MIN FOOD ABSENT	30 MIN FOOD AVAILABLE	100 MIN FOOD ABSENT
FLUID			ALL USED TAP WATER			
DEPRIVATION	23.5 HR WATER DEPRIVED	23.7 HR WATER DEPRIVED	23.7 HR WATER DEPRIVED	WATER DEPRIVED TO 80% OF ORIGINAL BODY WT	23.5 HR WATER DEPRIVED	22.3 HR WATER DEPRIVED
REFERENCE	KAPATOS & GOLD (1972)	GOLD ET AL (1973)	GOLD ET AL (1973)	RAMSAUER, MENDELSON & FREED (1974)	DEAUX (1973)	GOLD ET AL (1973)

Figure 2

baseline condition from which reduced or enhanced water intakes at cooler or warmer water temperatures deviate.)

Following our original demonstration of the cold (and hot) water suppression of water intake, we first asked whether this was an isolated phenomenon; in other words, whether it would replicate in a variety of testing situations and subjects. The phenomenon has replicated but with some notable exceptions.

3. Ponds, Puddles, and Dew Drops

One of the first variables that we explored was the nature of the water-delivery system. In the laboratory, most water-intake studies use sipper tubes (open-end), which is perhaps similar to licking tiny dew drops. Yet, in nature many animals obtain their water from open sources, such as ponds or puddles. In an attempt to come closer to the ponds of nature, we offered animals water in open bowls. The first thing that was evident was that rats obtain a great deal more water with each lick from an open bowl than from a sipper tube: 2 to 4 times as much water per lick. We were pleased to find that with the open-bowl situation, cool water suppressed water intake just as much as it had with the sipper tube. Figure 2 (B) shows the results for daily 20-min tests using sipper tubes (43.5% suppression). These results should be compared with Figure 2 (C), in which all the parameters were the same except that the water was in open bowls (40.7% suppression) (Gold, Kapatos, Prowse, Quackenbush, and Oxford, 1973).

4. Sexual, Developmental, and Interspecific Considerations

Figure 2 (A) shows data from our initial report (Kapatos and Gold, 1972). In that study, female albino rats were exposed for 30 min/day to water at a particular temperature. The temperature was varied from day to day. On the days that only 6°C (cool) water was available, water intake was suppressed 40.6%, as compared with the intake on days when only body-temperature (37°C) water was available.

Soon thereafter, both Ramsauer, Mendelson, and Freed (1974) and Deaux (1973) were successful in replicating the cool-water suppression of water intake. Their results are shown in Figure 2, (D) and (E), respectively. In contrast to the 40.6% suppression that Kapatos and Gold had found, Ramsauer *et al.* and Deaux obtained only 13.3% and 10.2% suppressions, respectively.

Ramsauer *et al.* used mostly male rats, and Deaux used only male rats, whereas Kapatos and Gold used female rats. Can this sex difference account for the greater suppression found by Kapatos and Gold? The likelihood of a sex difference in the response to water temperature is plausible in light of the existence of a parallel sex difference in the responsiveness to sweet-tasting solutions (Valenstein, Kakolewski, and Cox, 1967). Valenstein and his colleagues found that female rats are more responsive to sweet solutions.

Nevertheless, we (Gold, Laforge, and Fraser, unpublished) have recently found that males suppress at least as much as females. In our study, male and female rats had access to either cool (4°C) or warm (36°C) water for 20 min/day. Food was available only in the home cage. Male rats (n = 4) drank 36.7% less on the days when the water was cool. Females in the same study suppressed only 26.2%! When food was present both in the test cage and in the home cage, similar suppressions were also obtained (males, 33.3%; females, 29.8%). Thus, sex is not a factor in the cool-water suppression of water intake. Our direct comparisons between males and females should take precedence over between-laboratory sex comparisons. An explanation for the failure of some workers to obtain a robust cool-water suppression of water intake must reside in some other variable.

A sex difference, if it were found to exist, would have suggested that juvenile rats might respond differently than adults. As yet, there are no data with 10-min to 100-min daily drinking sessions for juvenile rats. However, we have explored a rather different aspect of development.

In response to the perplexed "What is it good for anyway?" response that our friends often gave us when we told them we were giving rats warm and cool water, we have recently explored the developmental consequences of prolonged *ad libitum* (24-hr) exposure to warm (3 rats housed together) and cool (3 rats housed together) water (Gold and Laforge, unpublished). The young female rats were followed for 4 weeks (age 33 days through to age 61 days). We regret to have to report that there were no discernible differences in food intake or growth as a consequence of the differential water temperatures. We also ran a condition designed to be analogous to the effect an inconsistent parent might exert. In this inconsistent-parent condition, the water was warm and cool on alternate days (6 rats housed in two groups of 3). Once again, there was no impact on food intake or on weight gain by rats treated inconsistently as compared to constant temperature conditions.

The absence of developmental differences suggests that water and perhaps milk temperatures are not an important factor in development— that it is not necessary to warm (or cool) a baby's bottle.

Very little work has been done on the temperature of baby's milk.

Opinions and custom vary as to whether it is desirable to warm up bottle-fed babies' milk to be more like mother's own. One of us (Gold) has explored this question with two human babies. One, a diminutive boy, had been breast-fed as a small infant. By the age of 7 months, when the data were collected, he had been weaned to a bottle. Three milk temperatures, refrigerator (6°C), ambient (22°C), and warm (36°C) were used, each on different days for a 50-min drink. Milk temperature had virtually no influence on his intakes. In contrast, a large girl, the first author's daughter, was tested at three ages: 3½, 8, and 26 months. At each age, warm (36°C) and cool (6°C) milk days were alternated. At the two younger ages, the first bottle in the morning was weighed. At 3½ months, milk temperature had almost no impact on milk intakes. At 8 months, however, significantly less cool milk was consumed. At 26 months, the only bottle she had was at bedtime. For that bottle, cool-milk suppression persisted. She also appeared calmer and fell asleep more rapidly when the milk was cool.

Perhaps coolness of ingested fluid is a cue to the offspring for weaning. Thus, it is to be avoided before a certain age; for example, it is not adaptive for an organism to start drinking cool fluids such as water until it is old enough to be weaned. Conversely, cool fluids might become preferable after the proper weaning age. It is not of adaptive value to nurse for too long because breast milk is low in iron, etc. Perhaps the female baby was at this postweaning stage at 26 months and perhaps was just getting there at 8 months. Therefore, cold milk was more satiating at these ages and so less was ingested. Consistent with this interpretation is the greater contentedness of the baby after drinking the cool milk. These limited data on only two human babies are offered in the hope that a reader with access to a greater population will do a more definitive study on milk temperature in children.

Although it would be folly to attribute the sometimes tenacious milk aversions seen in older children exclusively to their earlier orothermal experiences, it is conceivable that skillful manipulation of early milk temperatures could minimize such aversions. Perhaps cool milk is a more salient cue, hence easier to establish as a learned source of security or feeling of well-being (or aversion).

As well as in rats and babies, cool water also suppresses water intake in wild opossums and in Barbary doves. Eight wild opossums (M. I. Phillips, 1973, and personal communication) obtained all of their water in daily 1-hr drinking sessions. Food was available *ad libitum* but not during the daily drinking session. On days when the water was cool (5°C), water intakes were severely depressed as compared with days when the water was at body temperature. An intermediate water temperature (24°C) yielded intermediate intakes. Barbary doves also overdrink in a warm environment and underdrink in a cool environment (Budgell, 1970b).

5. Time Course of Water Intake

We have shown that for once-a-day drinking sessions taken as a whole, rats consume less when the water is cool, but, you may be wondering, when during the course of 20-min to 30-min daily sessions does the suppression occur? An early appearance—say, in the first minute—would suggest a neural or conceivably neural–hormonal mediation. A later appearance would suggest an osmotic or volemic postabsorptive mechanism for the suppression.

The answer turns out to be the former, see Figure 3. Cool-water intake is suppressed during the 1st min of 20-min drinks (Gold *et al.*, 1973).

Such short latency responses to oropharyngeal stimulation are not without precedent. Hot, thirsty humans can begin to sweat within 2–7 sec after starting to drink, and in rats, water in the mouth can produce diuresis within 1 min (Nicolaidis, 1969).

The cool-water suppression of drinking could be related to different surface-tension properties in cool water that slow the flow of water from sipper tubes or onto the tongue. With open bowls, we suspect that some rats do not insert their tongues as far into cool water as into warm water. Yet, differences in lick efficiencies are readily compensated for by rats. For example, the considerably greater lick efficiences from open bowls as compared with sipper tubes do not give greater fluid intakes, and the rats that

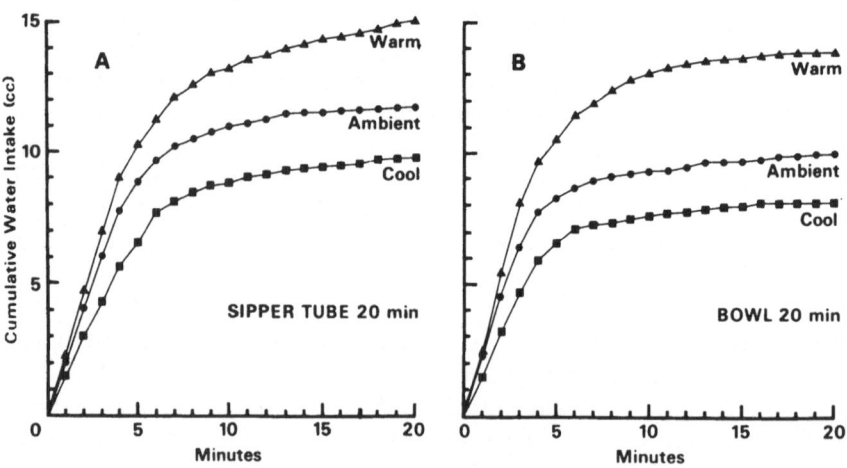

Figure 3. Minute-by-minute time course of water intakes by female rats at three water temperatures for (A) sipper tube and (B) open-bowl delivery. Warm = 37°C, Ambient = 23°C, Cool = 6°C. Only one temperature was available per day to any one rat (no-choice). (From Gold, Kapatos, Prowse, Quackenbush, and Oxford, 1973; copyright 1973 by the American Psychological Association. Reprinted by permission.)

do have equivalent lick efficiencies for cool versus warm water still drink less cool water (Gold *et al.*, 1973).

We suggest, instead, that the locus at which cool water exerts its uncompensated suppressant effect lies at the other end of the esophagus.

6. Time to Drink

6.1. Intakes on Restricted Schedules

The next parameter we explored was the duration of the daily drinking test. That is, is the cool-water suppression of water intake limited to a narrow range of drinking-session durations? On the contrary, cool-water suppresses water intake when water is available for discrete sessions of 10, 20, 30, or 100 min/day (Fig. 2).

6.2. Ad Libitum *(24-hr)* Intakes

In our 10- through 100-min studies, cool-water intakes were suppressed around 40%.

The water deprivation and the brief daily periods of water availability of even our daily 100-min tests are a far cry from the *ad libitum* situation found in most naturalistic settings. With some trepidation, we have recently measured water intakes at two water temperatures under *ad libitum* (24-hr) access conditions. The major methodological difficulties were (1) to maintain the water temperature and (2) to obtain reliable evaporative measures for the warm water, as the amount of evaporation over 24 hr is quite high (22 cc in this study). When the water temperature was changed daily, alternating between 36°C and 4°C, we obtained a reliable suppression of 21.7% (Figure 4A), which was quite consistent over our 5 adult female rats (Gold and Laforge, unpublished). When we then left these same rats with each water temperature for a whole week, however, the data became chaotic. Some drank twice as much warm as cool, and some vice versa. When we tried the same design with 6 group-housed weanling female rats (3 per cage), we obtained comparable results. With daily alternation, they drank 19.6% less 4°C water than 36°C water, whereas a group kept with 4°C water continuously did not drink any less than a group kept with 36°C water.

We are left looking for a mechanism that reacts to once-a-day changes in water temperature but adapts when the temperature changes come even

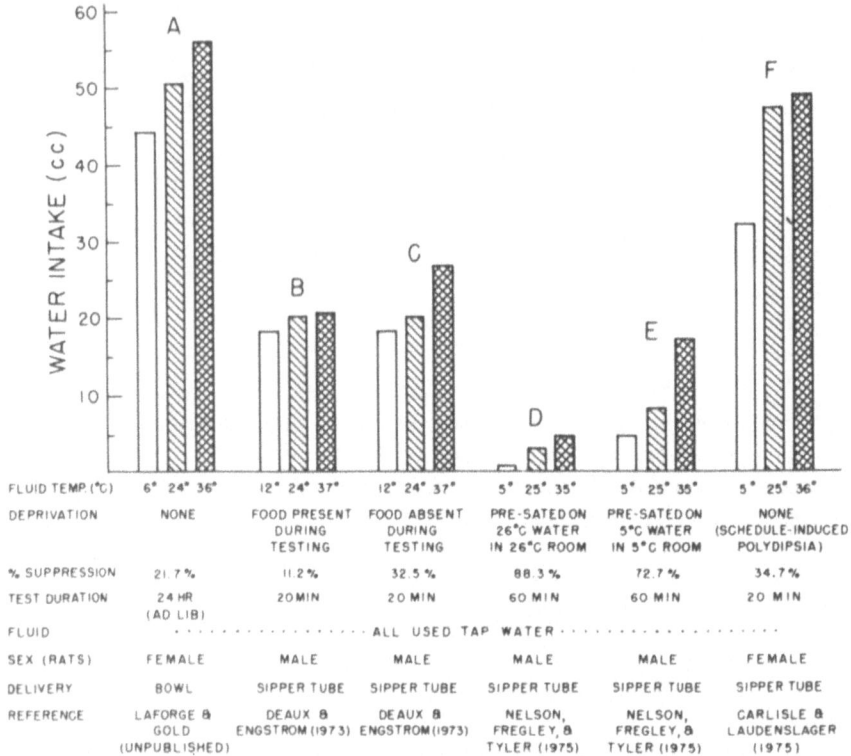

Figure 4

less frequently. It is no wonder, then, that the rats that we raised from weaning on cool water did not grow any less rapidly than those raised on warm water, as there were no water-intake differences.

6.3. Prior Exposure

The loss of cool-water suppression of water intake when the water temperature remains constant over many days appears to occur only in *ad libitum* testing. With daily 20-min testing, a series of 5 consecutive days of monothermic water does not diminish the suppressant effect of cool water (Gold, 1973).

This is not to say that experiential thermal variables are without impact in brief daily testing. Deaux and Engstrom (1973b) found that prior exposure to warm, ambient, or cool water enhanced intake of the familiar temperature on a subsequent two-bottle preference test. Whether such influences extend much beyond one day, however, remains to be demonstrated. In contrast, in all of our studies we have systematically discarded the first few (inconsistent) days' data when we exposed rats to new water temperatures. Thus, prior experiential factors can be important but could not have produced the results reported in our studies.

Furthermore, if a learned association between orolingual cooling and other, supposedly innate sensory aspects of water intake were behind the cool-water preference, then the converse—extinction—should be expected when the cooling stimulus is repeatedly presented alone. This prediction is tested and refuted by airlicking, which does not extinguish despite repeated exposure.

7. Thirst-Dependent Preferences

From the beginning of the water-temperature studies, we had wondered whether the appearance of cool-water suppression of water intake could somehow yield to a preference interpretation. This outcome seemed unlikely because the airlicking studies suggest that thirsty animals would prefer cool water, whereas we found in no-choice testing that thirsty rats drink less cool water. Meanwhile, Ramsauer, Mendelson, and Freed (1974) demonstrated that thirsty rats bar-press more, and at a higher rate, for cool water than for warm water, indicating that cool water is at least initially preferred over warm. It seemed paradoxical that cool water would be both more satiating and more preferred. Nevertheless, the preferences for cooling air, and cool metal, and the enhanced lever pressing for cool water might not translate directly into a preference for cool water. We finally measured water-temperature preferences in 1974 (Gold and Prowse). At the time that we did this preference study, we were set up to measure licks in 2-min time bins over a 20-min drinking session. We maintained this time-course analysis while we looked at two-bowl, two-temperature preferences. This time-course measure turned out to be very informative, as can be seen in Figure 5. We found that during the first few minutes of drinking, the thirsty rats almost exclusively preferred the cooler of two temperatures offered, whereas before the end of the 20-min drinking sessions, the rats switched to the opposite extreme and almost exclusively preferred the warmer of the two temperatures.

Ramsauer, Mendelson, and Freed also measured water-temperature preferences in 1974 but only reported net preferences over the entire drinking session, during which time the preferences shift from one extreme to the other.

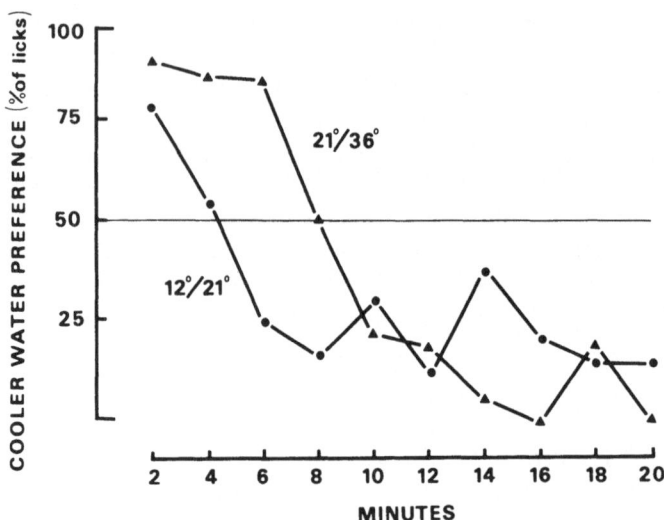

Figure 5. Time course of water-temperature preferences in two-bowl sessions. Each lick is considered an indicator of preference. When the choice is 12°/21°C, the 21°C water is the warmer, whereas when the choice is 21°/36°C the 21°C water is the cooler. (From Gold and Prowse, 1974; reprinted by permission.)

8. Mechanisms

8.1. Reflex Volemic Satiety

The initial cool-water preference in thirsty rats is exactly what was predicted based on the airlicking results. Recall, however, that in no-temperature-choice testing, the intakes on cool-water days are suppressed, and this cool-water intake suppression is apparent even during the first minutes of drinking. Since there is no warm-water preference at that time, we cannot use preference to explain the initial suppression of cool-water intake. We suggest instead that within seconds after the onset of cool-water drinking, the cooling of certain parts of the body reflexively satiates hypovolemic thirst. As a result, the rat drinking cool water is driven only by osmotic thirst and therefore drinks less rapidly. Oatley (1967) has shown that osmotic and volemic thirst stimuli are additive in their effect on thirst (see review by Fitzsimons, 1972). Subtraction of one of the two from water-deprivation thirst should therefore reduce the rate of water intake.

A likely mechanism for this reflexive satiety of volemic thirst is that orolingual, esophageal, and/or stomach cooling causes a reflexive constriction of vascular musculature, so that the available vascular space shrinks.

This mechanism would satiate hypovolemic thirst, as the existing fluid volume would more adequately fill its shrunken container.

Rapid neural satiation of volemic thirst could account for both the cool-water preference in two-choice testing and the suppression of cool-water intake in the early minutes of no-choice testing. Indeed, the cooler the water, the less rapidly it would have to be ingested in order to maintain the reflexive satiation of volemic thirst. As actual blood volume swells, the amount of shrinkage of the vascular container needed to achieve volemic satiety would gradually diminish. If, in addition, we posit that excessive (hyper)volemia is unpleasant, then we have explained the gradual shift in preference during a drink from cool to warm water.

Volemic thirst is probably detected in part via stretch receptors in the large-capacitance veins near the heart and in the right auricle. Signals from these stretch receptors probably travel in the vagus nerve (Gauer and Henry, 1963).

These vagal fibers should remain intact after subdiaphragmatic vagotomy, but not after cervical vagotomy. Thus, if vascular stretch receptors are at work, then cervical but not subdiaphragmatic vagotomy should reverse cool water's suppression. As will be seen in the subsequent section, subdiaphragmatic vagotomy does eliminate cool water's suppression, thus disputing vascular mediation.

8.2. Stomach Stretch

An alternate, or perhaps complementary, mechanism is that the temperature-preference switch occurs when the stomach becomes full. This information would need to be sent to the brain, perhaps via the vagus, in response either to stomach stretch or to the onset of gastric emptying onto duodenal stretch receptors or duodenal osmoreceptors.

Strong support for a stomach-fill mechanism contributing to the temperature-preference switch is provided by a comparison of our time course of the switch with Hatton and Bennet's (1970) data on the time course of changes in stomach volume, blood volume, and serum osmolarity following drinking. Our switch toward a warm preference is clearly evident as early as during the third 2-min time bin (5th and 6th min of drinking), whereas osmotic changes in serum are still not apparent as late as 8 min after the start of drinking. Plasma-volume changes are even slower, with no significant shift as late as 10 min after the start of drinking. What, then, does occur during the 5th and 6th min of drinking? It is then that the stomach reaches its greatest fill and the rate of drinking begins to slow. For 4 min thereafter, the reduced level of drinking matches the rate of stomach emptying, and the stomach maintains a steady level of fill (Hatton and Bennet, 1970).

In order to prevent ingested water from reaching possible duodenal stretch receptors or osmoreceptors, one of us (Gold) has used the duodenal clamp technique (Hall, 1973). Rats were implanted with remotely closable duodenal clamps. After their water-temperature switching behavior had stabilized, they were clamped just before the start of one day's preference session. With the clamp in place, the preference switch still occurred, ruling out mediation by duodenal receptors, and the switch occurred even earlier than on the preceding no-clamp days (Gold, unpublished). This finding suggests a stomach-stretch mechanism. The clamp prevented stomach emptying and so hastened stomach filling.

Based on these data, we (Gold, Sawchenko, Laforge, and Fraser, unpublished) predicted that subdiaphragmatic vagotomy would eliminate the cool-water suppression of water intake. Male and female water-deprived rats (n = 8) had 20-min access to warm (36°C) water on even days alternated with 20-min access to cool (4°C) water on odd days. Prior to surgery, on cool-water days water intake was 31.0% lower than on warm-water days. Following bilaterial vagotomy, this suppression fell to a nonsignificant 4.8%. The loss of the temperature difference, incidentally, was due primarily to a decrease in water intake on warm-water days, not to an increase on cool-water days. We (Sawchenko, Gold, and Ferrazano, unpublished) have further characterized the mechanism by measuring stomach emptying. After a drink of cool water the stomach empties more rapidly than it does after a warm drink. Gastric vagotomy selectively speeds up the emptying of warm water. These effects of vagotomy and water cooling are not additive, thus suggesting that in intact rats cool water suppresses vagal tone at the pylorus. Thus, both cool water and vagotomy would elicit dumping onto duodenal receptors.

Deaux (1973) has shown that cool water (12°C) is absorbed into blood more slowly than warm water (37°C). Deaux's data appear to suggest that cool water leaves the gut more slowly. Perhaps cooling causes localized vasoconstriction, which impedes absorption.

Enhanced stomach emptying has also been reported by Kraly and Blass (1976). These workers were studying the intake of liquid diet in food- but not water-deprived rats. They found *enhanced* intake of liquid food in a cool environment. They attribute the enhanced intake in part to a more rapid emptying of the stomach. In most of their experiments, ambient and food temperatures varied together, thus introducing a confound. It is not clear, therefore, whether a cool drink, a cool environment, or both enhance gastric emptying. If enhanced gastric emptying were at work in our cool-water studies and if rapid gastric emptying translates into rapid absorption, then gastric emptying could well account for some of the rapid satiety observed with cool water.

Paradoxically, however, when Kraly and Blass controlled the ambient temperature and manipulated only the temperature of the liquid diet, intakes were *not* enhanced. (Stomach emptying was not measured under these controlled conditions.) Despite numerous methodological differences, the contrast between our water-intake data and Kraly and Blass's food-intake data clearly emphasizes that temperature has vastly different effects on thirst as compared with hunger.

Fluid temperature is not the only thing that influences gastric clearance. Osmolarity is also a factor. Saline slows stomach clearance (Hall and Blass, 1975). In order to achieve and maintain equivalent levels of gastric fill, warm water or hypertonic saline would have to be consumed more slowly. In a 2 × 2 design (2 temperatures × 2 salt concentrations) Engstrom and Deux (1974) showed both the thermal and the osmotic influences on intake and also found that these influences are not additive. As compared with warm (37°C) hyposmotic fluid (0.44% NaCl), cooling (to 12°C) or increasing the salinity (to 1.32%) both decreased fluid intake about 17%, as expected. Combining these two influences (cool and hyperosmotic) did not increase the suppression any further. Thus, the mechanisms behind these two variables may somehow be redundant or even antagonistic. Additional data, including plain-water and isotonic (0.9% saline) groups, are needed to verify this strange interaction between temperature and salinity.

8.3. Hedonism

Finally, we have to account for the continuation of the suppression of cool-water intake during the later portions of no-choice testing sessions, for at that time the stomach is no longer full. To do this accounting, we suggest a hedonistic explanation. We suggest that as the drinking session begins, we have very thirsty rats whose rate of intake is guided mostly by how thirsty they are. The more thirsty they are, the faster they drink. When there is no choice, they readily drink the less-preferred warm water and drink it even more rapidly than the reflexively partially satiating cool water. As the drinking session progresses, however, the amount consumed becomes more and more sensitive to preference factors. As we have shown, after several minutes of drinking, rats (1) prefer warm water when given a choice and (2) drink more of it when there is no choice. In other words, preference factors assume greater importance toward the end of a drink.

Note that when rats have a choice between 12°C and 21°C water (the lower curve of Figure 5), they switch from a cooler preference to a warmer preference after only 4 min of drinking, whereas when the choice is between 21°C and 36°C water, the switch is delayed till after 8 min of drinking. (If Figure 5's abscissa were plotted as water intake instead of as time, then the difference between the two curves would be even greater, as the 6°C water was initially consumed more slowly in the 6–21°C condition than was the 21°C water in the 21–36°C condition.)

There is, as a result, an intermediate point at about 6 min into the drinking session where the 21°C water is preferred over both temperature extremes. It appears from this observation that an exhaustive series of two-choice tests, or cafeteria-style testing with an array of temperature choices, would reveal a gradual increase in the preferred water temperature as the animal satiates. It would follow also that if deprivation times were manipulated, the initially chosen water temperature would rise as deprivation time decreased. The end point of decreasing deprivation time is, of course, *ad libitum* 24-hr availability, during which the animal is in the middle of a never-ending drinking session (as defined by water availability).

Fitzsimons (1972) referred to the drinking that occurs under 24-hr availability as secondary. Primary drinking occurs in response to osmotic or volemic stimuli. All other drinking is secondary. In most natural settings, such as *ad libitum* 24-hr water, primary thirst rarely develops. It appears that in *ad libitum* (secondary thirst) testing, which incidentally probably also comes closest to the typical human situation, neither the cool (0–6°C) preference seen at the beginning of a daily 20-min drink nor the warm (36°C) preference seen at the end of that brief drink would be seen. The water-temperature preference on *ad libitum* (24-hr) preference testing should remain somewhere in between these two extremes. Surely, rats getting water for only 20 min per day are more thirsty at the beginning of their daily drinking session than they would ever be in a 24-hr *ad libitum* situation, and conversely, by the end of that daily 20-min session, they surely have stuffed themselves with so much water that they are more hydrated and distended with water than they would ever get under 24-hr *ad libitum* drinking. (Hall and Blass, 1975, show that this overhydration is only osmotic. Volemic deficits are not even fully compensated by drinks of as long as 5 hr).

With no temperature choice, *ad libitum* water intakes should be maximal at the most-preferred temperature. As yet, we have measured *ad libitum* no-temperature-choice water intake at only three temperatures: 6°C, 24°C, and 36°C. Intakes on 36°C days clearly exceed intakes on 24°C days. Thus, the preferred temperature under *ad libitum* conditions is probably quite high—closer to 36°C than it is to 24°C.

No-choice *ad libitum* testing at additional temperatures and direct 24-hr *ad libitum* preference testing (secondary drinking) are needed, but in the meantime, it appears reasonable to predict that the 24-hr preferred water temperature is a few degrees lower than body temperature. This is not to say, however, that water-temperature preferences should be expected to remain constant across the entire *ad libitum* day. More likely, there is a modest temperature preference shift, from cooler to warmer, associated with each individual bout of *ad libitum* drinking.

9. Fluids Other Than Water

So far, we have dealt only with the intake of water, but what about the intake of other fluids? One that has been studied somewhat systematically is saline. In Figure 6, we show data first for distilled water (Figure 6 (A); Kapatos and Gold, 1972), for comparison purposes, and then, from left to right, for increasing concentrations of saline. Unfortunately, some of the studies had food available during testing and others did not; some used males and and others used females. With 0.44% saline (Figure 6 (B); Engstrom and Deaux, 1974), with food available, the cooler solution caused only a 17.3% suppression of water intake in male rats. At approximately isotonic (0.9%) saline, with food absent and female rats, water intake was suppressed 41.3% in 20-min sessions [Figure 6 (C)] and 47.5% in 100-min sessions [Figure 6 (D); Gold *et al.*, 1973]. These suppressions were greater

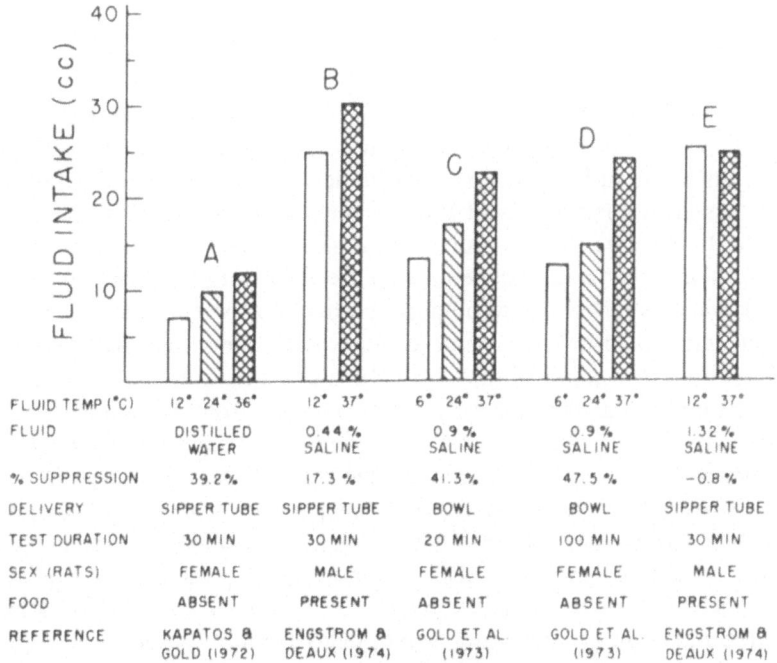

FLUID TEMP (°C)	12° 24° 36°	12° 37°	6° 24° 37°	6° 24° 37°	12° 37°
FLUID	DISTILLED WATER	0.44% SALINE	0.9% SALINE	0.9% SALINE	1.32% SALINE
% SUPPRESSION	39.2%	17.3%	41.3%	47.5%	-0.8%
DELIVERY	SIPPER TUBE	SIPPER TUBE	BOWL	BOWL	SIPPER TUBE
TEST DURATION	30 MIN	30 MIN	20 MIN	100 MIN	30 MIN
SEX (RATS)	FEMALE	MALE	FEMALE	FEMALE	MALE
FOOD	ABSENT	PRESENT	ABSENT	ABSENT	PRESENT
REFERENCE	KAPATOS & GOLD (1972)	ENGSTROM & DEAUX (1974)	GOLD ET AL. (1973)	GOLD ET AL. (1973)	ENGSTROM & DEAUX (1974)

Figure 6

than those depicted in Figure 6 (B) with the less-concentrated (0.44%) solution. Note that the temperature spreads used also varied. Nevertheless, with all of these concentrations, there was a clear cool-water suppression of fluid intake. When we move along to hypertonic (1.32%) saline solutions [Figure 6 (E)], the cool fluid no longer suppresses the intake. Unfortunately, these authors (Engstrom and Deaux, 1974) had previously found [Figure 2 (E)] only a 10.2% suppression of water intake using tap water. The impact of hypertonic saline on the suppression of cool-fluid intake should be replicated with parameters known to yield large differences for warm versus cool tap water.

These between-laboratory comparisons suggest that the sex of the rats and the availability of food during the drinking session are powerful variables. The impact of these two variables may be more apparent than real, however, as we (Gold, Laforge, and Fraser, unpublished) recently compared males with females each with and without food present during 20-min daily drinking tests. We found that in all four testing conditions, cool water (0°C) suppressed water intake by at least 30% as compared with warm water (36°C). Thus, neither maleness nor the presence of food diminishes the intake-suppressant effect of cool water.

10. Osmotic, Volemic, or Gastric?

One impetus for using saline solutions is to dissect what are believed to be the two aspects of primary thirst—that is, osmotically induced thirst and volemic thirst—in order to determine whether cool water satiates primarily one of these two components of thirst or the other, or both. The rationale would be that in the ingestion of an isotonic or slightly hypertonic saline solution, the volemic component of water-deprivation thirst would be satiated, whereas the osmotic component would not be satiated because the ingested fluid would have the same osmolarity as the serum and therefore could not relieve osmotic thirst. Unfortunately, the interpretation of this effect is complicated, as saline may not be absorbed from the gut as rapidly as water. As a result, both osmotic and volemic satiation would be delayed. Engstrom and Deaux (1974) have suggested that the rate of absorption may also vary with fluid temperature, with cool water being absorbed more slowly than warm water. This mechanism cannot account for the behavioral data, in that if cool water were absorbed more slowly, it would not suppress water intake but rather would enhance it by slowing osmotic satiation. While the retention of cool water would delay osmotic satiation, it might, however, produce gut distention, which could suppress further water intake. Following this argument one might expect that saline solutions would also be very satiating, as the absorption of saline may be delayed as compared with water, but of course saline solutions are consumed in much greater volumes than water (see e.g., Gold *et al.*, 1973). These arguments are further complicated by the recent demonstration that cooling can stimulate more rapid gastric emptying (Sawchenko *et al.*, unpublished).

We (Gold and Pytko, unpublished) have recently extended our two-bowl (two-temperature) preference testing to isotonic and hypertonic saline solutions. With isotonic (0.9%), the initial preference is for cool saline. The preference switches to warm saline during the drinking session, but the switch with saline is delayed as compared with the switch from cool to warm water (Figure 7). Our initial interpretation of this result was that since the 0.9% saline should not satiate osmotic thirst, then by default the temperature-preference shift was due to volemic satiation. We attributed the delay in the shift to a slower absorption of isotonic saline from the gut, as compared with water.

Subsequent two-bowl testing with warm and cool 1.5% saline has forced us to reevaluate this conclusion. With the more concentrated 1.5% saline, the rats eventually profit from their drink by excreting an even more concentrated urine, but in the short run, the 1.5% saline increases both blood volume and blood osmolarity. In other words, volemic thirst is sated while osmotic thirst is enhanced (Hall and Blass, 1975).

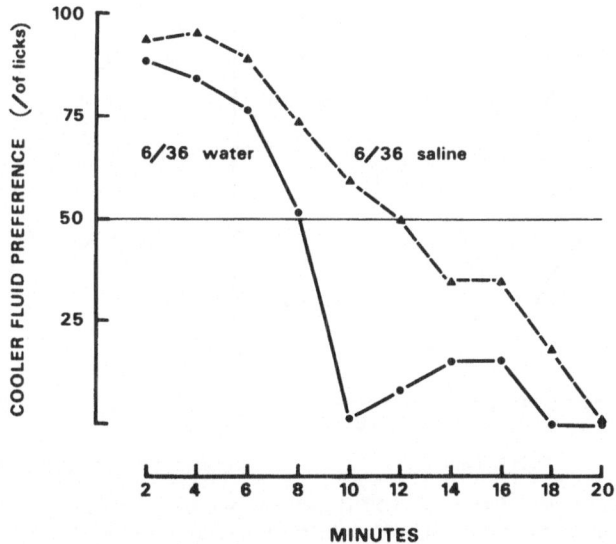

Figure 7. Time course of water and 0.9% (isotonic) saline temperature preferences. On half the days, the choice was 6°C versus 36°C water; on the other days, the choice was 6°C versus 36°C saline. The temperature-preference switch occurred with the 0.9% saline, but it was delayed. (Gold and Prowse, unpublished.)

Because of large between-rat differences, we present these two-bowl (warm versus cool) 1.5% saline data separately for each rat. The rats tended to drink the 1.5% saline more slowly than they did water. When we plot temperature preference against time, the water temperature preference curves and the saline preference curves do not agree with one another (Figure 8, left side). However, when we correct for the slower saline intake by plotting preference against cumulative amount consumed (Figure 8, right side), we find that for each rat, the saline and water curves are more com-

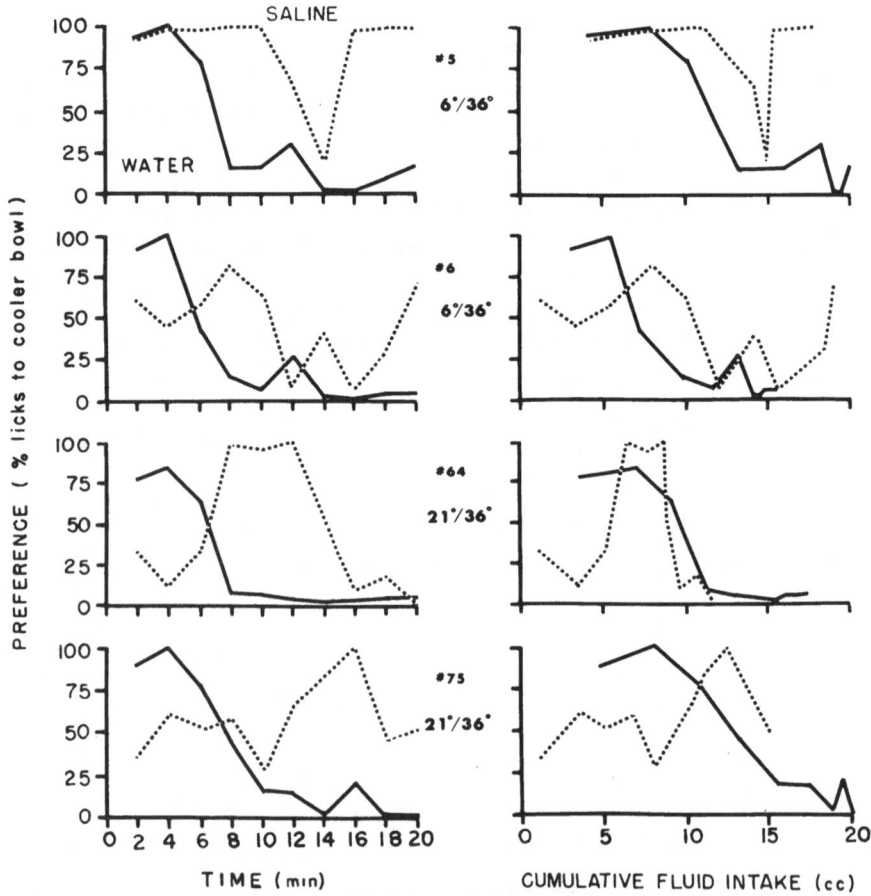

Figure 8.   Time course of water and 1.5% (hypertonic) saline temperature preferences. On half the days, the choice was 6°C versus 36°C water; on the other days, the choice was 6°C versus 36°C saline. Each curve is data from one female rat. At least four days' data are averaged for each data point. On the left-hand curves, the preferences are plotted against time. Saline intakes tended to be slower than those of water. On the right-hand curves, the same preferences are plotted against cumulative intake. (Gold and Pytko, unpublished.)

parable. The rats started out with (or acquired) a cool-fluid preference and then, after a 9–13cc fill (which equals stomach capacity), they all began to prefer the warm saline. This warm preference was not maintained throughout the remainder of each saline day. The two rats, #5 and #6, with a 6–36°C choice (the other two had a 21–36°C choice) reversed their preferences back to cool saline. We believe that this reversal back to the cooler fluid occurred in response to the predicted increase in serum osmolarity following hypertonic saline ingestion.

Rats #64 and #65 started their saline preference sessions with a mild warm preference (21–36°C choice). After 8–12 min, enough time for serum osmolarity to begin to rise, these rats switched to the coolest (21°C) temperature offered. Only after 9–10 cc had been consumed did they again prefer the warmer fluid (gut filling, blood volume increasing).

We conclude that fluid-temperature preferences do not correlate exclusively with any one aspect of thirst or satiety. Gut stretch, serum osmolarity, and blood volume may all play a role. The temperature preference then becomes simply a correlate of thirst.

11. Pagophagia

An extreme version of the deprivation-induced preference for oral cooling may be the phenomenon of pagophagia (ice eating) exhibited by severely anemic rats (Woods and Weisinger, 1970) and severely anemic humans (Reynolds, Binder, Miller, and Chang, 1968) and occasionally indulged in by apparently normal humans and rats (Woods and Weisinger, 1970). In 20-min preference testing (ice versus cool water), anemic rats select crushed ice for 96% of their H_2O intake versus only 45% for controls.

In anemia, there is an inadequate supply of hemoglobin, which is needed to transport oxygen from lungs to tissues. Constriction of the total blood space should get the limited supply of hemoglobin moving more quickly on its endless shuttle between lungs and tissues. The anemic rat or human would only have to learn through experience that mouth cooling makes him feel better.

Vasoconstriction through the sipping of cold water would not alleviate the sluggishness of anemia as well as ice would because (1) water does not cool as much as ice, and (2) the high water intake that would be required to maintain a cooling stimulus via cold water would also tend to swell the blood volume and thereby dilute the hemoglobin concentration, defeating the object of the cooling. The best relief would be obtained from a rapidly circulating low blood volume. Pagophagia seems uniquely suited to accomplish the desired end.

12. Thermoregulation and Water Temperature

The consumption of cool water lowers body temperature (Deaux and Engstrom, 1973a). An obvious interpretation of both the suppressed intake of cool water when there is no temperature choice and the switch away from cool water when there is a choice is that the rat is thermoregulating. In order to test this interpretation, we set up a situation in which a rat would do one thing if it was using water temperature to thermoregulate and another if it was choosing a water temperature according to its level of thirst. We (Gold and Prowse, 1974) did this by offering rats a choice between warm and cool water in a very warm (32°C) environment. The rats started, as usual, with the cool water but then switched to warm water even more rapidly than controls. They maintained their warm-water preference throughout the remainder of the 20-min test, despite elevated colonic temperatures. We conclude that thirsty rats do not select a water temperature for its thermoregulatory value. A more conservative concluding statement would be that warm, partially satiated rats do not select cool, thermoregulatory water. Whether cold thirsty rats will display a cold preference based on thirst or a warm preference based on thermoregulation has not been investigated.

13. Thermogenic Drinking

Some have suggested that drinking occurs in rapid response to increases in ambient temperature. Exposure to an ambient-temperature increase for as little as one-half hour will elicit a drinking response in the rat (Budgell, 1970b; Fregly and Waters, 1966; Grace and Stevenson, 1971; Nelson, Fregly, and Tyler, 1975). This drinking response, which occurs before any significant osmotic or volemic effects due to water loss (Hainsworth, Stricker, and Epstein, 1968), is usually referred to as *thermogenic drinking*. Thermogenic drinking anticipates the water loss that would occur with continued exposure to the increased ambient temperature. In this section, we examine these studies in light of our finding that water temperature alone affects water intake. We propose that thermogenic drinking is merely the expression of a preference for warm water.

Budgell (1970a) reported that water-deprived rats inside a temperature-controlled Skinner box that bar-press for 0.1-cm^3 volumes of water increased their water intake as the ambient temperature increased from 5°C to 40°C. This increase was interpreted as a direct result of the ambient temperature, since dehydration (as measured by body-weight loss during exposure) proved negligible. The rats in this study were acclimated in the Skinner box to the various ambient temperatures for 30 min before the water reinforcer was made available. No mention was made of the temperature of the water nor of any attempts at controlling water temperature during the acclimation and testing periods. Lacking specific knowledge to the contrary, we surmise that increases in ambient temperature were accompanied by increases in the water temperature. As we have shown, increases in water temperature alone increase water intake.

Grace and Stevenson (1971) also reported thermogenic drinking, this time for satiated rats. Their rats were satiated at 22°C. Then they were shifted to a warm (32°C) environment. During the first hour of exposure to 32°C, they drank 65% more water than during the same hour on control (22°C) days. The authors note that in their experiments the water was allowed to equilibrate with the environmental temperature before testing began. Grace and Stevenson did not appear to consider this increase in water temperature to be a potent variable.

In conclusion, we challenge all of the data purporting to demonstrate thermogenic drinking. In no case can we find an instance of thermogenic drinking in which water temperature was controlled. Several studies, cited by Fitzsimons (1972), failed to obtain thermogenic drinking. Perhaps these studies placed the water in the heated chamber at the same time as the rats went in, so that there was insufficient time for the water to warm up.

14. Prandial versus Nonprandial Drinking

The possible confounding effects of prandial drinking during *ad libitum* water intake versus the nonprandial drinking imposed during most of the brief (20–30 min) daily testing regimes was explored by Deaux and Engstrom (1973). They used daily 20-min drinking tests at 12°C, 24°C, or 37°C with or without food present during the drinking sessions. They had a small *N* (3 per cell) and the possibly bothersome presence of rectal temperature probes. They found that the presence or absence of food had no effect on water intake on days when the water was at 12°C or 24°C, but the presence of food did reduce the intake of 37°C water [compare Figure 4 (B) with Figure 4 (C)]. In the presence of food, the intake of 37°C water fell to the level of intake obtained with 24°C water. In our laboratory, this interaction has not replicated. We (Gold, Laforge, and Fraser, unpublished) have, as previously mentioned, found that the presence of food does not diminish the suppressant effect of cool water. Additionally, we will see presently that in schedule-induced drinking, in which food is unavailable most of the time, despite the fact that the drinking is food-associated, more warm water is consumed.

Thus, in our laboratory, the sex of the subjects and food availability during drinking sessions do not seem to be potent variables so far as the cool-water suppression of water intake is concerned. This within-one-laboratory study should, we believe, take precedence over between-laboratory comparisons that give the appearance of sex- and food-related blunting of the suppression. The factor(s) behind the between-laboratory discrepancies remain unresolved.

15. Schedule-Induced Drinking

The effect of water temperature on water intake also extends to an extreme form of prandial drinking: schedule-induced polydipsia (see Chapter 9). Carlisle (1973) had animals lever-press for small pellets of food on a VI 1-min schedule that produced polydipsia. When the freely available water was warm, the rats drank more than when the water was cool. Using a wider range of temperatures, Carlisle and Laudenslager (1976) have replicated their finding. They obtained a 34.7% suppression of water intake for 5°C water as opposed to 36°C water [Figure 4 (F)]. Independent manipulation of the ambient temperature had no effect. Weijnen (1975) has concurred in this interpretation of Carlisle's data.

16. Intake by Sated Animals

A variant on *ad libitum* testing is the work of Nelson, Fregly, and Tyler (1974). Their animals were exposed to 26°C water on an *ad libitum* basis and, in addition, were given 1-hr daily tests at different temperatures of water. In this situation, the only things that might have induced drinking during the 1-hr tests were (1) the stimulation of being transferred to the testing cage; (2) the new temperature of the water being offered during the test; and (3) thirst that might have developed during the hour of testing. In this situation, all intakes were quite small, but cool water suppressed water intake 88.3% [Figure 4 (D)], one of the biggest suppressions found. Incidentally, these were male rats.

The very high (88.3%) suppression of cool-water intake obtained by Nelson *et al.* can be explained by their use of previously sated rats. For example, take our typical 40% suppression of water intake, which would be expected if rats obtained all of their water in a daily 1-hr test. Reasonable intakes would be on the order of 20 cc of 36°C water versus 12 cc of 6°C water, for a 40% suppression. However, if you subtract from both of those intakes an arbitrary 11 cc of water to achieve hydration comparable to the situation at the start of Nelson *et al.*'s testing, then you are left with 9 cc of 36°C water intake and only 1 cc of 6°C water intake for an 88.9% suppression.

In the same Nelson *et al.* study, other rats were kept for 24 hr in a cool (5°C) environment with *ad libitum* 5°C water and then were brought to a 25°C environment, where they were offered no-temperature-choice water at either 5°C, 25°C, or 35°C. In this situation, just as when there was no prior cold-room experience, the animals drank less cool water than warm water during the 1-hr test—in this case, 72.7% less (Figure 4E).

Nelson *et al.*'s animals drank considerably more water following the day of cold exposure than following the control condition in which there was no cold exposure. We interpret this as a rebound from suppressed water intake during the cold exposure. Nelson *et al.* did not measure water intake during the day of cold exposure, presumably because they were thinking of their manipulation in terms of the cold environment rather than in terms of the cold water that the animals encountered in the cold environment. Nevertheless, under all of these different testing conditions, cool water suppressed water intake, replicating our initial finding.

17. Speculation

Orolingual and/or gastric cooling may account for a variety of heretofore mysterious phenomena, such as airlicking, pagophagia, thermogenic drinking, post-cold-exposure drinking, reflexive sweating, reflexive diuresis, and the initial preference for cool water. The importance of fluid temperature as a variable in the laboratory and as an explanation for these curious phenomena seems convincing. Whether fluid temperature responses have significant survival value in naturalistic settings remains to be demonstrated. It does, however, appear to be adaptive to drink more (warm water) in the summer in anticipation of needs for evaporative cooling and to drink less (cool water) in the winter to conserve body heat.

Another possible explanation is that orally naïve young animals use mouth cooling to identify drinking water (Mendelson and Chillag, 1970). If this were the sole cue, then body-temperature water would be refused. In our experience with mature rats, even body-temperature isotonic saline solutions, which offer neither thermal nor osmotic signals, are readily consumed. However, we have not tried a more balanced electrolyte than straight 0.9% NaCl, and we have not closely observed the initial responses of saline-naïve animals. We have also found that body-temperature water, which does not cool, is preferred by thirsty rats to ice, which offers much cooling but little water per lick (Gold, unpublished data).

It is nevertheless attractive to speculate that the instant reward of mouth cooling initiates water drinking in the newborn bird or weanling mammal and that the textural, taste, visual, and olfactory properties of water acquire rewarding properties through their learned association with the cooling fluid and with the delayed osmotic and volemic consequences thereof. Consistent with this suggestion, Mendelson and Chillag (1970) have shown that no prior experience is required to establish a cold-licking response.

To attack this nature–nurture question in mammals, it would be necessary to raise them from birth with tube feeding and then observe their initial responses when the first water they encounter is at body temperature. We are not aware of any such work. Observations on newly hatched chicks might fit this bill, with considerably less trouble.

In conclusion, the temperature of what enters our mouths appears to be an important hedonic and regulatory variable, but experimentally it has, until recently, been neglected. Prior to 1972, when Fitzsimons's exhaustive 93-page review of the thirst literature (including 433 references) appeared, not one paper had dealt with water temperature's influence on water intake. Fitzsimons devoted only one sentence to the topic: "Temperature may also be important; as is well known, a cool drink may quench thirst more effectively than a warm one" (p. 498). We assume that "as is well known" means that despite the volume of literature on thirst, no citation could be found at that time to support the conclusions of personal oral experience.

18. Summary

The temperature of what enters our mouths is an important hedonic and regulatory variable. Until recently, we knew this only because of our anecdotal personal experiences. The effects of varying the temperature of drinking water have now come under experimental investigation.

Cooling drinking water suppresses water intake in tests as short as 20 min/day and as long as 24 hr. The suppressant effect of cool water appears to be mediated via the gut, as after subdiaphragmatic vagotomy, water temperature no longer influences water intake.

Orolingual and gastric cooling appear to account for a variety of heretofore-mysterious phenomena, including airlicking, pagophagia, and thermogenic drinking.

ACKNOWLEDGMENTS

We are indebted to Earl Simson for preparation of the figures and to Ed Hirsch and Paul Sawchenko for critical reading of the manuscript.

References

Banet, M., and Seguin, J. J., 1970 Effects of preoptic cooling in rats acclimated to 21° and 4°C, *J. Appl. Physiol.* **29**:385–388.

Box, B. M., Montis, F., Yeomans, C., and Stevenson, J. A. F., 1973, Thermogenic drinking in cold-acclimated rats, *Am. J. Physiol.* **225**:162–165.

Budgell, P., 1970a, The effect of changes in ambient temperature on water intake and evaporative water loss, *Psychon. Sci.* **20**:275–276.

Budgell, P., 1970b, Modulation of drinking by ambient temperature changes, *Anim. Behav.* **18**:753–757.

Carlisle, H. J., 1973, Schedule-induced polydipsia: Effect of water temperature, ambient temperature, and hypothalamic cooling, *J. Comp. Physiol. Psychol.* **83**:208–220.

Carlisle, H. J., and Laudenslager, M. L., 1976, Separate effects of water and ambient temperature on polydipsia, *Physiol. Behav.* **16**:121–124.

Deaux, E., 1973, Thirst satiation and the temperature of ingested water, *Science* **181**:1166–1167.

Deaux, E., and Engstrom, R., 1973a, The temperature of ingested water: Its effect on body temperature, *Physiol. Psychol.* **1**:152–154.

Deaux, E., and Engstrom, R., 1973b, The temperature of ingested water: Preference for cold water as an associative response, *Physiol. Psychol.* **1**:257–260.

Engstrom, R., and Deaux, E., 1974, Stomach distention as a regulation of fluid intake, *Physiol. Psychol.* **2**:337–340.

Epstein, A. N., 1967, Oropharyngeal factors in feeding and drinking, *in* "Handbook of Physiology (Alimentary Canal, Vol. I, Sec. 6)" Washington, D.C., American Physiological Society, pp. 197–218.

Fitzsimons, J. T., 1972, Thirst *Physiol. Rev.* **52**:468–561.

Fregly, M. J., 1971, Effect of a low bulk diet on water and electrolyte exchange in rats exposed to cold, *Can. J. Physiol. Pharmacol.* **49**:959–966.

Fregly, M. J., and Waters, I. W., 1966, Water intake of rats immediately after exposure to a cold environment, *Can. J. Physiol. Pharmacol.* **44**:651–662.

Gauer, O. H., and Henry, J. P., 1963, Circulatory basis of fluid volume control, *Physiol. Rev.* **43**:423–481.

Gold, R. M., 1973, Cool water suppression of water intake: One day does not a winter make, *Bull. Psychon. Soc.* **1**:385–386.

Gold, R. M., Kapatos, G., Prowse, J., Quackenbush, P. M., and Oxford, T. W., 1973, Role of water temperature in the regulation of water intake, *J. Comp. Physiol. Psychol.* **85**:52–63.

Gold, R. M., and Prowse, J., 1974, Water temperature preference shifts during hydration, *Physiol. Behav.* **13**:291–296.

Grace, J. E., and Stevenson, J. A. F., 1971, Thermogenic drinking in the rat, *Am. J. Physiol.* **220**:1009–1015.

Hainsworth, F. R., Stricker, E. M., and Epstein, A. N., 1968, Water metabolism of rats in the heat: Dehydration and drinking, *Am. J. Physiol.* **214**:983-989.

Hall, W. G., 1973, A remote stomach clamp to evaluate oral and gastric controls of drinking in the rat, *Physiol. Behav.* **11**:897-901.

Hall, W. G., and Blass, E. M., 1975, Orogastric, hydrational, and behavioral controls of drinking following water deprivation in rats, *J. Comp. Physiol. Psychol.* **89**:939-954.

Hamilton, C. L., 1963, Interactions of food and temperature preference shifts during hydration, *J. Comp. Physiol. Psychol.* **56**:476-488.

Hatton, G. I., and Bennet, C. T., 1970, Satiation of thirst and termination of drinking: Roles of plasma osmolarity and absorption, *Physiol. Behav.* **5**:479-487.

Hendry, D. P., and Rasche, R. H., 1961, Analysis of a new nonnutritive positive reinforcer based on thirst, *J. Comp. Physiol. Psychol.* **54**:477-483.

Kapatos, G., and Gold, R. M., 1972, Tongue cooling during drinking: A regulator of water intake in rats, *Science* **176**:685-686.

Kraly, F. S., and Blass, E. M., 1976, Mechanisms for enhanced feeding in the cold in rats, *J. Comp. Physiol. Psychol.* **90**:714-726.

McFarland, D., and Wright, P., 1969, Water conservation by inhibition of food intake, *Physiol. Behav.* **4**:95-99.

Mendelson, J., and Chillag, D., 1970, Tongue cooling: A new reward for thirsty rodents, *Science* **1970**:1418-1419.

Mendelson, J., Chillag, D., and Paramesvaran, M., 1973, Effects of airstream temperature and humidity on airlicking in the rat, *Behav. Biol.* **8**:357-365.

Mendelson, J., and Zec, R., 1972, Effects of lingual denervation and desalivation on airlicking in the rat, *Physiol. Behav.* **8**:711-714.

Nelson, E. L., Fregly, M. J., and Tyler, P. G., 1974, Effects of water temperature on post-cold-exposure drinking response of rats, *Am. J. Physiol.* **227**:977-980.

Nelson, E. L., Fregly, M. J., and Tyler, P. E., 1975, Factors affecting thermogenic drinking in rats, *Am. J. Physiol.* **228**:1875-1879.

Nicolaidis, S., 1969, Early systemic responses to orogastric stimulation in the regulation of food and water balance: Functional and electrophysiological data, *N.Y. Acad. Sci.* **157**:1176-1203.

Oatley, K., 1967, A control model of the physiological basis of thirst, *Med. Biol. Eng.* **5**:225-237.

Ramsauer, S., Mendelson, J., and Freed, W. J., 1974, Effects of water temperature on the reward value and satiating capacity of water in water deprived rats, *Behav. Biol.* **11**:381-393.

Reynolds, R. D., Binder, H. J., Miller, M. B., and Chang, W. W. Y., 1968, Pagophagia and iron deficiency anemia, *Ann. Intern. Med.* **69**:435-440.

Valenstein, E. S., Kakolewski, J. W., and Cox, V. C., 1967, Sex differences in taste preference for glucose and saccharin solutions, *Science* **156**:939-942.

Weijnen, J. A. W. M., 1975, Lingual stimulation and water intake, *in* "Control mechanisms of drinking" (G. Peters, J. T. Fitzsimons, and L. Peters-Haefeli, eds.), Springer-Verlag, Berlin, pp. 9-13.

Woods, S. C., and Weisinger, R. S., 1970, Pagophagia in the albino rat, *Science* **169**:1334-1336.

Note: Experiments cited as unpublished will appear in Gold, R. M., Laforge, R. C., Sawchenko, P. E., Fraser, J. C., and Pytko, D., 1977, Cool water's suppression of water intake: Persistence across deprivation conditions, ages, sexes, and osmolarities, *Physiol. Behav.*, and in Sawchenko, P. E., Gold, R. M., Ferrazano, P. A., 1977, Abolition by selective gastric vagotomy of the influence of water temperature on water intake: Mediation via enhanced gastric clearance, *Physiol. Behav.*, in press.

7

Taste Modulation of Fluid Intake

Douglas G. Mook and Nancy J. Kenney

1. Introduction

In any discussion of "nonregulatory" or "nonhomeostatic" determinants of behavior, the influence of taste on ingestion is certain to receive at least a mention. And at first glance, the phenomena observed when animals are offered various sapid materials are impressive to one accustomed to think of animals as "finely tuned regulatory machines" (Ernits and Corbit, 1973).

Taste factors can guide ingestive behavior in ways opposed to homeostatic needs, sometimes catastrophically so. Rats with lateral-hypothalamic damage may refuse to drink, and accept a lethal voluntary dehydration, when their drinking water is made slightly bitter (Teitelbaum and Epstein, 1962). Even intact rats, maintained in a cafeteria-feeding situation, may grow poorly or even die through simple refusal to ingest the amounts of casein they need, probably because they dislike its taste (Kon, 1931). Taste can also influence intake very powerfully in ways that are simply irrelevant to need. For example, rats made thirsty by hypertonic saline injection will drink a dilute glucose solution in preference to the plain water that is at least as adequate for repairing the hydrational deficit (Burke, Mook, and Blass, 1972). Finally, it is well known that taste factors can elicit a bout of ingestion even when no need at all has been imposed (Young, 1955; Ernits and Corbit, 1973).

Douglas G. Mook ● Department of Psychology, University of Virginia, Charlottesville, Virginia. *Nancy J. Kenney* ● Department of Psychology, University of Washington, Seattle, Washington. This chapter was completed in February, 1975.

Despite the force of these examples, the "nonregulatory" aspect of the responses to taste are not emphasized in this review. There are several reasons. Falk (1973) has argued cogently that it is a mistake to play off "regulatory" and "nonregulatory" factors in fluid intake; there are many factors that affect ingestion, and we ought not to permit a preoccupation with homeostasis to lead us to prejudge their relative importance in the animal's life.

The instances in which taste seems to pose a challenge to regulation should not blind us to the vital role it plays in the service of regulation—even if instances of the two cases were always easy to distinguish, and we shall see that they are not. Ingestion takes place in the environment. Therefore, its control requires a system that takes into account both the internal state of the animal and the commodities available in the environment, so that the demands of the one are met by appropriate selections from the other. A hungry animal seeks food; a thirsty one, water; and it is by sensory input such as taste that the system is able to identify these. Even writers who have emphasized the dispensability of mouth factors in the quantitative control of ingestion (Epstein, 1967) have been careful to restrict their discussion to instances in which identification and selection are problems that the animal does not face.

So close is the cooperation between internal and external factors in regulation—between what the animal needs and what it knows—that such factors as taste may not only serve but even determine the regulatory system that is in control of an animal's behavior in a given instance. Thus, a hungry but nonthirsty rat will refuse to drink water but will accept a saccharin solution avidly; and the hungrier it is, the more it will drink and the more dilute is the solution it will accept (Teitelbaum and Epstein, 1962; Mook & Cseh, in preparation). Presumably, it is treating the sweet solution as food and "eating" it because of its taste. If so, what this means is that taste can recruit a controlling system that would not be operative if taste were not there. The sweet taste, once it is registered by the brain, can switch the rat's lapping from control by one homeostatic system, concerned with water balance, to control by another concerned with energy balance.

Finally, some identifiable factors control intake in ways that cut across regulatory-nonregulatory distinctions. As we shall see later, even in cases in which drinking can be considered to be initiated by taste *per se,* it is terminated by the same factors that would terminate it if it were a response to need. It seems to us most significant that this is so; it is not at all obvious that it has to be that way; and it underscores our doubts that a regulatory-nonregulatory polarity will be part of the classification scheme that the analysis of mechanisms will eventually require.

2. Methods in the Study of Taste Modulation

Although much has been learned from the study of sapid materials added to complex foods, in this paper we focus on the response to solutions. (This limitation is partly in recognition of the editors' intentions and partly in self-defense. If we even tried to say all that there is to say about solutions, we would be in trouble, and we shall make no such attempt.) Adding sapid solutes to water has the definite advantage that it simplifies matters on the input side. In addition, it provides baselines that are sensitive to the animal's physiological state, that result from manipulation of single variables (such as solute concentration) that can be specified with precision and varied over a wide range, and that, finally, are remarkably reliable and stable. The well-known "preference-aversion function"—a rise, then a decline in intake with increasing solute concentration—characterizes the response to a variety of taste substances; and it does so, at least in our experience, in virtually every rat, unless abnormalities have been deliberately introduced into the situation.

The methods used to study fluid preferences and aversions differ with respect to how many commodities are offered, for how long, and what the animal's hydrational and/or nutritional state is when they are offered. In the most typical case, a test solution is offered along with the food and water the rat would have anyway, and the measure of interest is how the animal divides its spontaneous fluid intake between water and the solution. This is the familiar 24-hr, two-bottle procedure associated with Richter (1942-1943). Alternatively, a test solution may be offered *ad libitum* as the animal's only source of fluid for a while (see, e.g., Beebe-Center, Black, Hoffman, and Wade, 1948; Stellar, 1961). This procedure obviously simplifies the analysis. Its disadvantage is that it couples the intake of commodities whose ingestion may actually be controlled by quite different factors. For example, an adrenalectomized rat offered only isotonic saline to drink is forced to take on a water load in order to meet its salt requirements.

Ad libitum procedures may be regarded as most closely approximating physiological condition, in the sense that the animal is faced with no barriers to the satisfaction of its needs. Brief-exposure experiments, on the other hand, seemed initially to offer more convenience (the data are gathered over a brief period) and, especially when combined with deprivation, an increment in experimental control (the animal begins each session in a relatively constant state of readiness to drink). Actually, modern automatic techniques (for recording, infusion, and the like) have brought both convenience and control to the study of ingestion under *ad libitum* conditions. Examples of what can be done are Thomas and Mayer's (1968) studies of the factors affecting meal size and meal frequency and Kissileff's (1973) finding that infusions of water concurrent with feeding affect the frequency but not the size of spontaneous water drafts.

Nevertheless, brief-exposure methods are widely employed and have been fruitful. A frequently employed method is the one invented by Wiener and Stellar (1951). A rat is deprived of water overnight and offered a test solution to drink for a brief period (e.g., 1 hr) the next morning. Food and water are offered during the afternoon "maintenance" period, following which water is withdrawn and the cycle repeated. Variants of this method include offering the animal a choice among fluids during the morning's test (see, e.g., Stellar and McCleary, 1952) or manipulating its state prior to the test in some other way, as by adrenalectomy (Epstein and Stellar, 1955).

Since the animal is ready to drink when each test begins, what is of interest is how various factors modulate the drinking bout that ensues. The study of satiety mechanisms fits in naturally here, and various postingestive influences on intake can be studied if they are incremented or decremented artificially, as by gastric preloads (Stellar, Hyman, and Samet, 1954; McCleary, 1953) or concurrent injections (Kenney, 1974; Campbell and Davis, 1974b) or if the stomach contents are emptied along with the rat's drinking (Davis and Campbell, 1973). The role of taste can be assessed by variation of the material presented to the mouth in circumstances in which postingestive events either are held constant artificially (Mook, 1963, 1969; Rabe and Corbit, 1973; Smith, Holman, and Fortune, 1968) or can be shown to exercise negligible control over intake, as is the case with quinine (Teitelbaum and Epstein, 1962) or saccharin (McCleary, 1953; Bryner and Mook, in preparation).

Another way of isolating mouth factors is to limit still further the animal's contact with the solutions. Such "miniexposure" procedures, which allow the animal only a few laps at the fluid, have been used to assess the role of taste in guiding an animal's choices (Young and Falk, 1956) or in initiating or maintaining appetitive behavior (Young and Shuford, 1954). They also have been used to explore the trade-offs among positive and negative gustatory inputs (cf. Young, 1967). Closely related, of course, is the use of small quantities of solution as reinforcers in operant experiments. Here, especially if intermittent reinforcement schedules are used, large samples of behavior can be taken with very little ingestion occurring. Moreover, one can assess the strength of "motivation" (Teitelbaum, 1966) under various conditions uncontaminated by changes in amount taken in, since with interval-defined reinforcement schedules the rate of ingestion is nearly independent of operant rate.

3. Taste and the Peripheral Control of Ingestion

3.1. The Response to the Salt Taste

3.1.1. Preference-Aversion Drinking

A good example of the reliability of preference phenomena is the rat's response to sodium chloride. In Figure 1, we see intake of water and of saline solutions at varying concentrations, each paired with water for 48 hr in rats fed *ad libitum*. The three rats shown (arbitrarily selected) display preference-aversion patterns that are strikingly similar from animal to animal and are in good agreement with averaged group curves presented by other investigators (e.g., Bare, 1949). Saline intake rises as concentration increases up to about the isotonic point (0.15 *M* or 0.9% w/v) and declines progressively as concentration increases beyond that point. Changes in water intake are roughly reciprocal.

A variety of controlling factors operate at different saline concentrations and, for a given concentration, at different stages of the ingestive sequence. The apparent simplicity of the pattern, then, is deceptive.

Access to a preferred saline solution does not increase the total number of daily drafts (Kissileff, 1973; Mook and Kozub, 1968). Given that the animal is going to drink, however, it will choose a dilute saline solution in preference to water (Chiang and Wilson, 1963; Kissileff, 1973). This selection is guided by taste: it is seen when taste is isolated, either surgically (Mook, 1969) or by limited access, which minimizes postingestive events

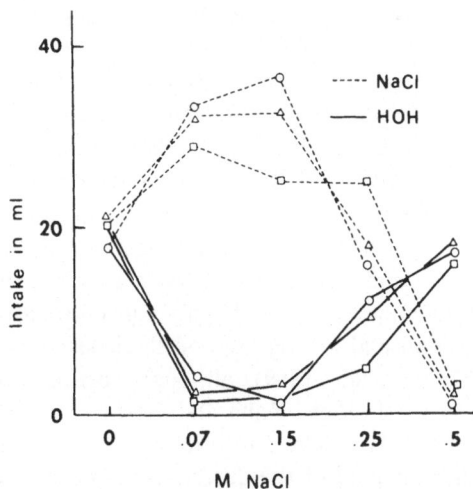

Figure 1. Two-choice *ad libitum* preference-aversion functions for saline in three normal female rats. (Unpublished observations.)

(Young and Falk, 1956). Conversely, it is not seen when the solutions are ingested without being tasted (Borer, 1968).

Even the notion of "selection," however, turns out to be unexpectedly complex. Chiang and Wilson (1963) suggested that rats continue drinking if they encounter the saline cylinder in an *ad libitum* situation, whereas if they encounter water, they will then switch to the saline cylinder. A similar phenomenon—a readiness to switch from water to saline but not the reverse—has been shown directly by Fisher (1965). On the other hand, Kissileff (1973) did not see such switching in his experiments. He suggested rather that his rats quickly learned the positions of the water and saline cylinders and thereafter must have been guided by the memory of the salt taste in approaching the saline cylinder when thirsty. This conclusion is based on the fact that a greater number of drafts was taken in a 24-hr period from the saline cylinder than from the water cylinder. Some other considerations also make the notion of "memory" as a guiding factor in such situations entirely plausible (see Section 3.1.2.).

Thus, on or even before the animal's contact with the drinking fluid, and before any consequences of its drinking come into play, there are at least three possible influences of taste to consider. A sapid solution might, simply by being available, be sufficient to elicit drinking that would not otherwise occur (although this seems not to be true of the salt taste). It, or its memory, may guide the animal's approach to the fluid cylinder. Or it may lead the animal to correct an "erroneous" approach and switch to a fluid more preferred than the one encountered. The difference between Kissileff's conclusions (1973) and Fisher's (1965) suggests that specific situational factors affect the relative importance of these three aspects of the gustatory control of choice.

Once drinking has begun, bouts of ingestion are longer when saline in the isotonic range is drunk than when water is. This aspect of saline drinking, however, is primarily under postingestive control (Mook, 1963; Mook and Kozub, 1968; Myer and Van Hemel, 1969). It appears that such solutions are less effective in relieving the animal's thirst (or, better, in inhibiting its thirst; cf. Stricker, 1969) than water is, and it is likely that the rat drinks such solutions in larger quantities than water for this reason.

Thus far, we have considered only "preferred" saline solutions in the hypo- and isotonic range. As higher concentrations are presented, other factors come into play. Highly concentrated salt solutions are aversive as taste stimuli and are rejected on the basis of taste alone (Young and Falk, 1956; Mook, 1963, 1969), although postingestive influences can also be shown (Kissileff, 1969; Rabe and Corbit, 1973). Mildly hypertonic solutions also tend to be rejected in brief-exposure choice tests, but in 24-hr experiments, there is part of the descending limb of the preference–aversion function in which solution intake is about the same as the intake of water (Figure 1; see also Bare, 1949). Presumably the taste of these solutions is not aversive, or at least not aversive enough to discourage intake when the rat encounters them.

If the solutions in this range are not aversive—and given that while they do hydrate the rat eventually (for they are well within the concentrating power of the kidney), they do so less rapidly than water does—why are they ingested in such small amounts? Probably because of the acute postingestive consequences of drinking them. In thirsty rats, gastric preloads of mildly hypertonic saline inhibit the subsequent ingestion of such solutions (Stellar, Hyman, and Samet, 1954). Since the same preloads augment water intake, it seems that the postingestive factor implicated here must operate in conjunction with mouth factors to inhibit further drinking of saline solutions. In the *ad libitum* case, presumably, such a mechanism limits the rat's drinking even of a mildly hypertonic solution on each encounter with it, so that total daily intake from the saline cylinder is not great.

There is a bit more, however. Thirsty rats, offered a mildly hypertonic solution, take a draft and then stop, returning later for brief drafts. This pattern might reflect the inhibitory postingestive effects of the solution, followed by the excretion of solutes, which removes the inhibition and releases further drinking. But this is not so, for the pattern persists in nephrectomized rats, which cannot excrete solutes (Blass, 1974). Some unidentified factor must account for the episodic nature of the drinking that occurs in such a case.

3.1.2. The Internal Environment

The saline preference–aversion pattern is sensitive to long-term drifts in internal state. It takes special intervention to produce the phenomenon in laboratory-maintained animals, but conditions producing "salt appetite" do develop in the wild and can assume considerable significance. According to Denton (1967), herds of ungulates have migrated great distances fo find salt, and men have fought wars over salt deposits.

Salt appetite, which can be induced in a variety of ways (Wolf, McGovern, and DiCara, 1974), can express itself as a large increase in intake of saline solutions in the preferred range; but more dramatic is the avidity with which such rats ingest solutions so concentrated that they are rejected under normal conditions (Bare, 1949; Epstein and Stellar, 1955). This change from rejection to acceptance can occur as a response to the salt taste alone, for it occurs when taste is isolated (Mook, 1969; Smith, Stricker, and Morrison, 1969; Smith, Holman, and Fortune, 1968). Since such procedures reduce or eliminate the physiological benefit that normally follows salt ingestion, these data also show that the salt appetite does not require the support of such benefit; it is not learned on the basis of relief from deficiency symptoms.

In discussing these findings, Mook commented (1969): "sodium deficiency can extend [the selection of saline in preference to water] to concentrations that are normally avoided" (p. 1171). It now seems to us that this is not the best way of looking at the matter. More is involved than a widening of the preferred range of saline concentrations. Indeed, it is likely that the mechanisms underlying salt appetite are quite different from those involved in salt preference by nondeficient rats. In the first place, hydration of the animal by intragastric water suppresses salt drinking by non-salt-deficient rats (Mook, 1963; Mook and Kozub, 1968), but it can actually enhance salt appetite in deficient rats (Jalowiec and Stricker, 1970). Moreover, salt preference, as we discussed it earlier (Section 3.1.1.), seems a rather casual matter. As we noted then, bouts of drinking are no more frequent in normal rats when a preferred solution is offered than when it is not, though they are directed to the solution in the former case (Kissileff, 1973). It is as if the rat said, "As long as I'm going to take a drink anyway, I might as well drink from the saline cylinder." Salt appetite, on the other hand, is focused and purposeful. Rats needing salt will press a lever (Quartermain, Miller, and Wolf, 1967), tolerate quinine adulteration in otherwise discouraging amounts (Gaines, 1965), or overcome a learned aversion (Stricker and Wilson, 1970) to get it. Finally, salt preference and salt appetite can be dissociated by damage to the brain (Wolf, 1964; Antunes-Rodrigues, Gentil, Negro-Vilar, and Covian, 1970). For example, Antunes-Rodrigues *et al.* showed that medial hypothalamic damage, which greatly depressed spontaneous saline intake, did not block its elevation following adrenalectomy.

Therefore, sodium deficiency (or its correlates or consequences) must act through some specific mechanism to cause an abnormally vigorous response to salt taste. What precisely has changed? It almost certainly does not involve the peripheral part of the gustatory pathway (Pfaffmann and Bare, 1950; Nachman and Pfaffmann, 1963; but see Wilcove, 1973). Indeed, the change need not be in the sensory system at all. The ingenious experiments of Krieckhaus and Wolf, 1968, (Krieckhaus, 1970; Wolf, 1969) have done much to clarify this matter and have provided us with yet another example of the rat's remarkable capabilities. In one experiment, thirsty rats were trained to lever-press for either water or $0.15\,M$ sodium chloride. Then they were tested with the apparatus turned off, after half of each group had been made sodium-deficient. The rats that were deficient, and had been trained with saline, made two or three times as many responses as did the other groups to the lever that had, in the past, delivered saline.

This difference, of course, could not reflect responsiveness to the salt taste as a stimulus. It wasn't there. An appeal to "reinforcement history" is no help here, for the depleted and nondepleted groups had had the same history prior to the tests in extinction, when no reinforcers were delivered. It appears rather that the rats were guided by a memory, or "search image" (Uexküll, 1934). To sacrifice rigor for clarity: the rats knew quite well what they wanted and where it should be found. Thus, the effect of sodium depletion must have been to give that image a positive "valence" (Tolman, 1959) or, in more modern terminology, to bring it into play as a target in the control system that guided the animals' actions. If that sort of thing is what happens here, it certainly is not unique to taste and ingestion. In thermoregulation, for instance, there is reason to think that some challenges may call corrective action into play by biasing a target or goal state, rather than by acting as stimuli that recruit the effector systems directly (Hammel, 1965).

In view of our previous discussion, it seems likely that the "image" is a gustatory one in some sense that we cannot now specify. But it clearly would be futile to look for changes in the sensory system, as it is driven by peripheral stimulation, if a change in value of the "image" is what is essential. Unfortunately, our ideas as to what to do instead are not, at this point, very clear.

3.1.3. The Oral Environment

Richter's original suggestion was that "salt appetite" might come about through a change in salivary composition, which, in turn, would affect the adaptation level of the taste receptors. While such a factor probably is of little or no importance in this particular context (cf. Halpern, 1967), concern with the oral environment has provided valuable information about the phenomenology of taste (see Chapter 8) and has led to preference experiments that, if nothing else, are instructive for the problems of interpretation they raise.

A reduction or disappearance of salt preference following desalivation in the rat has been reported by several investigators (Vance, 1965; Catalanotto and Sweeney, 1972; Kissileff, 1967; Lawson, Hagstrom, and Walter, 1974). Of these, some (Vance, 1965), but not others (Lawson *et al.,* 1974), have seen a transient enhancement in preference before it disappeared. This would seem to be a case in which the local environment of the taste receptors can have powerful effects on the intake patterns that they, in turn, control.

There is a difficulty, however. Salivaless or saliva-deficient rats are faced with the mechanical problem of swallowing dry food with a dry mouth. To solve it, they typically adopt a "prandial-drinking" pattern, taking a few laps of water after every morsel of food, as if to wash it down. The normal rat is quite different. It does most of its drinking in long, sustained drafts, usually before or after a meal rather than during one (Kissileff and Epstein, 1969). Kissileff (1967) suggested that preference loss in salivaless rats might reflect a simple failure to taste the solution because of the masking taste of food in the mouth during drinking. He found that in desalivate "prandial drinkers" that had lost the salt preference, it could be restored at once by the withdrawal of food. A similar dependence of salt preference on the absence of food was found by Lawson (1969) in intact rats whose salivation was blocked by atropine. Lawson and Hagstrom (1972) went on to show that preference thresholds for sucrose and saccharin, and the aversion threshold for quinine, were elevated following desalivation—but again, only if food was present. These findings, all consistent with the taste-masking idea, raise difficulties for the other obvious interpretations of the effect of salivary loss. That the findings apply to sucrose makes it unlikely that the adapting properties of salivary electrolytes are a critical factor; and the restoration of preferences and aversions by removal of food argues against a change in receptor morphology or function (Catalanotto and Sweeney, 1972) as the factor responsible for the preference deficits in desalivate rats.

The question comes to this: Does a preference deficit in a desalivate rat reflect the effect of desalivation on taste, or is it a secondary consequence of the abnormal feeding pattern that develops? Lawson and Hagstrom (1972) synthesized the prandial-drinking pattern in intact rats suffering no interference with the salivary system by a brilliantly simple tactic: they required the rat to drink a few drops of water in order to obtain a small pellet of food, and the rats were required to earn all their food this way. Drinking, then, was forced into close and intermittent association with feeding. Would imposition of this situation interfere with normal avoidance of quinine? It did.

The prandial-drinking pattern, then, must be reckoned with as a source of artifact in preference work. This problem is especially worrisome for students of the neural control of preference phenomena. Since brain damage at a number of sites along the trajectory of the medial forebrain bundle and related structures can produce the prandial-drinking syndrome (see e.g., Kissileff and Epstein, 1969; Toth, 1973; Dickinson, 1973), the problem may lurk in the paths of many investigators.

3.2. The Response to the Sweet Taste

In Figure 2, we see another example of the reliability of basic preference–aversion phenomena: three arbitrarily selected rats showed virtually identical preference–aversion functions for glucose in the Richter 24-hour, two-bottle paradigm. Functions for sucrose look very similar to these. So do the saline preference–aversion functions shown in Figure 1 above, and

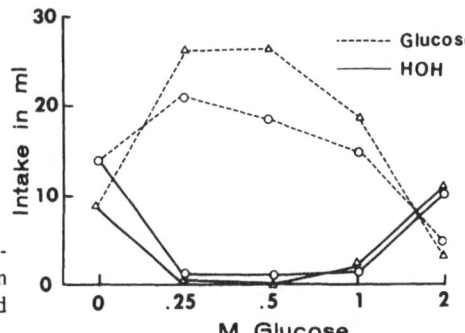

Figure 2. Two-choice, *ad libitum* glucose preference-aversion functions in two normal female rats. (Unpublished observations.)

what this shows is the danger of parallels. The intake patterns characterizing the response to glucose and saline are virtually identical. The controlling mechanisms are not.

3.2.1. Preference-Aversion Drinking

Most of the analysis of the sugar preference-aversion pattern has been conducted with brief-exposure experiments, which give rise to functions similar to the *ad libitum* one. Figure 3 shows 1-hr intake in hungry and thirsty rats, and under normal drinking conditions (lower curves) they show the typical rising and falling preference-aversion pattern.

The upper curves in Figure 3 show what intake looked like when post-ingestive influences were removed: These rats were sham-drinking the various solutions through esophageal fistulas. Although the curves are dis-

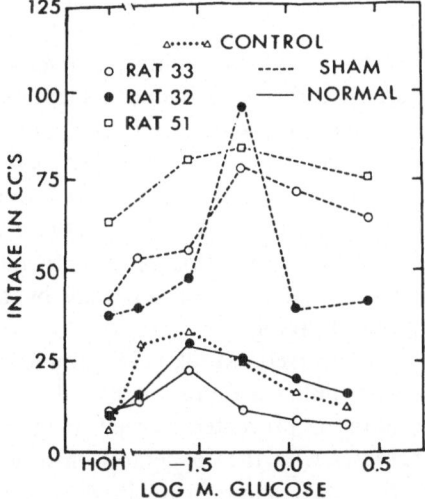

Figure 3. Single-bottle intake of glucose at various concentrations. One of these rats was intact ("control"); the others were esophagostomized. In the "sham" condition, the operated rats were simply passing the fluid offered through the mouth and out the fistula, so that all postingestive influences on intake were removed. In the "normal" condition, a fluid identical to the one presented to the mouth was intubed into the stomach as the rat drank, thus reinstating the conditions of normal drinking. (From Mook, 1963. Copyright 1963 by The American Psychological Association. Reproduced by permission.)

placed upward and to the right, their form remains: intake rose over the low-concentration range, falling off at higher ones. Consideration of these data led Mook (1963) to speak of "positive" and "negative" mouth factors affecting intake at low and high concentrations, respectively.

Subsequent work has made us highly suspicious of sham-drinking data. In the first place, such data are quite variable. That effect might be corrected by better control, but more serious is the copious salivation that is evoked by the sweet solutions and then is lost through the fistula. We have not measured salivary loss directly, but we have seen rats lose as much as 10 g in body weight over an hour's period of sham drinking. If we assume that the loss is mostly water, this means that the rat incurs a water loss nearly equal to its total blood volume. Since dehydration can attenuate drinking of concentrated glucose solutions (see below), the "negative mouth factor" may actually reflect systemic dehydration.

Life certainly would be simpler if there were no negative mouth factor, for a wealth of other evidence would render its interpretation problematic. In two-choice, brief-exposure experiments, rats at least initially select the higher of two glucose concentrations, even if it is at a concentration well above the peak of the preference–aversion function (Stellar and McCleary, 1952; Young and Greene, 1953). In operant situations using intermittent reinforcement, the response rate increases with the concentration of sugar used as incentive material, up to very high levels (Guttman, 1953). Lick rate also increases with the sugar concentration offered (Davis, 1973). If there is a negative mouth factor, it is something other than aversiveness of the sugar taste, even at very high concentrations.

A more important set of factors that inhibit intake of sugar at high concentrations, giving rise to the descending limb of the preference–aversion function, are the postingestive consequences of drinking. Guttman's bar-pressing data are compatible with this idea: with periodic reinforcement, which minimizes ingestion, he found the response rate to increase with increasing concentration, as noted above. But with continuous reinforcement, which permits appreciable ingestion, the response rate fell off at the high concentrations, so that the plot of rate against sugar concentration took on a bowed appearance similar to the preference–aversion function. McCleary (1953) and Jacobs (1961, 1962) have found that stomach preloads of glucose inhibit subsequent glucose intake, thus showing that the solutions may bypass the taste receptors and still limit further drinking.

Conversely, Mook (1963) found that if the normal consequences of glucose drinking are prevented and the dehydration that accompanies sham drinking is prevented as well, then intake (like choice and bar-press rate) increases with the concentration of glucose offered to the mouth. Figure 4 (solid lines) shows 1-hr intake as a function of glucose concentration

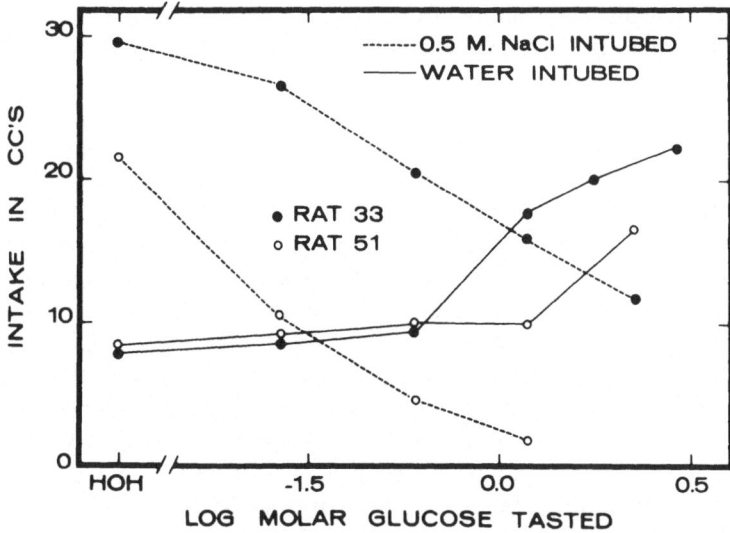

Figure 4. Single-bottle, 1-hr intake by esophagostomized rats passing glucose solutions through the mouth. The rats received concurrent injections of water (solid line) or hypertonic saline (dashed line) into the stomach as they drank. (Some of the data are from Mook, 1963.)

"tasted" but passed out through the esophageal fistula, under conditions in which water was injected into the stomach as the animal drank. The resulting intake curve is flatter than the normal preference–aversion curve, suggesting that postingestive factors play some part in its normal rise; but intake continued to increase as concentration rose, well into the range where in rats drinking normally it would be sharply limited (compare the lower curves in Figure 3).

Postingestive inhibition of glucose drinking does not require that the nutritive action of glucose come into play (McCleary, 1953; Mook, 1963). The broken lines in Figure 4 show the effect of hypertonic saline injections, as opposed to water injections, delivered into the stomach as the rat drank. Here, intake at the high concentrations was markedly depressed by the dehydrating stomach loads. We also see that such loads must exert their suppressant effects in cooperation with taste input, since the same loads markedly enhanced intake when coupled with the taste of water. This finding also means that such factors as malaise cannot account for the suppression of glucose drinking.

Dehydration is only one factor that can inhibit glucose drinking, however, and it probably is not the major one to operate under physiological conditions. Jacobs (1961, 1964) has pointed out that even concentrated glucose solutions, once swallowed, do not produce appreciable dehydration. He found, for instance (1964), that whereas gastric preloads of hypertonic saline depress glucose drinking and enhance the intake of water, glucose preloads have only the former effect and not the latter, even at high concentrations. These and other findings led him to argue that the postingestive effects of glucose on intake are expressed through an "extraosmotic route."

Further work has supported this conclusion and has brought us closer to identification of the system(s) involved. Kenney (1974) showed that intragastric and intraduodenal glucose infusions were more effective than acaloric infusions in suppressing concurrent glucose intake, and Campbell and Davis (1974a) found them more effective in suppressing the subsequent rate of licking at a glucose solution. Campbell and Davis also found that intraportal injections had similar effects but that intrajugular ones had none, suggesting the participation of hepatic glucoreceptors (see Russek, 1970) in the control of sugar-solution drinking. It is clear, then, that short-term sugar intake is responsive to a chemospecific "satiety" system, triggered by receptors proximal to the general circulation in the route that ingested sugars follow—in the gut or the liver, or both.

How do these findings, all from brief-exposure experiments, bear on the *ad libitum* preference-aversion patterns shown in Figure 2? The element of choice is added, of course, and with the sugars as with dilute saline solutions, the choice of sugar over water has been shown to be taste-dependent (Borer, 1968; Mook, 1969). Of course, the simple notion of "selection," here as in the salt case, needs to be specified further (see Section 3.1.1).

Beyond that, it is tempting to regard the total daily intake as reflecting the cumulative effect of the short-term controls as they operate on each bout of ingestion. It is plausible to look at the saline case that way (as we did earlier), since availability even of a preferred saline solution does not increase drinking frequency. But though direct evidence is lacking, we doubt that this would be true of a sugar solution. And if it is not—that is, if the presence of the solution evokes frequent approaches to the cylinder containing it—then again, it must not be the taste itself, but its memory, that triggers the approach (unless such factors as aftertaste are involved).

There is direct evidence of another sort that long-term exposure to such solutions brings factors into play that the brief-exposure experiments do not tap. Hammer (1968) found that sucrose solutions were preferred to equimolar solutions of the less-sweet disaccharide, maltose, in brief-exposure tests under a variety of conditions. But in the single-bottle 24-hr tests, intake of maltose was the greater. The simplest explanation, though it was not directly tested, is that bouts of ingestion may have been more frequent when maltose was offered than when sucrose was, in the *ad libitum* case. Why this should be so is not clear. It would, however, fit in with the evidence that meal size and meal frequency are controlled by different factors (Le Magnen and Tallon, 1966; Thomas and Mayer, 1968).

3.2.2. Food or Fluid?

It makes sense to speak of "meals" in this connection because sugar solutions are food as well as fluid, and rats are fully sensitive to this fact: they will "drink" such solutions when thirsty, or they will "eat" them when hungry (Teitelbaum and Epstein, 1962; Collier and Bolles, 1968).

Here, then, is another interaction into which taste enters—not with the internal consequences of drinking this time, but with the antecedent state of need. That taste is involved is most clearly shown by the fact that these dependencies apply to nonnutritive saccharin solutions as well as to the sugars (Teitelbaum and Epstein, 1962; C. L. Hamilton, 1969; Strouthes, Volo, and Unger, 1974).

That sugar solutions are foods is of course well known; indeed, they have been used as stimulus material by many investigators whose interests lie in the control of feeding rather than in the taste modulation of drinking. Nevertheless, at least some of us who have addressed questions of fluid preference did not have this idea clearly in mind. If we had, we would not have needlessly complicated our experiments by using rats that were both food- and water-deprived (Mook, 1963, 1969) or that were tested after water deprivation (Mook, Walshe, and Farris, 1975), which, of course, causes hunger as well as thirst (Adolph, 1947; Verplanck and Hayes, 1953).

Philip Farris, in our laboratory, showed that perfectly normal brief-exposure preference–aversion functions, identical to those shown in Figure 3 (lower curves), could be produced in rats tested not after water deprivation (which produces both hunger and thirst) but after food deprivation alone (which produces only hunger). Mook and Cseh (in preparation) found that rats that were mildly hungry, but were actually overhydrated by automatic water intubations and never drank water, accepted sweet solutions (including saccharin) avidly. And the hungrier they were, the more saccharin they drank, and the lower was their acceptance threshold for saccharin solutions.

The fact that sweet solutions are foods, or are treated by the rat as if they were, introduces some worrisome second thoughts about the Richter two-bottle procedure as applied to them. What we really are offering the rat in such an experiment is a choice between two fluids (water and a sweet solution), but also a choice between two foods (the solution and the dry food typically offered). Thus, the rat, in drinking the solution, may be expressing not a preference for one fluid over another but for one food over another— especially inasmuch as solid-food intake is depressed when sugar solutions are offered, as if in compensation for the calories ingested when they are drunk (Jacobs, 1962, and our own observations). Such an idea by itself might not affect our interpretations very much. But sweet solutions differ from dry foods along a number of dimensions, of which sweetness is only one. In particular, it takes less effort to "eat" a liquid than to eat a dry food (Morrison, 1968).

The "effort" factor can be eliminated if the rat is offered only a liquid diet, and then the effect of sweetness in isolation can be assessed when a choice is offered between an unsweetened diet and the same diet sweetened with saccharin. When this was done, there was no trace of a saccharin preference, at any concentration, in the majority of the animals (Mook, 1975). This was so even when the rats were starved, a condition that markedly lowers preference thresholds for sweet solutions in water (Campbell, 1958; Mook and Cseh, in preparation), and even though it could be shown that the rats did detect the saccharin additive and that it tasted sweet to them (Mook, 1975). Apparently sweetness, when it is "redundant"—that is, when other cues already identify the sweetened commodity as food—has in itself no effect on the choices most animals make.

It seems quite possible that the preference for sweet solutions, when offered with water and dry food, depends not so much on the "palatability" of the sweet taste—if this were so important, we should have seen a saccharin preference in the liquid-diet experiments—but on the fact that the commodity is identified as food. In other words, a purported "hedonic" function of the sweet taste may reduce to an "identificational" or "releasing" function. Some ramifications of this idea, and some problems with it (of which perhaps the most serious is the failure of rats to reduce solid-food intake when offered saccharin), are discussed elsewhere (Mook, 1975).

While these experiments were being done, and while we were rethinking the relative roles of hunger and thirst in preferences for sweet fluids, Ernits and Corbit (1973) were well ahead of us. About the time Phil Farris found that thirst is not necessary to produce normal preference–aversion functions, they showed that hunger is not necessary either. Rats that were simply offered a sweet solution to drink for 1 hr, with no significant deprivation and at a time of day when spontaneous ingestion would otherwise be minimal, drank copious amounts and generated preference–aversion patterns indistinguishable from those seen in deprived animals. Indeed, the "dipsogenic" properties of taste were so potent as to occlude the effect of moderate thirst; imposition of 16-hr water deprivation caused very little increase in the amount drunk. The effect seems to be specific to solutions that taste sweet. Sodium chloride is drunk in minimal amounts by nondeprived rats, a finding that fits well with Kissileff's report (1973) that access to saline does not increase draft frequency in the *ad libitum* case.

Since Ernits and Corbit's rats would have eaten and drunk very little if the sweet solution had not been offered, it seems reasonable to regard the drinking that did occur as evoked by the taste stimulus. Certainly, if any data require the concept of palatability, theirs do. Yet, there are other considerations that make one wonder whether taste as a dipsogenic stimulus really operates in isolation even here.

In the first place, even if the initiation of such drinking is taste-determined, its termination is another matter. Intake patterns of the various sweet solutions showed the usual rising and falling preference–aversion pattern; relatively little ingestion occurred at the high concentrations. If taste were operating alone, one would expect it to produce intake patterns more closely approximating those seen in other cases in which taste is presumed to act alone; and in such cases, as we saw earlier (Section 3.2.1), intake varies directly with concentration even up to very high levels. The turndown in the function is probably the result of postingestive influences. Therefore, its occurrence in the Ernits and Corbit situation suggests that even if a bout of drinking is initiated by taste, its termination can remain sensitive to postingestive events. Kenney (1974) showed this directly, finding that nutritive solutions infused into the gut during drinking suppressed intake and were more effective in doing so than nonnutritive ones, even in nondeprived rats adapted to the Ernits and Corbit schedule. Therefore, intake in response to "taste" is susceptible to depression by nutrient loads, even if no nutrient "need" was present at its inception.

A second problem arises when we look again at the matter of initiation of drinking. To speak of drinking as taste-evoked implies that it should have a certain invariance in the face of changes in the animal's state. In fact, Ernits and Corbit showed that the amount drunk is invariant, to a surprising degree, as the rat's hydrational status is varied. But we recall that rats also treat sweet solutions as foods; what about nutritional status?

Mook and Cseh (in preparation) arranged for rats to be fed and watered by automatic intubation and overhydrated them so that they could be said to be "eating," not "drinking," the test solutions offered for 1 hr each day. Then they manipulated body weight by varying the size or the caloric density of the meals intubed (it made no difference which way it was done). As shown in Figure 5, hungry rats, which were severely underweight, accepted saccharin solutions avidly. They drank less and less as weight was incremented by frequent infusion of a calorically dense liquid diet (overhydration continuing throughout). Most important for present purposes, we see that as weight was elevated beyond the animals' normal weight, intake dropped to low or immeasurable levels. The rats would approach the cylinder each time it was offered (habit? memory?) and lap once or twice but then withdrew; no sustained drinking was evoked by the sweet taste. Therefore, the Ernits and Corbit effect is not seen if the animal is obese as compared to the weight it would maintain if feeding freely.

We may have reservations, then, about the interpretation of the Ernits and Corbit findings. But at a descriptive level, Ernits and Corbit have pro-

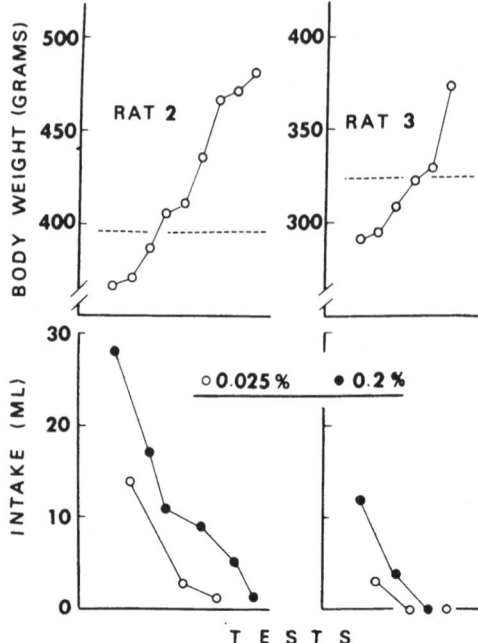

Figure 5. Single-bottle, 1-hr saccharin intake by rats otherwise fed and watered through intragastric cannulas. The two intake curves in each lower panel represent different concentrations of saccharin. The rats were administered large amounts of water at frequent intervals by automatic pumping apparatus; in addition, a liquid diet was infused in quantities sufficient to induce progressive weight gain. The horizontal dashed lines represent the animals' body weights before experimentation began. (From Mook and Cseh, in preparation.)

vided us with valuable information and, quite likely, some valuable tools for further analysis. They have shown that the response to taste can be invariant across deprivation conditions to a surprising extent. And if we admit that there was no acute deprivation state operating in their experiments, the continued dependence on postingestive control may give us a means for studying the effect of nutrient loads without the complication of a preexisting, corresponding nutrient depletion. At the very least, they have shown us how we can drastically simplify the brief-exposure experiment, attaining the experimental control such a procedure promises without the interference with the animal by imposed deprivations that it once seemed to require. The student of preference may still be driven to drink, but his subjects need not be.

3.2.3. The Nature of Oral–Systemic Interaction

At several points, we have referred to the limiting effects on intake of postingestive "satiety" factors as ones that must cooperate with mouth factors to exert their effects. Hydration, for example, suppresses water drinking but not drinking in response to a sweet taste (Mook, 1963; Mook and Cseh, in preparation). Dehydration conversely inhibits drinking when coupled with the taste of concentrated glucose but enhances drinking when coupled with the water taste (Figure 4). What sorts of interactive processes are involved here?

Let us take as a model case the selective depression of glucose drinking (but not of water drinking) by a postingestive load. There seem to be two ways in which this inhibition could come about.

First, it may be a direct subtractive process. Perhaps taste alone grows more and more potent as a stimulus to ingestion as concentration increases, but the postingestive inhibition that results from drinking becomes stronger, too. Then, since total intake is progressively lower over the descending limb of the preference–aversion function, perhaps the inhibitory postingestive effect increases with concentration more sharply than the stimulatory taste effect does, and overrides it.

Such inhibition, of course, cannot be exerted on the systems controlling the lapping response itself. Otherwise, water drinking would be suppressed along with glucose drinking. Perhaps, if we think of the rat as "eating" the glucose solution, the inhibition is applied to the systems mediating hunger; this would leave the thirst system ready to drive lapping, if engaged by contact with water. Then, if this were so, the continuous descending limb of the preference–aversion function could be accounted for in another way: perhaps the higher the concentration, the greater the extent to which the fluid is identified as food (the better it fits a "search image"?), and hence the greater the extent to which the lapping response is affected by the inhibition of a hunger system, which may be driving it concurrently with the thirst system.

A second possibility is that the effect of the postingestive load is to modify the stimulatory or the "hedonic" properties of taste, not just to oppose them. Maybe as satiety accrues, the sweet taste changes from "pleasant" to "neutral" or even "aversive."

In man, an interaction of the second type has been reported (see Cabanac, 1971). The rated hedonic tone of sweet solutions varies more or less directly with concentration before a glucose load, inversely following one. But as Booth, Lovett, and McSherry have pointed out (1972), it is uncertain what relationship this phenomenon bears to normal satiety as it accrues during a meal.

Attempts to demonstrate such a hedonic shift as a factor in the control of feeding have not been convincing. For instance, Hammer (1968) reasoned that if sweetness becomes aversive as satiety develops, one might see a shift in preference toward the less sweet of two isocaloric solutions, in rats offered the choice between them, as a bout of feeding progressed. She did not see such a shift, and we know of no one else who has looked for one in precisely this context. Something like it has been reported over long time periods: offered two solutions of different sweetness, rats may reverse an initial preference for the sweeter one after a few days' access to the choice.

But again, how this relates to short-term satiety is problematic (see Booth *et al.*, 1972, for a detailed examination of this problem).

Some other lines of evidence seem hard to reconcile with a hedonic-shift interpretation. Campbell and Davis (1974b) showed that infusion of very dilute glucose into the gut was sufficient to depress oral intake of dilute glucose, but much more concentrated loads were needed to affect intake of sweetened condensed milk. If the former loads were enough to shift the glucose taste toward aversion, one would expect them to have a greater effect on the sweeter fluid, milk. Of course, milk and glucose solutions differ in ways other than sweetness.

The effect of induced obesity on saccharin "eating" in rats raises similar considerations. Looking again at Figure 5, we see that as body weight was incremented, intake of saccharin was progressively depressed at both concentrations used. Their relative positions remained unchanged, however; in particular, notice that intake at the lower concentration was driven down to very low values at levels of obesity that still permitted appreciable intake at the higher concentration. If the taste of the dilute solution had "turned the corner" and become aversive at this point, the more intense taste of the concentrated one should have been more aversive still. All this assumes, of course, that the effect of satiety should be to reverse the slope of the function relating hedonic tone to stimulus intensity, not just to raise or lower it. In the latter case, the hedonic-shift model would be only a special case of the subtraction one, with a special (and to be sure, significant) assumption as to where the opposition occurs: what is "subtracted from" is the hedonic value of the stimulus.

For one who would investigate an interaction of this kind, perhaps the best tactic is to start with a case in which the interaction is absent. Such a case exists (Bryner and Mook, in preparation). In performing experiments like the one shown in Figure 5, we were puzzled by the fact that the rats, especially when underweight, would drink moderate amounts of a saccharin solution and then stop. These rats were not thirsty but hungry, and the saccharin provided no repletion. Why then did they stop? Because their stomachs were full? No, for concurrent injections of an identical solution into the stomach, milliliter for milliliter as the rat drank, had no effect on the amount passed through the mouth before drinking ended (although of course it doubled the rate of stomach filling).

This seems to be a case, then, in which the normal postingestive consequences of drinking play no part in its termination. Yet drinking does stop, and it may be quite strictly limited, especially if the rat is overweight. Apparently, both the initiation and the termination of drinking, in this case, are under oropharyngeal control.

Why this should be—in particular, what the "stop" mechanism is—we do not know, though we do know that the controlling mechanism must be sensitive to body weight. Possibly sensory adaptation plays a role in the rat as in the blowfly (Dethier, Evans, and Rhoades, 1956), though in the rat such an adaptation would probably have to be central rather than peripheral (Wilcove, 1973) and might better be called habituation. Body weight might then act by setting the apparent stimulus intensity that is sufficient to support continued ingestion. There may come a point, as the effectiveness of taste fades through continued exposure, at which a given concentration is no longer "acceptable" as food. That point would be reached earlier for a weak solution than for a stronger one. And since the hungry rat accepts dilute solutions that the not-hungry one refuses, the cutoff point may be set lower for a hungry rat than for a less-hungry one (relaxation of "search image" criteria?). The observed results would follow.

For present discussion, however, the important point is that these phenomena may permit us to introduce postingestive controlling factors into a situation in which we can see their effects against a blank background. Would intake of saccharin be driven below the baseline set by mouth factors alone, if nutrient material were injected along with drinking? If so, and once intake of a weak solution were so terminated, what would happen if a stronger one were substituted? If vigorous drinking then resumed, it would be strong evidence for a trade-off, rather than an interaction, between inhibition arising from postingestive events and the excitatory properties of the taste input.

3.3. The Response to Noxious Tastes

At first glance, the role of aversive tastes in the peripheral control of intake seems simple enough, though its neurology is mysterious (see Section 4.3). Some tastes are unpleasant, and given the choice, the animal avoids these. This seems straightforward, but questions arise when we look at the matter more closely.

Indeed, to convince himself that there are problems here, one need only examine the raw data typically obtained in aversion-threshold studies. An arbitrary but reasonable criterion of "aversion," in the two-bottle case, is that a solution is aversive if less than 25% of the animal's total intake is from the solution when it is paired with water. This means that if a rat drinks, say, 5 ml of a quinine solution and 16 ml of water (as often happens), the solution is to be called aversive. But if it is aversive, why does the rat drink even 5 ml? It doesn't have to, after all; and 5 ml represents a substantial amount of lapping—on the order of 1000 laps (cf. Stellar and Hill, 1952).

The possibilities that come to mind here closely parallel those that could enter into positive choices based on taste, as discussed earlier (Section

3.1.1). One can imagine a rat that has become thirsty and encounters the quinine cylinder; then, if the taste is not too aversive, the animal simply may not take the trouble to switch to the water cylinder. *Trouble*, of course, is the operative word here. Perhaps the concentration of quinine, over the aversion-threshold range, affects the likelihood of such switching. Alternatively, it may affect the probability of the rat's approaching the quinine cylinder rather than the one containing water; if so, the resulting data would measure the effectiveness of the memory of the bitter taste in controlling subsequent approaches. Or maybe draft size is affected (cf. Levitsky, 1970). As we wondered earlier when a rat is "hungry enough" to take in the offered sweet fluid, perhaps in the quinine case we are seeing the point at which it becomes "satiated enough" to stop drinking the fluid it has encountered.

Detailed records of drinking pattern, along the lines of Kissleff's (1973) analysis of salt preference, could do much to clarify these issues. To our knowledge, no such data exist, though Booth (1972) has made a start on a related issue. His provocative data, which cannot be summarized here, provide fair warning as to the complexities we may expect to face.

The effectiveness of the bitter quinine taste as a barrier to fluid ingestion is coupled, for reasons we do not understand, to the specific nature of the hydrational deficit that is responsible for the ingestive sequence. Burke, Mook, and Blass (1972) found that rats drinking for reasons of intracellular dehydration refused solutions of quinine that were readily accepted by the same rats when they were drinking because of a reduction in blood volume. Such finickiness does not mean that cellular dehydration evokes a less vigorous thirst, for when cellular dehydration was added to a preexisting water deprivation, intake of water was augmented, but intake of a quinine solution was actually depressed, as compared with the response to water deprivation alone. Thus, it seems that something about cellular dehydration intensifies the barrier to ingestion that the bitter taste poses.

In later (unpublished) explorations, Burke has shown that the effect is not specific to quinine or even to bitter solutions. The same effect is seen when noxious solutions of urea or hydrochloric acid are offered the rat: solutions that are avidly accepted after water deprivation are refused following intracellular dehydration. More confusing still, the solution need not be aversive in itself. Solutions of alcohol that are actually preferred to water in choice tests are refused, whereas water is drunk copiously, after cellular dehydration.

One wonders at this point whether the rat may treat any solution that has an intense taste as if it were hypertonic and therefore threatened further cellular dehydration. If so, the finickiness may simply represent an unwillingness on the rat's part to incur further cellular dehydration, given that cellular dehydration is its problem to begin with (cf. Stellar *et al.*, 1954).

The concentrations of quinine used in such experiments are very hypotonic, but then quinine is very potent as a taste stimulus (Pfaffmann, 1961), so perhaps a quinine solution, because of its intense taste, is treated as if it were potentially dehydrating even though it is not.

Such an account would make good sense, but it is false (Blass, 1974). Nephrectomized rats were intubed a mildly hypertonic solution of sodium chloride by gavage, then offered the same solution to drink by mouth. The gastric load in itself produced a marked hypernatremia with plasma expansion, so that the drinking that ensued was surely driven by cellular dehydration. Yet the rats drank copiously and persistently, even though the solution offered was in fact hypertonic and, since they could not excrete solutes, intensified their hydrational problem.

Thus, the explanation for the special properties of intracellular thirst eludes us. The solutions that rats reject when drinking for that reason do not fall into any obvious class such as "hypertonic" or even "aversive," and why such rats should reject *any* solution that is otherwise acceptable is not obvious. In the absence of a guiding theory, this phenomenon will simply have to be explored as an empirical enigma.

Finally, we have an exciting breakthrough on the technical side to report. A number of investigators, the present authors included, have tried to measure the aversiveness of solutions directly by training rats to make an arbitrary response, such as lever pressing, to escape or avoid infusions of noxious solutions into the mouth. We have had no success whatsoever. But Kissileff (1974) has done it. He has reported that infusion of concentrated saline solutions into the mouth will support escape lever-pressing and that the response rate varies in the expected way with such variables as solute concentration of the infusate. He has suggested that others may have failed to achieve such control because they used quinine solutions (as we did). Quinine's bitter aftertaste can persist for long times, even if the mouth is rinsed with water, and it may be that the termination of quinine infusion into the mouth is an ineffective reinforcer for that reason.

A number of problems lend themselves to study by Kissileff's method, as he noted in his presentation. For example, the problem we discussed earlier, whether the sweet taste of food becomes aversive as "satiety" develops, may now be accessible to direct experiment.

4. Central Control of Preference and Aversion

Our discussion here of brain mechanisms of preference and aversion is highly selective, a sample rather than a survey. The directly pertinent literature is not very large, though it is rapidly growing, and much of it is exceedingly difficult to interpret. This difficulty is not for lack of technical sophistication on the part of students of the problem. On the contrary, much work has been done that is technically superb on the surgical, anatomical, or pharmacological side.

Rather, the problem as we see it is that with few exceptions, the work on brain mechanisms of preference and aversion has not kept pace with our developing appreciation of the multiplicity of factors that control a bout of fluid ingestion. A couple of examples must suffice. Many investigators of the salt-preference phenomenon have used, as test material, saline solutions in the mildly hypertonic range—on the order of 2% w/v, or about 0.3 M. Unfortunately, saline solutions in that concentration range represent about the most complicated case imaginable. They are above the maximally preferred range but are not strongly aversive (see Section 3.1.1); they are hypertonic but within the concentrating power of the kidney. Thus, a change in amount ingested may reflect changes in the response to an attractive taste component, to an aversive taste component, or to postingestive osmotic transients; or it might be an adaptive reaction to alterations in the internal environment produced by changes in renal function (cf. Covian, 1974). This web of influences could be disentangled, but attempts to disentangle it have thus far been few.

Yet another difficulty attends the investigation of sweet solutions, which, as we have emphasized, may be treated as foods as well as fluids. Thus, if intake of a sweet solution responds differently to some cerebral manipulation from intake of water or even a "preferred" saline solution, this may mean that quite different mechanisms are controlling its ingestion in the first place. Conversely, we really should not expect sweet-solution intake to be affected in the same way by water deprivation (which produces both thirst and self-imposed hunger) as by, say, cholinergic stimulation of the brain (which can evoke drinking but whose effects on feeding are less well explored). For the same reason, the often-used tactic of using a sweet solution to anchor a "hedonic dimension" (sweet solution positive, water intermediate, a bitter or salty solution negative) ignores the identificational role of the sweet taste and may be seriously misleading.

All this is in addition to the staggering complexity of the brain itself, a point we need hardly belabor. For a particularly thoughtful and instructive example of the sorts of problems that arise from that source, see Wolf and DiCara (1974).

And there is one more problem, for the reviewers this time. As our previous discussion implies, we often are uncertain where to place the phenomena of preference in our more inclusive conceptual categories. For example, to the extent that intake of sweet solutions is a feeding response, a full account of it must consider the vast literature dealing with the control of food intake. To the extent that it reflects a hedonic process, one might expect the literature on emotion or "reward" mechanisms to be pertinent. And inasmuch as memory and search are involved in preference phenomena, what we learn about preference may make full sense only in the context of what can be learned about the neurology of search, memory, attention, and the like.

What this means is that we cannot hope to discuss fully the context and implications of the studies we survey—even if we always knew what context is most appropriate, and we don't. Of necessity, we simply point the reader to some representative instances of what is going on in the area and refer him to the individual writers' ideas as to how the facts might best be organized.

4.1. Salt Solutions

Since about the middle of the last decade, with the work of Novakova and Cort (1966), Wolf (1964), Covian and Antunes-Rodrigues (1963), and their respective collaborators, interest in the neurological basis of salt preference and aversion has become substantial and a sizable literature has accumulated. We cannot review the details here. The picture that seems to be emerging is that salt intake can be affected, in ways not dependent on changes in salt excretion, by appropriately placed hypothalamic lesions, and that the systems so implicated are in turn subject to modulatory influences from the limbic system, in particular the septal nuclei (Gentil, Mogensen, and Stevenson, 1971; Grace, 1968; Saad, Antunes-Rodrigues, Gentil, and Covian, 1972), the cingulate cortex (Grace, 1968), and the amygdaloid complex (Chiaraviglio, 1971; Gentil, Antunes-Rodrigues, Negro-Vilar, and Covian, 1968). The interactions of facilitation and inhibition are complex (see, e.g., Grace, 1968; Gentil *et al.,* 1968).

Unfortunately for our present purposes, these studies do not tell us what role gustation plays in the effects reported or whether it plays any role at all. The problems of interpreting changes in the intake of hypertonic solutions have already been mentioned. When intake of a "preferred" saline solution (in the isotonic range or below) is augmented after brain damage, this still could reflect a release from postingestive inhibition, to which intake even of an isotonic solution must be subject (Rolls and Jones, 1972). Or, of course, it could implicate a change in the response to taste, which might be expressed in a number of ways. Perhaps the influence of taste on selection has been made more potent, so that the rat selects saline in preference to water (in the two-choice case) more consistently than before. Were this the case, one would expect a corresponding decline in water intake; this sometimes happens, but not always (see e.g., Grace, 1968). Where it does not, perhaps an influence of taste on draft size and/or frequency, weak or absent in the intact rat (see Section 3.1.1), has been made visible by the interference in the central nervous system (CNS). Something of the sort almost surely is involved in the effects of mesencephalic stimulation by implants of $FeCl_3$ (Chiaraviglio, 1972). This implantation caused substantial intake of isotonic saline, in a brief period, by nondeprived rats that otherwise drank very little.

When intake of "preferred" saline solutions is depressed, the converse problems in interpretation arise. To them we may add the possible taste-masking effect of the "prandial-drinking" pattern, as discussed earlier (Section 3.1.3). Indeed, this issue arose in the first place through investigations of preference in rats recovered from lateral hypothalamic lesions. Such rats are prandial drinkers, and they sometimes show no saline preference (Kissileff and Epstein, 1969; but see Wolf, 1964; Wolf and Quartermain, 1967). The possible role of prandial drinking in such preference deficits can be checked in neurologically intact rats by tests for preference in the absence of food. But in the lateral-hypothalamic rat, which does not drink if it cannot eat (Epstein and Teitelbaum, 1964), such a test cannot meaningfully be made, and the question goes unresolved.

In the case of salt appetite, as opposed to "spontaneous" salt preference, the role of gustation has been addressed more directly. Wolf (1968) has found that damage in or posterior to the gustatory thalamic relay causes deficits in salt appetite, though rats with such damage showed the normal avoidance of strong salt solutions, suggesting (but not proving) that they could taste. In view of our earlier discussion, one possible line of speculation is that such damage may interfere with the rats' "recognition" of an encountered saline solution as the substance needed, rather than with their appreciation of sodium deficit or of taste quality. One wonders whether rats so prepared would work at a lever for a saline solution that is "expected" but not delivered (and therefore need not be recognized), as in the Krieckhaus and Wolf (1968) experiment.

Hypothalamic involvement in salt appetite has also been reported. Wolf (1964) found that the enhancement of isotonic-saline intake normally seen after Formalin injection did not occur in rats with damage to the lateral hypothalamus, even though they did show the usual "spontaneous" preference. The effects of adrenalectomy are less clear (Wolf and Quartermain, 1967). When lateral hypothalamic rats were offered a milk diet, isotonic saline, and water, they did not increase their saline intake after adrenalectomy as unlesioned rats did. But then, they took little or nothing from the two drinking fluids in any case and may simply not have approached them in the absence of dry food (cf. Epstein and Teitelbaum, 1964). When offered dry food, all rats, lesioned and not, preferred saline to water, but when saline concentration was increased into the normally aversive range, the preference dropped out in the brain-damaged adrenalectomized rats.

Here, however, the problem of specificity arises (Epstein, 1971). Lateral-hypothalamic rats are very finicky about their drinking fluids and refuse to drink if the fluid is even mildly aversive. Perhaps, assuming that the strong salt solution was unpalatable, the failure of salt appetite was a manifestation of such finickiness.

In light of these uncertainties, it is encouraging to note that evidence of quite a different sort suggests that something is wrong with the lateral rat's response to the salt taste *per se*. When a neurologically intact, anesthetized rat's tongue is stimulated with a saline solution, there is an increase in spinal-reflex output recorded from the ventral root. This response does not occur after lateral-hypothalamic damage (Wayner, Kahan, and Stoller, 1966).

4.2. Sweet Solutions

To the extent that intake of a sweet solution is a feeding response, one would expect it to be affected by manipulations that influence food intake. The most obvious starting place here has led to disappointing results. Ventromedial-hypothalamic lesions are well known to produce vigorous hyperphagia, often amounting to a two- or threefold increase in food intake. But their effects on sweet-solution intake are subtle, if demonstrable at all. Using a brief-exposure procedure, Mook and Blass (1970) found no differences between hyperphagic and control rats in response to glucose or sucrose at various concentrations. One might relate this to the finding that hyperphagic rats do not overeat when offered food for only brief periods (Panksepp, 1971). With *ad libitum* tests, Sclafani (1973) found that hyperphagic rats did take in larger-than-normal amounts of concentrated (but not dilute) glucose solutions. Since limitations on the intake of concentrated glucose solutions are primarily postingestive (see Section 3.2.1), these findings fit in well with the notion that ventromedial cells are concerned with postingestive "satiety" effects. Sclafani reported other data, however, that suggest that matters are not so simple. For instance, intact rats increased their intake of sweet solutions when deprived of solid food, but hyperphagic rats did not. These data can be made sense of, but only in ways that remain *ad hoc* at present.

The situation is especially puzzling in light of such rats' reported hyperreactivity to the sweetening of solid food (Teitelbaum, 1955) and what seems to be the corresponding hyperreactivity to bitter tastes. In our own unpublished observations, we have often seen hyperphagic rats that were extremely finicky about quinine in their food or water respond quite normally to sweet solutions, showing neither rejection nor exaggerated acceptance. There seems to be an asymmetry in the finickiness phenomenon, one that we encounter again later.

Other CNS interference can produce more profound effects on preference for sweet solutions, but ones that are just as hard to interpret. Mook, Lindsey, Pace, and Graeber (1972) found that rats with lesions in the lateral preoptic area (LPO) of the forebrain, tested after deprivation of water (but not of food; this is important), showed no preference for glucose solutions and often actively rejected them in favor of water. We later showed (Mook, Walshe, and Farris, 1975) that the same loss of preference under these conditions extends to sucrose and saccharin. And in all three cases, the normal preference patterns are restored in such rats if they are deprived of food, as well as of water, prior to the preference test (Figure 6). The aberration in preference produced by LPO lesions, therefore, has something to do with the availability of food during water deprivation.

A strikingly similar deficit was reported by Murphy and Brown (1970; Brown and Murphy, 1973) in rats with hippocampal damage. After water

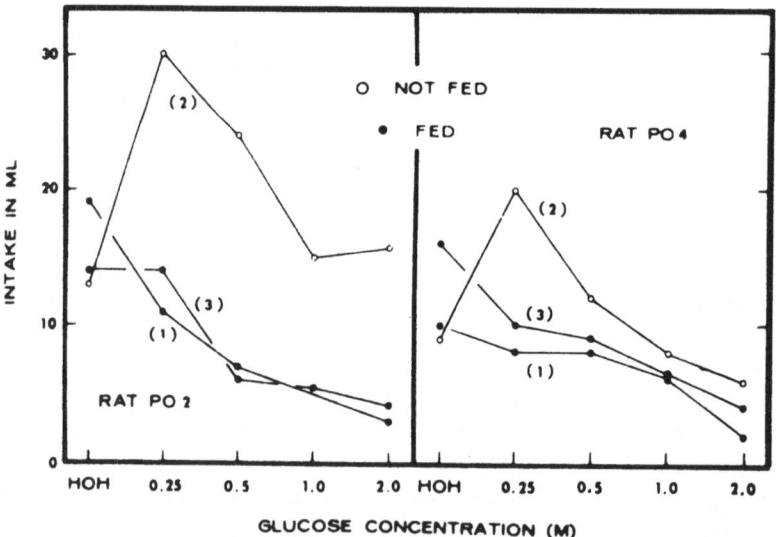

Figure 6. Single-bottle, 1-hr intake of glucose at various concentrations in rats with damage to the lateral preoptic area. Such rats showed no glucose preference when tested after water deprivation in the presence of food (the "fed" condition). When deprived of both food and water prior to the test ("not fed"), they showed the typical rising and falling preference-aversion function. Normal rats (not shown) displayed the preference-aversion pattern under either condition. The numbers in parentheses represent the order in which the various functions were determined. (From Mook, Lindsey, Pace, and Graeber, 1972. Copyright 1972 by the Psychonomic Society. Reproduced by permission.)

deprivation (with food present), such rats showed no sucrose preference; after food deprivation, they did show this preference. These authors considered, but later rejected, the possibility that hyperactivity could have interfered with the expression of preference by causing frequent shifts from one to the other drinking cylinder. Data on food intake during deprivation were not reported.

Food intake may be critical here. Our own first thought was that the failure of preference in water-deprived LPO rats might occur because of the high intake of food, in the absence of water, that we observed them to display (Mook *et al.*, 1972). Our reasoning was as follows. Intact rats reduce their food intake when water-deprived and thus are both hungry and thirsty. When offered a sweet solution, they may then be drinking it, eating it, or both. LPO rats, since they ate more during deprivation, may have been thirsty but not hungry and may have refused sweet solutions (especially when water was offered as an alternative) for that reason.

Some of the data are compatible with this idea (Mook *et al.*, 1972; Mook, Walshe, and Farris, 1975), but later observations not yet published are less clearly in its favor. We have seen a number of LPO rats display the preference deficit even when they ate no more than normal amounts during water deprivation. One other possible factor is suggested by some preliminary observations of the feeding pattern in such rats. Deprived of water, intact rats eat a few meals of nearly normal size and then virtually cease to eat until water is restored. LPO rats, in contrast, eat very small and frequent meals and persist in this pattern even throughout a night of water privation. Perhaps the subsequent failure of preference occurs not because such rats eat *more* but because they eat *later* than do intact rats on a similar deprivation and testing regimen and therefore have fed more recently when the preference test is conducted.

In any case, it is likely that the preference deficits we first observed in such rats are secondary to something that happens during the period of water deprivation prior to test, since it occurs only if food is available then. Other deficits, however, may be primary. As we noted earlier, intact rats seek and ingest sweet solutions offered for brief periods, even if they are not deprived at all (Ernits and Corbit, 1973). In LPO rats, such a response is abolished or greatly diminished (Mook, Walshe, and Farris, 1975). Thus, the need for food can "push" such rats to ingest sweet solutions in normal amounts, but they are unresponsive to the "pull" of the sweet taste in the absence of need. Perhaps damage to the LPO area or elsewhere in the medial-forebrain-bundle complex will provide a way of dissociating the identificational from the hedonic function of the sweet taste. This aspect of the syndrome, incidentally, seems not to be produced by hippocampal damage, which has been reported not to depress sweet-solution seeking by nondeprived rats (Beatty and Schwartzbaum, 1968a).

Whereas damage to certain brain areas may depress the "hedonic" response to the sweet taste, damage to other areas may release it. This at least is one way of interpreting the effects of damage to the septal area, which enhances the intake of sweet solutions under a variety of conditions (Beatty and Schwartzbaum, 1967, 1968a,b). Both lick rate and bar-pressing rate—that is, both the response to the taste and the vigor of seeking it—are affected (Beatty and Schwartzbaum, 1968a,b).

The effect on lick rate is independent both of concentration and of preloads, and it occurs largely because control rats reduce their rate of licking over the course of a session, but septals do not (Beatty and Schwartzbaum, 1968b). Flaherty and L. W. Hamilton (1971) found that whereas in control rats more licks are taken at a concentrated sucrose solution than at a dilute one, septals treat both solutions as if they were of the higher concentration. Finally, the effect is seen whether or not the rats are hungry; in fact, when shifted from a deprivation schedule to *ad libitum* feeding, Beatty and Schwartzbaum's control rats reduced their rate of licking, but septals did not.

These last findings in particular are intriguing in their suggestion that septal damage may interfere with factors that normally impose a certain selectivity on the response to sweets, making it sensitive to a variety of internal and external factors, for example, deprivation, concentration, exposure duration, momentary probability of receiving the fluid (Beatty and Schwartzbaum, 1968a), and even environmental novelty (L. W. Hamilton and Flaherty, 1971). This idea is the more attractive in view of the finding that amygdaloid damage depresses lapping in response to dilute, but not concentrated, sucrose solutions (Kemble and Schwartzbaum, 1969; Rolls and Rolls, 1973). That pattern of findings might be seen as an increase in selectivity, so that only powerful stimuli evoke the full-blown normal response.

The notion of *selectivity* is ambiguous, of course; we mean it to be so, for we can do little to make it precise. One might think of reduced selectivity as a matter of increased "persistence," so that having begun to drink (or bar-press), the septal rat continues to do so, heedless of suppressant factors that affect the intact rat.

If such "persistence" is thought to apply to emission of the response itself, this idea is the same as McCleary's (1961) response–suppression interpretation of septal function. The problem here is that the effect of septal damage on intake of sweets is separable from its effects on perseveration in reversal tasks, a situation that would also be expected to tap "persistence" in this sense (Beatty and Swartzbaum, 1968b). Alternatively, the persistence might be on the perceptual–attentional side, affecting the rat's tendency to keep its attention on drinking (cf. McFarland and McFarland, 1968) or, what may amount to the same thing, the tenacity with which it holds to certain "search-image" specifications (cf. Andrew, 1972). In the converse case, Kemble and Swartzbaum (1969) advanced the same idea in suggesting that their rats with amygdaloid damage were more distractable than controls. This possibility raises complex issues that cannot be dealt with here. If it is true, the response to distraction experiments (and reversal and other transfer tests) depends heavily on what aspects of the situation are included in "search-image" specifications (cf. Archer, 1974).

On the other hand, one might think of reduced selectivity as a general hyperreactivity, so that stimuli minimally effective in intact rats are treated as potent ones by septals (see Beatty and Schwartzbaum, 1967, for discussion). But here again, the septal effect on sweet-solution intake is dissociable from its effects on reactivity to bitter tastes (see, e.g., Burright and Zuromski, 1970) or to handling and foot shock (see, e.g., Carey, 1972).

Thus, this aspect of the septal syndrome stands coyly aloof from the categories in which we have tried to place it. It is an intriguing phenomenon in search of a context.

4.3. Noxious Solutions

A refusal to tolerate quinine solutions at concentrations normally accepted by rats can be produced by damage to the septal area (Beatty and Schwartzbaum, 1967) and by damage to the ventromedial hypothalamic area (Corbit, 1965). The two syndromes have a great deal in common.

In both cases, preoperative experience may be important (Carey and Procopio, 1974; Singh, 1974). In both cases, quinine consumption tends to rise as thirst accrues (Corbit, 1965; Donovick, Burright, and Zuromski, 1970) and water consumption is high when water is restored, suggesting that neither a disorder in thirst, nor a failure to "recognize" the bitter solution as a source of fluid is responsible for the finickiness. Nevertheless, "a considerable number of rats with septal . . . lesions would rather die than drink 0.025% quinine" (Gittelson, Donovick, and Burright, 1969, p. 293), and we have seen hyperphagic rats make the same decision (unpublished). Whatever is released by the injury to the brain, it is a powerful factor.

In both cases, the finickiness is unaccompanied by a decrease in aversion thresholds (Carey, 1971; Mook and Blass, 1968). Rather than an increased sensitivity to the bitter taste, then, we are seeing an exaggerated response to aversiveness that the intact rat also appreciates but will tolerate if it must. One is reminded here of cases in which the affective component of pain is exaggerated or suppressed by damage to the brain without change in threshold for detection of the sharpness, prickiness, etc., of potentially painful stimuli (cf. Melzack and Casey, 1968). The two disorders are surely not the same, for finickiness is dissociable from hyperreactivity to pain (Carey, 1972). But their mechanisms might be similar.

Finally, in both cases, the effect is dissociable from other changes that at first look like aspects of the same phenomenon: hyperreactivity to saccharin and hyperemotionality in the case of septal damage (Carey, 1972; Donovick, Burright, and Zuromski, 1970) and hyperphagia and obesity in the case of ventromedial damage (Graff and Stellar, 1962, and our own observations). This last point is worth emphasizing, for it is often lost sight of, especially in discussions of the ventromedial syndrome.

Whereas intake of sweet solutions is augmented by anterior or posterior septal damage, only posterior septal lesions exaggerate the response to quinine, as reported, for example, by Donovick, Burright, and Zuromski (1970). These authors have also reported a similar effect of habenular lesions (but see Carey and Procopio, 1974) and opposite effects of damage to the interpeduncular nucleus of the midbrain, which actually reduces quinine responsiveness. They have sketched out a hypothetical circuit that might mediate the avoidance of bitter tastes—one that bears little relation, by the way, to the circuits offered by other writers for the avoidance of "noxious" salty tastes (cf. Grace, 1968)—and they suggest some ways in which the system might in turn be overridden as thirst becomes intense.

The fact that the interplay of excitation and inhibition extends into the midbrain is compatible with other evidence that we are dealing with a multilevel system here. According to Peiper (1963), an anencephalic human infant will make characteristic grimaces when a bitter solution is applied to the tongue. Perhaps the role of forebrain systems is to elaborate and implement a "decision" to approach or to avoid, which itself may be made at a level far down the neuraxis. Since the sequence of approach and ingestion (or its converse) has many components, as we emphasized in our discussion of peripheral control, it is unfortunate that we have so little detailed information about what various preparations actually do when faced with a bitter taste. We do know (Gittelson et al., 1969) that rats with septal damage, unlike controls, refused even to sample a quinine solution during the last half hour of an hour's access to it. Unless aftertaste is a factor, this response would imply that not only the suppression of ingestion by the bitter taste but also the suppression of approach by its memory has been exaggerated by the injury to the forebrain.

A number of investigators have found that the artificial induction of thirst by CNS manipulations—for example, by cholinergic stimulation of the brain—is accompanied by an intolerance for bitter solutions much greater than that observed after deprivation-induced thirst (see Krikstone and Levitt, 1974, for review). These studies have been criticized on methodological grounds (Krikstone and Levitt, 1974). Franklin and Quartermain (1970) have pointed out that whereas water deprivation evokes the multiple controls over thirst concurrently, CNS manipulations may evoke them selectively, thus giving rise to less intense motivation. To this we would add one further consideration: the controls over drinking activated by various procedures may differ in nature as well as in number. We recall that cellular-dehydration thirst is characterized by hyperreactivity to quinine, whereas thirst evoked by blood-volume reduction is not (Burke *et al.*, 1972). If drinking evoked by CNS stimulation is easily suppressed by the addition of a bitter taste to the drinking fluid, this may be because the manipulation has selectively activated a system involved in intracellular thirst, one whose action is characterized by such finickiness. Even if all this were true, of course, its effect would only be to reduce by one the number of questions that face us—not to answer any of them.

5. Retrospect and Prospects

We began this chapter with an almost apologetic disclaimer of intent to discuss the "nonhomeostatic" aspects of taste responsiveness that have been the major focus of theoretical concern. By now, we trust that the reason for our lack of attention to this problem is clear: we simply have had few occasions to refer to it.

Even where we have referred to it, we have often done so with distinct twinges of doubt. Several times in this discussion, we have made an implicit distinction between drinking as a "response to taste" and as a "response to need." It has been a convenient distinction at times, but we have wondered even as we wrote whether that convenience may bear an unacceptable cost in accuracy. A hungry rat, which "needs" food, may drink a sweet solution and refuse water. In such a case, the response to need *is* to respond to taste—that is, to the taste that identifies the needed commodity. States of deprivation and repletion can then be seen as affecting the parameters of that response (its threshold, its persistence, its susceptibility to inhibition by other factors, etc.). These states may do so by affecting the gustatory system, but in the case of salt appetite, we have direct evidence that they may do it in quite a different way: by affecting the properties of the "search image" with which gustatory input is to be compared. As to the "dipsogenic properties" of taste itself, the Ernits and Corbit (1973) experiments have shown us that the response to taste can be invariant across a wide range of internal variation in certain commodities. This is important to know. But with variation in other aspects of the animal's internal economy, the plasticity of the response to taste remains clearly visible, as we have seen.

In addition to the interaction of taste responsiveness with antecedent state, we must consider its interaction with the consequences of drinking.

We have seen that the taste of concentrated glucose, in conjunction with water intubation, leads to very high intake, whereas intake terminates much sooner if water is presented to the mouth. Combined with a dehydrating stomach load, in contrast, the same glucose taste leads to sharply limited intake, whereas the water taste drives intake to very high levels. What this means is that the taste of concentrated glucose (like the systemic effects of a hypertonic load) is not, in itself, either an augmentor or an inhibitor of ingestion. Rather it enters into a system that takes account both of the commodity offered and of the internal state of the animal and then, in turn, augments or inhibits.

Running through the above examples, we see that we have come very close to arguing ourselves out of a topic. In a sense, there probably is no such thing as "taste modulation of fluid intake"; the causal factor in the instances covered is not taste itself but the resultant of the interactions into which it enters.

Thus, on the causal side, we have been compelled to broaden our analytic units. The role of taste cannot be studied in isolation; what we must understand are the interactions themselves. At the same time, what is affected—the actual ingestive sequence—is being examined in ever finer detail, and the separate phenomena we must study are proliferating at an alarming rate. Even in the simple case of the selection of one fluid over another, for instance, we must ask how the presence of a sapid solution guides the animal's approach to the container; how it leads to a decision whether to ingest or to sample further in the environment, as in the "switching" demonstrations; and how it drives the consummatory response once the decision to ingest is made. The distinction between the first two questions and the last corresponds to the distinction between appetitive and consummatory behavior that has been so useful in other contexts. The distinction between the first question and the other two reminds us of the distinction between the effects of taste once it is encountered as a stimulus and the effects of the "memory" or the "image" that can operate before the taste itself is encountered during the bout of ingestion under study. This last aspect of the problem, which links the study of ingestion with the study of learning and memory, has been neglected by rodentologists (except for students of conditioned taste-aversion and the like), though the early reward–expectancy studies might have pointed the way (cf. Tolman, 1959). Investigators of avian behavior have confronted problems of a similar kind and have addressed them in imaginative ways (cf. Andrew, 1972; Archer, 1974; Tinbergen, 1960).

As to how preference and aversion are mediated by the brain, the state of the art is more difficult to assess. Quite literally, we are not sure how much we know. The literature that is directly and demonstrably pertinent to the problem is not very large (yet), but the literature that *might* be relevant is almost limitless; it encompasses the work on feeding, on drinking, on attention, on memory, on reward and punishment—and more.

Of course, the problems of solution intake would be of little interest unless they did bear on broader issues, and their relations to other phenomena will sooner or later have to be specified. But to a surprising extent, the work especially on peripheral control of solution intake has proceeded without regard to these problem areas (and others we could list). Perhaps this is just as well. It is quite likely that the rapid progress the area has seen would only have been impeded by embroilment in the theoretical difficulties that beset the work on attention, reinforcement, and even feeding.

Most important, we have at least the right to hope that the rather isolated empiricism of this research area has taken us well along the path of identification and classification of phenomena, on which investigation of their mechanisms depends. The identification of various stages in the ingestive sequence, and the multiple factors controlling them, has presented us with an appalling number of factors to consider, but it has done so only by showing us ways of isolating them and investigating the mechanism of each. And it is these specific mechanisms that will fall into their respective places in the broader scheme of things. For example, the mechanisms by which an animal seeks a sweet taste are certain to be not only quite different from the mechanisms by which such a taste drives the lapping response but also related to a quite different class of phenomena. Perhaps, recognizing this, we are now in a position to ask questions of the brain that are specific enough to permit of intelligible replies.

ACKNOWLEDGMENTS

This paper was written in large part while the first author was visiting the University of Sussex, and he is grateful to Dr. R. J. Andrew and the other members of the School of Biology for their hospitality. We also thank Dr. Andrew and Frank W. Finger, Harry R. Kissileff, and Nella D. Mook for their comments on the chapter. We have not always taken their advice, however, and they bear no responsibility for what it says.

References

Adolph, E. F., 1947, Urges to eat and drink in rats. *Am. J. Physiol.* **151**:110–125.

Andrew, R. J., 1972, Recognition processes and behavior with special reference to the effects of testosterone on persistence, *in* "Advances in the Study of Behavior" (D. S. Lehrman, R. A. Hinde, and E. Shaw, eds.), Academic Press, London, pp. 175–208.

Antunes-Rodrigues, J., Gentil, C. G., Negro-Vilar, A., and Covian, M. R., 1970, Role of adrenals in the changes of sodium chloride intake following lesions in the central nervous system, *Physiol. Behav.* **5**:89–93.

Archer, J., 1974, The effects of testosterone on the distractability of ducks by irrelevant and relevant novel stimuli, *Anim. Behav.* **22**:397–404.

Bare, J. K., 1949, The specific hunger for sodium chloride in normal and adrenalectomized white rats, *J. Comp. Physiol. Psychol.* **42**:242–253.

Beatty, W. W., and Schwartzbaum, J. S., 1967, Enhanced reactivity to quinine and saccharin solutions following septal lesions in the rat, *Psychon. Sci.* **8**:483–484.

Beatty, W. W., and Schwartzbaum, J. S., 1968a, Commonality and specificity of behavioral dysfunction following septal and hippocampal lesions in rats, *J. Comp. Physiol. Psychol.* **66**:60–68.

Beatty, W. W., and Schwartzbaum, J. S., 1968b, Consumatory behavior for sucrose following septal lesions in the rat, *J. Comp. Physiol. Psychol.* **65**:93–102.

Beebe-Center, J. G., Black, P., Hoffman, A. C., and Wade, M., 1948, Relative per diem consumption as a measure of preference in the rat. *J. Comp. Physiol. Psychol.* **41**:239–251.

Blass, E. M., 1974, The physiological, neurological and behavioral basis of thirst, *in* "Nebraska Symposium on Motivation, 1974" (M. R. Jones, ed.), University of Nebraska Press, Lincoln.

Booth, D. A., 1972, Taste reactivity in starved, ready to eat and recently fed rats, *Physiol. Behav.* **8**:901–908.

Booth, D. A., Lovett, D., and McSherry, G. M., 1972, Postingestive modulation of the sweetness preference gradient in the rat. *J. Comp. Physiol. Psychol.* **78**:485–512.

Borer, K. T., 1968, Disappearance of preferences and aversions for sapid solutions in rats ingesting untasted fluids, *J. Comp. Physiol. Psychol.* **65**:213–221.

Brown, T. S., and Murphy, H. M., 1973, Factors affecting sucrose preference behavior in rats with hippocampal lesions, *Physiol. Behav.* **11**:833–845.

Burke, G. H., Mook, D. G., and Blass, E. M., 1972, Hyperreactivity to quinine associated with osmotic thirst in the rat, *J. Comp. Physiol. Psychol.* **78**:32–39.

Cabanac, M., 1971, Physiological role of pleasure, *Science* **173**:1103–1107.

Campbell, B. A., 1958, Absolute and relative preference thresholds for hungry and satiated rats. *J. Comp. Physiol. Psychol.* **51**:795–799.

Campbell, C. S., and Davis, J. D., 1974a, Licking rate of rats reduced by intraduodenal and intraportal glucose infusion, *Physiol. Behav.* **12**:357–365.

Campbell, C. S., and Davis, J. D., 1974b, Peripheral control of food intake: Interaction between test diet and postingestive chemoreception, *Physiol. Behav.* **12**:377–384.

Carey, R. J., 1971, Quinine and saccharin preference-aversion threshold determinations in rats with septal ablations, *J. Comp. Physiol. Psychol.* **76**:316–326.

Carey, R. J., 1972, A neuroanatomical investigation of enhanced cutaneous and gustatory responsivity associated with septal forebrain injury, *J. Comp. Physiol. Psychol.* **80**:449–457.

Carey, R. J., and Procopio, G., 1974, Differential effects of septal, preoptic, and habenula ablations on thirst-motivated behaviors in rats, *J. Comp. Physiol. Psychol.* **86**:1163–1172.

Catalanotto, F. A., and Sweeney, E. A., 1972, The effects of surgical desalivation of the rat upon taste acuity, *Arch. Oral Biol.* **17**:1455–1465.

Chiang, H. M., and Wilson, W. A., 1963, Some tests of the diluted-water hypothesis of saline consumption in rats, *J. Comp. Physiol. Psychol.* **56**:660–665.

Chiaraviglio, E., 1971, Amygdaloid modulation of sodium chloride and water intake in the rat, *J. Comp. Physiol. Psychol.* **76**:401–407.

Chiaraviglio, E., 1972, Mesencephalic influences on the intake of sodium chloride and water in the rat, *Brain Res.* **44**:73–82.

Collier, G. H., and Bolles, R. C., 1968, Hunger, thirst, and their interaction as determinants of sucrose consumption. *J. Comp. Physiol. Psychol.* **66**:633–641.

Corbit, J. D., 1965, Hyperphagic hyperreactivity to adulteration of drinking water with quinine HCl, *J. Comp. Physiol. Psychol.* **60**:123–124.

Covian, M. R., 1974, Septal control of sodium balance. Paper read at 26th international Congress of Physiological Sciences, Jerusalem, 1974.

Covian, M. R., and Antunes-Rodrigues, J., 1963, Specific alterations in sodium chloride intake after hypothalamic lesions, *Am. J. Physiol.* **205**:922–926.

Davis, J. D., 1973, The effectiveness of some sugars in stimulating licking behavior in the rat, *Physiol. Behav.* **11**:39–45.

Davis, J. D., and Campbell, C. S., 1973, Peripheral control of meal size in the rat: The effect of sham feeding on meal size and drinking rate, *J. Comp. Physiol. Psychol.* **83**:379–387.

Denton, D. A., 1967, Salt appetite, *in* "Handbook of Physiology," Vol. 1, Sec. 6 (C. F. Code, ed.), American Physiological Society, Washington, D.C., pp. 433–459.

Dethier, V. G., Evans, D. R., and Rhoades, M. V., 1956, Some factors controlling the ingestion of carbohydrates by the blowfly, *Biol. Bull.* **111**:204.

Dickinson, A., 1973, Prandial drinking following septal lesions in rats, *Physiol. Behav.* **10**:335–338.

Donovick, P. J., Burright, R. G., and Zuromski, E., 1970, Localization of quinine aversion within the septum, habenula, and interpeduncular nucleus, *J. Comp. Physiol. Psychol.* **71**:376–383.

Epstein, A. N., 1967, Oropharyngeal factors in feeding and drinking, *in* "Handbook of Physiology" Vol. 1, Sec. 6 (C. F. Code, ed.), American Physiological Society, Washington, D.C., pp. 197–218.

Epstein, A. N., 1971, The lateral hypothalamic syndrome: Its implications for the physiological psychology of hunger and thirst, *in* "Progress in Physiological Psychology," Vol. 4 (E. Stellar and J. M. Sprague, eds.), Academic Press, New York, pp. 263–317.

Epstein, A. N., and Stellar, E., 1955, The control of salt preference in the adrenalectomized rat, *J. Comp. Physiol. Psychol.* **48**:167–172.

Epstein, A. N., and Teitelbaum, P., 1964, Severe and persistent deficits in thirst produced by lateral hypothalamic damage, *in* "Thirst" (M. J. Wayner, ed.), Pergamon Press, New York, pp. 395–406.

Ernits, T., and Corbit, J. D., 1973, Taste as a dipsogenic stimulus, *J. Comp. Physiol. Psychol.* **83**:27–31.

Falk, J. L., 1973, Comments on Dr. Kissileff's chapter, *in* "The Neuropsychology of Thirst" (A. N. Epstein, E. Stellar, and H. R. Kissileff, eds.), V. H. Winston & Sons, Washington, D.C., pp. 225–228.

Fisher, G. L., 1965, Saline preference in rats determined by contingent licking, *J. Exp. Anal. Behav.* **8**:295–304.

Flaherty, C. F., and Hamilton, L. W., 1971, Responsivity to decreasing sucrose concentrations following septal lesions in the rat, *Physiol. Behav.* **6**:431–437.

Franklin, K. B. J., and Quartermain, D., 1970, Comparison of the motivational properties of deprivation-induced drinking with drinking elicited by central carbachol stimulation, *J. Comp. Physiol. Psychol.* **71**:390–395.

Gaines, S., 1965, Tolerance for quinine in adrenalectomized rats. Unpublished M.A. thesis, University of Virginia.

Gentil, C. G., Antunes-Rodrigues, J., Negro-Vilar, A., and Covian, M. R., 1968, Role of amygdala complex in soldium chloride and water intake in the rat, *Physiol. Behav.* **3**:981–985.

Gentil, C. G., Mogenson, G. J., and Stevenson, J. A. F., 1971, Electrical stimulation of septum, hypothalamus and amygdala and saline preference, *Am. J. Physiol.* **220**:1172–1177.

Gittelson, P. L., Donovick, P. J., and Burright, R. G., 1969, Facilitation of passive avoidance acquisition in rats with septal lesions, *Psychon. Sci.* **17**:292-293.

Grace, J. E., 1968, Central nervous system lesions and saline intake in the rat, *Physiol. Behav.* **3**:387-393.

Graff, H., and Stellar, E., 1962, Hyperphagia, obesity and finickiness. *J. Comp. Physiol. Psychol.* **55**:418-424.

Guttman, N., 1953, Operant conditioning, extinction and periodic reinforcement in relation to concentration of sucrose used as reinforcing agent, *J. Exp. Psychol.* **46**:213-224.

Halpern, B., 1967, Some relationships between electrophysiology and behavior in taste, *in* "The Chemical Senses and Nutrition" (M. R. Kare and O. Maller, eds.), Johns Hopkins Press, Baltimore, pp. 213-242.

Hamilton, C. L., 1969, Ingestion of nonnutritive bulk and wheel running in the rat, *J. Comp. Physiol. Psychol.* **69**:481-484.

Hamilton, L. W., and Flaherty, C. F., 1971, Behavioral patterns associated with water intake in normal and septal rats, *J. Comp. Physiol. Psychol.* **76**:165-174.

Hammel, H. T., 1965, Neurons and temperature regulation, *in* "Physiological Controls and Regulations" (W. S. Yamamoto and J. R. Brobeck, eds.), Saunders, Philadelphia, pp. 71-97.

Hammer, L., 1968, Relationship of reinforcement value to consummatory behavior, *J. Comp. Physiol. Psychol.* **66**:667-672.

Jacobs, H. L., 1961, The osmotic postingestion factor in the regulation of glucose appetite, *in* "Physiological and Behavioral Aspects of Taste" (M. R. Kare and B. P. Halpern, eds.), University of Chicago Press, Chicago, pp. 16-27.

Jacobs, H. L., 1962, Some physical, metabolic, and sensory components of appetite for glucose, *Am. J. Physiol.* **203**:1043-1054.

Jacobs, H. L., 1964, Evaluation of the osmotic effects of glucose loads in food satiation, *J. Comp. Physiol. Psychol.* **57**:309-310.

Jalowiec, J. E., and Stricker, E. M., 1970, Restoration of body fluid balance following acute sodium deficiency in rats, *J. Comp. Physiol. Psychol.* **70**:94-102.

Kemble, E. D., and Schwartzbaum, J. S., 1969, Reactivity to taste properties of solutions following amygdaloid lesions, *Physiol. Behav.* **4**:981-985.

Kenney, N. J., 1974, Postingestive factors in the control of glucose intake by satiated rats, *Physiol. Psychol.* **2**:433-434.

Kissileff, H. R., 1967, Loss of salt preference in rats drinking prandially, *Am. Zool.* **2**:116.

Kissileff, H. R., 1969, Aversion for hypertonic saline solutions in rats ingesting untasted fluids, *in* "Olfaction and Taste III" (C. Pfaffman, ed.), Rockefeller University Press, New York.

Kissileff, H. R., 1973, Nonhomeastatic controls of drinking, *in* "The Neuropsychology of Thirst" (A. N. Epstein, E. Stellar, and H. R. Kissileff, eds.), V. H. Winston & Sons, Washington, D.C., pp. 163-198.

Kissileff, H. R., 1974, Instrumental responding terminates aversive gustatory stimuli. Paper read at the Eastern Psychological Association, Philadelphia.

Kissileff, H. R., and Epstein, A. N., 1969, Exaggerated prandial drinking in the "recovered lateral" rat without saliva, *J. Comp. Physiol. Psychol.* **67**:301-308.

Kon, S. K., 1931, The self-selection of food constituents by the rat, *Biochem. J.* **25**:473-481.

Krieckhaus, E. E., 1970, "Innate recognition" aids rats in sodium regulation, *J. Comp. Physiol. Psychol.* **73**:117-123.

Krieckhaus, E. E., and Wolf, G., 1968, Acquisition of sodium by rats: Interaction of innate mechanisms and latent learning, *J. Comp. Physiol. Psychol.* **65**:197-201.

Krikstone, B. J., and Levitt, R. A., 1974, Comparisons between drinking induced by water deprivation or chemical stimulation, *Behav. Biol.* **11**:547-559.

Lawson, W. B., 1969, Effects of atropine-induced desalivation on salt preference. Paper read at the Eastern Psychological Association, Philadelphia.

Lawson, W. B., and Hagstrom, E. C., 1972, Desalivation and taste preference. Paper read at Psychonomic Society, St. Louis.

Lawson, W. B., Hagstrom, E. C., and Walter, G. F., 1974, Salt preference in desalivate rats, *Physiol. Behav.* **12**:733-739.

Le Magnen, J., and Tallon, S., 1966, La périodicité spontanée de la prise d'aliments ad libitum du rat blanc, *J. Physiol.* (Paris) **58**:323-349.

Levitsky, D. A., 1970, Feeding patterns of rats in response to fasts and changes in environmental conditions, *Physiol. Behav.* **5**:291-300.

McCleary, R. A., 1953, Taste and postingestion factors in specific-hunger behavior, *J. Comp. Physiol. Psychol.* **46**:411-421.

McCleary, R. A., 1961, Response specificity in the behavioral effects of limbic system lesions in the cat, *J. Comp. Physiol. Psychol.* **54**:605-613.

McFarland, D. J., and McFarland, F. J., 1968, Dynamic analysis of an avian drinking response, *Med. Biol. Eng.* **6**:659-668.

Melzack, R., and Casey, K. L., 1968, Sensory, motivational and central control determinants of pain, *in* "The Skin Senses" (D. R. Kenshalo, ed.), Charles C Thomas, Springfield, pp. 423-443.

Mook, D. G., 1963, Oral and postingestional determinants of the intake of various solutions in rats with esophageal fistulas, *J. Comp. Physiol. Psychol.* **56**:645-659.

Mook, D. G., 1969, Some determinants of preference and aversion in the rat, *Ann. N.Y. Acad. Sci.* **157**:1158-1175.

Mook, D. G., 1975, Saccharin preference in the rat: Some unpalatable findings, *Psychol. Rev.* **81**:475-490.

Mook, D. G., and Blass, E. M., 1968, Quinine aversion thresholds and "finickiness" in hyperphagic rats, *J. Comp. Physiol. Psychol.* **65**:202-207.

Mook, D. G., and Blass, E. M., 1970, Specific hungers in hyperphagic rats, *Psychon. Sci.* **19**:34-35.

Mook, D. G., and Kozub, F. J., 1968, Control of sodium chloride intake in the nondeprived rat, *J. Comp. Physiol. Psychol.* **66**:105-109.

Mook, D. G., Lindsey, G. P., Pace, E. K., and Graeber, R. C., 1972, Preference for glucose disrupted by lateral preoptic lesions in the rat, *Psychon. Sci.* **28**:266-268.

Mook, D. G., Walshe, D. Z., and Farris, P. R., 1975, Some observations on the preference deficit produced by lateral preoptic lesions in the rat, *J. Comp. Physiol. Psychol.* **88**:785-795.

Morrison, S. D., 1968, The constancy of the energy expended by rats on spontaneous activity, and the distribution of activity between feeding and non-feeding, *J. Physiol.* (London) **197**:305-323.

Murphy, H. M., and Brown, T. S., 1970, Effects of hippocampal lesions on simple and preferential consummatory behavior in the rat, *J. Comp. Physiol. Psychol.* **72**:404-415.

Myer, J. S., and Van Hemel, P. E., 1969, Saline as a reinforcer of bar pressing by thirsty rats, *J. Comp. Physiol. Psychol.* **68**:455-460.

Nachman, M., and Pfaffmann, C., 1963, Gustatory nerve discharge in normal and sodium-deficient rats, *J. Comp. Physiol. Psychol.* **56**:1007-1011.

Novakova, A., and Cort, J. H., 1966, Hypothalamic regulation of spontaneous salt intake in the rat, *Am. J. Physiol.* **211**:919-925.

Panksepp, J., 1971, A re-examination of the role of the ventromedial hypothalamus in feeding behavior, *Physiol. Behav.* **7**:385-394.

Peiper, A., 1963, Cerebral Function in Infancy and Childhood. Consultants Bureau, New York.

Pfaffmann, C., 1961, Sensory and motivating properties of the sense of taste, *in* "Nebraska Symposium on Motivation," 1961 (M. R. Jones, ed.), University of Nebraska Press, Lincoln, pp. 71-108.

Pfaffmann, C., and Bare, J. K., 1950, Gustatory nerve discharges in normal and adrenalectomized rats, *J. Comp. Physiol. Psychol.* **43:**320–324.

Quartermain, D., Miller, N. E., and Wolf, G., 1967, Role of experience in relationship between sodium deficiency and rate of bar pressing for salt, *J. Comp. Physiol. Psychol.* **63:**417–420.

Rabe, E. F., and Corbit, J. D., 1973, Postingestional control of sodium chloride solution drinking in the rat, *J. Comp. Physiol. Psychol.* **84:**268–274.

Richter, C. P., 1942-1943, Total self-regulatory functions in animals and human beings, *Harvey Lect.* **38:**63-103.

Rolls, B. J., and Jones, B. P., 1972, Cessation of drinking following intracranial injections of angiotensin in the rat, *J. Comp. Physiol. Psychol.* **80:**26–29.

Rolls, B. J., and Rolls, E. J., 1973, Effects of lesions in the basolateral amygdala on fluid intake in the rat, *J. Comp. Physiol. Psychol.* **83:**240–247.

Russek, M., 1970, Demonstration of the influence of an hepatic glucosensitive mechanism on food intake, *Physiol. Behav.* **5:**1207-1209.

Saad, W. A., Antunes-Rodrigues, J., Gentil, C. G., and Covian, M. R., 1972, Interaction between hypothalamus, amygdala, and septal area in the control of sodium chloride intake, *Physiol. Behav.* **9:**629-636.

Sclafani, A., 1973, Deficits in glucose appetite and satiety produced by ventromedial hypothalamic lesions in the rat, *Physiol. Behav.* **11:**771–780.

Singh, D., 1974, Role of preoperative experience on reaction to quinine taste in hypothalamic hyperphagic rats, *J. Comp. Physiol. Psychol.* **86:**674–678.

Smith, D. F., Stricker, E. M., and Morrison, G. R., 1969, NaCl solution acceptability by sodium-deficient rats, *Physiol. Behav.* **4:**239–243.

Smith, M. H., Holman, G. L., and Fortune, K. H., 1968, Sodium need and sodium consumption, *J. Comp. Physiol. Psychol.* **65:**33-37.

Stellar, E., 1961, Drive and motivation, *in* "Handbook of Physiology," Vol. 3, Sec. 1 (J. Field, ed.), American Physiological Society, Washington, D.C., pp. 1501-1527.

Stellar, E., and Hill, J. H., 1952, The rat's rate of drinking as a function of water deprivation, *J. Comp. Physiol. Psychol.* **45:**96-102.

Stellar, E., Hyman, R., and Samet, S., 1954, Gastric factors controlling water and salt-solution drinking, *J. Comp. Physiol. Psychol.* **47:**220-226.

Stellar, E., and McCleary, R. A., 1952, Food preference as a function of the method of measurement, *Am. Psychol.* **7:**256.

Stricker, E. M., 1969, Osmoregulation and volume regulation in rats: Inhibition of hypovolemic thirst by water, *Am. J. Physiol.* **217:**98-105.

Stricker, E. M., and Wilson, N. E., 1970, Salt-seeking behavior in rats following acute sodium deficiency, *J. Comp. Physiol. Psychol.* **72:**416-420.

Strouthes, A., Volo, A. M., and Unger, T., 1974, Hunger, thirst and their interactive affects on the rat's drinking in a saccharin-water choice, *Physiol. Behav.* **13:**153-158.

Teitelbaum, P., 1955, Sensory control of hypothalamic hyperphagia, *J. Comp. Physiol. Psychol.* **48:**156-163.

Teitelbaum, P., 1966, The use of operant methods in the assessment and control of motivational states, *in* "Operant Behavior: Areas of Research and Application" (W. K. Honig, ed.), Appleton, New York, pp. 565-608.

Teitelbaum, P. and Epstein, A. N., 1962, The lateral recovery syndrome: Recovery of feeding and drinking after lateral hypothalamic lesions, *Psychol. Rev.* **69:**74-90.

Thomas, D. W., and Mayer, J., 1968, Meal taking and regulation of food intake by normal and hypothalamic hyperphagic rats, *J. Comp. Physiol. Psychol.* **66:**642-653.

Tinbergen, L., 1960, The natural control of insects in pine woods: I. Factors influencing the intensity of predation by song birds, *Arch. Néer. Zool.* **13:**265-343.

Tolman, E. C., 1959, Principles of purposive behavior, *in* "Psychology: A Study of a Science," Vol. 2 (S. Koch, ed.), McGraw-Hill, New York.

Toth, D. M., 1973, Temperature regulation and salivation following preoptic lesions in the rat, *J. Comp. Physiol. Psychol.* **82**:480–488.

Uexküll, J. V., 1934, Streifzüge durch die Unwelten von Tieren und Menschen, *translated in* "Instinctive Behavior" (C. H. Schiller, ed.), Methuen, London.

Vance, W. B., 1965, Observations on the role of salivary secretion in the regulation of food and fluid intake in the white rat, *Psychol. Monogr.* 79 (Whole No. 598).

Verplanck, W. S., and Hayes, J. R., 1953, Eating and drinking as a function of maintenance schedule, *J. Comp. Physiol. Psychol.* **46**:327–333.

Wayner, M. J., Kahan, S., and Stoller, W., 1966, Lateral hypothalamic mediation of spinal reflex facilitation during salt arousal of drinking, *Physiol. Behav.* **1**:341–350.

Wiener, I. H., and Stellar, E., 1951, Salt preference in the rat determined by a single-stimulus method, *J. Comp. Physiol. Psychol.* **44**:394–401.

Wilcove, W. G., 1973, Ingestion affected by the oral environment: The role of gustatory adaptation on taste reactivity in the rat, *Physiol. Behav.* **11**:297–312.

Wolf, G., 1964, Effect of dorsolateral hypothalamic lesions on sodium appetite elecited by desoxycorticosterone and by acute hyponatremia, *J. Comp. Physiol. Psychol.* **58**:396–402.

Wolf, G., 1968, Thalamic and tegmental mechanisms for sodium intake: Anatomical and functional relations to lateral hypothalamus, *Physiol. Behav.* **3**:997–1002.

Wolf, G., 1969, Innate mechanisms for regulation of sodium intake, *in* "Olfaction and Taste: III" (C. Pfaffmann, ed.), Rockefeller University Press, New York.

Wolf, G., and DiCara, L. V., 1974, Impairments in sodium appetite after lesions of gustatory thalamus: Replication and extension, *Behav. Biol.* **10**:105–112.

Wolf, G., McGovern, J. F., and DiCara, L. V., 1974, Sodium appetite: Some conceptual and methodologic aspects of a model drive system, *Behav. Biol.* **10**:27–42.

Wolf, G., and Quartermain, D., 1967, Sodium chloride intake of adrenalectomized rats with lateral hypothalamic lesions, *Am. J. Physiol.* **212**:113–118.

Young, P. T., 1955, The role of hedonic processes in motivation, *in* "Nebraska Symposium on Motivation, 1955" (M. R. Jones, ed.), University of Nebraska Press, Lincoln, pp. 193–238.

Young, P. T., 1967, Palatability: The hedonic response to foodstuffs, *in* "Handbook of Physiology," Vol. 1, Sec. 6, (C. F. Code, ed.), American Journal of Physiology, Washington, D.C., pp. 353–366.

Young, P. T., and Falk, J. L., 1956, The relative acceptability of sodium chloride as a function of concentration and water need, *J. Comp. Physiol. Psychol.* **49**:569–575.

Young, P. T., and Greene, J. T., 1953, Quantity of food ingested as a measure of relative acceptability, *J. Comp. Physiol. Psychol.* **46**:288–294.

Young, P. T., and Shuford, E. H., Jr., 1954, Intensity, duration, and repetition of hedonic processes as related to acquisition of motives, *J. Comp. Physiol. Psychol.* **47**:298–305.

8

Water Taste in Mammals

Linda M. Bartoshuk

1. Early Beliefs about the Intrinsic Taste of Water

Opinions about the taste of pure water can be traced back as far as Aristotle (384–322 B.C.). He believed that "Water in its own nature has no flavour" (Hammond, 1902, p. 164). We do not know how pure Aristotle's water sources were, but the process of distilling water is known to be very old. Sarton (1931) tentatively attributed a treatise on distilled water to the first half of the 12th century. Avicenna (980–1037), one of the most influential of the Arabian physicians, mentioned the distilling of water as a technique of purifying it in *The Canon of Medicine* (Gruner, 1930). Avicenna, like Aristotle, believed that pure water is tasteless. This belief was not seriously challenged until the 19th century. Henle (1880) and Öhrwall (1891) concluded that distilled water is flat or insipid rather than truly tasteless. About this same time, other investigators began to notice that some subjects reported that pure water tasted bitter (Camerer, 1870; Kiesow, 1894; Skramlik, 1922). Brown (1914) obtained evaluations of water from 100 subjects. About 50% of the judgments were "no taste," 25% were "bitter," and the remaining 25% were varied. Brown concluded, "The upshot of all of these observations seems to be that water is not tasteless (in the broader sense of the word), that it tastes more like bitter than anything else" (p. 253).

In the 19th century, the taste of water was also of interest in another context. Water was known to take on sweet tastes if the tongue was first exposed to substances like sulfuric acid (Adducco and Mosso, 1886) or

Linda M. Bartoshuk • John B. Pierce Foundation Laboratory, New Haven, Connecticut.

potassium chlorate (Nagel, 1896); however, these phenomena were not believed to be related to the intrinsic taste of water.

Despite Henle, Öhrwall, and Brown, the prevailing opinion about the taste of distilled water was that it is tasteless. Well into the 20th century, the occasional subjects reporting that water tasted bitter were considered to be anomalous (Anderson, 1959).

2. The "Water" Fiber of Zotterman and His Colleagues

In 1949, Zotterman reported that the activity in the glossopharyngeal nerve of the frog increased when distilled water was flowed across the tongue. He concluded that this water response "reflexly keeps the mouth closed, thus reducing an otherwise obvious increase of the intake of water and a subsequent greater loss of salts due to an increased diuresis" (p. 188). Some investigators have challenged the frog's water response on the grounds that Zotterman's water may have contained small amounts of calcium, to which frogs are very sensitive (Nomura and Sakada, 1965). However, additional experiments have supported the existence of a true water response in the frog (Akaike and Yamada, 1965; Nomura and Ishizaki, 1972). Liljestrand and Zotterman (1954) found that activity in the chorda tympani nerves of the cat, the dog, and the pig also increased when water was flowed across the tongue. This report was followed by a detailed analysis of the characteristics of the water fiber that they believed was responsible for the water responses observed from the whole nerve (Cohen, Hagiwara, and Zotterman, 1955). Table 1 shows the characteristics of the water-fiber type as well as the other fiber types they found in the cat chorda tympani. Liljestrand and Zotterman concluded that "most likely in all mammals there are specific nerve endings responding to water which thus may constitute a mechanism for discrimination of water containing only very small amounts of electrolytes or none" (p. 302).

The belief that water fibers would be found in all mammals differed

Table 1.   Types of Fibers Found in the Cat Chorda Tympani by Cohen et al. (1955)

Stimulus	"Water" fiber	"Salt" fiber	"Acid" fiber	"Quinine" fiber	Sensation evoked
H_2O (salt < 0.03 M)	+	0	0	0	Water
NaCl (> 0.05 M)	0	+	0	0	Salt
HCl (pH 2.5)	+	+	+	0	Sour
Quinine	+	0	0	+	Bitter

dramatically from some results of Pfaffmann and Beidler. When Pfaffmann did the first detailed analysis of the responses of cat chorda tympani fibers in 1941, he found fibers sensitive to HCl, to HCl and NaCl, and to HCl and quinine but did not find any of the water fibers that Cohen *et al.* (1955) later described. He did note that some fibers produced a transient response to water, particularly if the tongue had just previously been stimulated with acid. Pfaffmann and Beidler and their students added detailed analyses of the responses of rat single fibers (Pfaffmann and Bare, 1950; Pfaffmann, 1955; Beidler, 1953, 1955). They did not find "water fibers" as described by Liljestrand and Zotterman (1954) and Cohen *et al.* (1955). However, Beidler (1955) noted that water produced a transient response when it followed sodium benzoate.

The single-fiber studies so important to the water-fiber issue also provided the data that led to an important conclusion about coding. The fibers studied differed considerably from one another, casting doubt on the validity of the concept of fiber types.

Zotterman was initially skeptical about the failure to find water fibers in the rat (Liljestrand and Zotterman, 1954) but later recorded from rats himself and found no water fibers. He then concluded that the rat is so sensitive to weak NaCl that it can discriminate weak NaCl from water without the help of water fibers.

Man was added to the list of species tested for water fibers with dramatic recordings from the human chorda tympani nerve. These recordings were possible because the chorda tympani crosses the eardrum before entering the skull and proceeding to the medulla. During certain kinds of ear surgery, the chorda tympani can be cut and placed on an electrode. The whole-nerve recordings that resulted showed only slight decrements in neural response when water was applied to the tongue. This led Zotterman and Diamant (1959) to conclude, "the specific effect of water on taste in man can thus be looked upon as being of the same nature as that of blackness upon vision" (p. 192).

3. The Contingent-Water-Taste View

3.1. The Taste of Water to Man Is Contingent on the Substance Preceding the Water

The resolution of the arguments over the intrinsic taste of water in man turned out to depend on an understanding of taste adaptation. Hahn initiated a series of classic studies on taste adaptation in the 1930s. He showed that if the tongue is exposed to a constant concentration of a tastant, the taste intensity decreases until it disappears and the absolute threshold for that tastant is then located at a concentration just above the adaptation concentration (Hahn, 1949). McBurney and Pfaffmann (1963) reasoned that Hahn's conclusions about NaCl could be extended to saliva. That is, the NaCl in saliva might act as an adapting solution just like the NaCl rinses in Hahn's studies. McBurney and Pfaffmann removed saliva with a distilled-water rinse and found that the absolute threshold for NaCl dropped by a factor of 30.

The importance of saliva as an adapting solution for the salty taste led us to wonder whether or not saliva was related to the arguments over the intrinsic taste of water. We asked subjects to judge the taste of water after adaptation to water and after adaptation to NaCl (Bartoshuk, McBurney, and Pfaffmann, 1964). Water after adaptation to water was tasteless, while water after adaptation to NaCl of about the same concentration as that in saliva was bitter-sour. Figure 1 illustrates these results as well as showing the relationship of NaCl and water taste thresholds to the adaptation concentration. Note that after adaptation to NaCl, the bitter-sour taste is evoked not only by water but by all concentrations of NaCl lower than the adapting concentration. These data suggest that the intrinsic taste of water can be related to variation in salivary electrolytes. High concentrations of salivary electrolytes would make water taste bitter-sour, while low concentrations would make water tasteless. At intermediate salivary electrolyte concentrations, water may be detected as different from saliva but may fail to have a recognizeable taste. This phenomenon may be the origin of the "flat" water tastes of Henle and Öhrwall.

We now know that complete adaptation is not necessary for the production of a water taste. Even a few seconds' exposure is often enough. In addition, the effects of adaptation to a tastant on the taste of water are not limited to NaCl. Adaptation to other tastants produces a variety of water tastes. In fact, all of the four basic tastes can be produced by water if the correct tastants precede it (Bartoshuk, 1968; McBurney, 1969; McBurney and Bartoshuk, 1973). Table 2 shows the predominant tastes of water following exposure of the tongue to several tastants. The taste quality produced by water is predicted to some extent by the taste quality of the preceding stimulus (McBurney and Shick, 1971). However, there are obviously exceptions. Since the part of a molecule that gives it its characteristic taste need not be the part of the molecule that is responsible for producing the water taste, the exceptions are not surprising.

In general, as the concentration of the adapting solution increases, the intensity of the water taste increases. Figure 2 shows an example with an adapting solution of potassium chlorogenate, one of the components in globe artichokes. The intensity of the water may also begin to decrease above a particular adapting concentration. Figure 3 shows an inverted-U function for the taste of water following NaCl.

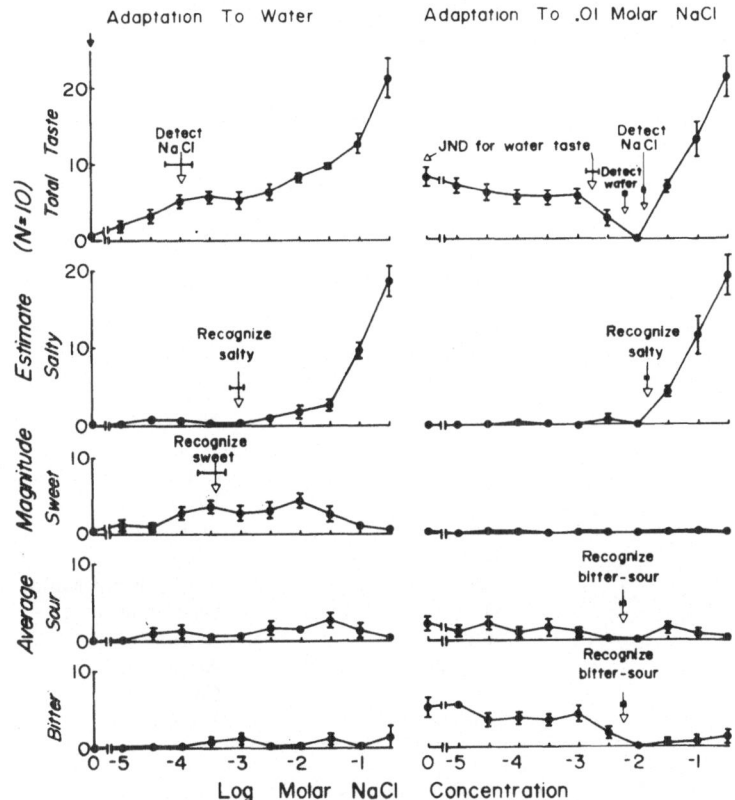

Figure 1. The taste of water and NaCl under two adaptation conditions: water and 0.01 *M* NaCl (approximately equivalent to the Na and Cl content of resting human saliva). The top functions show the total taste intensity. The lower functions show how the total is divided into salty, sweet, sour, and bitter components. The detection thresholds for water and NaCl are shown on the total functions. The JND for water taste represents the concentration of NaCl just noticeably less bitter-sour than water. The recognition thresholds are shown on the appropriate quality functions. The ± standard error of the mean is shown around each data point. (From Bartoshuk, 1974, Copyright 1974 by the American Psychological Association. Reprinted by permission.)

3.2. Electrophysiological Responses to Water Are Also Contingent on the Substance Preceding the Water

The contingent nature of the taste of water in man suggested to us that Zotterman's water fiber needed reevaluation. We designed experiments on the rat, the cat, and the squirrel monkey similar to those done on man. That is, responses of single chorda tympani fibers to water were evaluated after four different adapting solutions (Bartoshuk, 1965; Bartoshuk and Pfaffmann, 1965; Bartoshuk, Harned, and Parks, 1971; Pfaffmann, Frank,

Table 2.   Water Tastes Following Various Stimuli[a]

Taste of water following stimulus[b]	Taste of stimulus[b]	Stimulus
Bitter	Salty	NaCl
Sour	Salty–bitter	KCl
	Salty	KBr
	Sweet	Saccharin
	Bitter	KNO_2
Salty	Bitter	Urea
Sweet	Salty–sour–bitter	$NaNO_3$
	Salty–bitter	Na_2SO_4
	Salty–bitter	NH_4Br
	Bitter	$CaBr_2$
	Bitter	KNO_2
	Bitter	$MgSO_4$
	Bitter	Urea
	Bitter	Quinine hydrochloride
	Bitter	Quinine sulfate
	Bitter	Caffeine
	Sour	Acetic acid
	Sour	Citric acid
	Sour	Tartaric acid

[a] Modified from McBurney and Shick, 1971.
[b] Qualities listed were those reported on at least half the trials with that particular stimulus.

Bartoshuk, and Snell, 1975). The results were similar to those obtained with human subjects. Responses to water depended on the nature of the preceding stimulus. Figures 4, 5, and 6 show examples of responses to water in each species. The cat fiber in Figure 4 responded to water following NaCl, the squirrel monkey fiber in Figure 5 responded to water following HCl, and the rat fiber in Figure 6 responded to water following sucrose. Responses to water that are contingent on the adapting solution have also been observed in the rat by Smith and Frank (1972) and Yamamoto and Kawamura (1974) and in the macaque monkey by Ogawa, Yamashita, Noma, and Sato (1972). Hellekant (1965a) noted a particularly interesting contingency in rats, dogs, and cats. Whole chorda tympani nerve recordings showed responses to water following alcohol. A single-fiber analysis in the cat suggested that the fibers sensitive to water following alcohol were the same fibers described as "water fibers" by Zotterman (Hellekant, 1965b).

Some of our single fibers did not respond to water under any adapting solution tested; some responded to water after one adapting condition, some

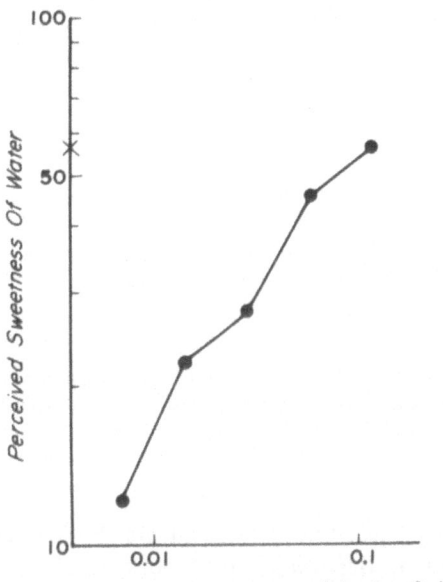

Figure 2. The sweet taste of water following potassium chlorogenate as a function of the concentration of potassium chlorogenate. The X represents the sweetness of two teaspoons of sucrose in six ounces of water (0.16 M sucrose). (Modified from Bartoshuk, Lee, and Scarpellino, 1972.)

after two, some after three and some after all four. One of the most important differences across species was the relative number of fibers responding to water after a particular adapting solution. For example, water following NaCl was a much more effective contingency for the cat than for the rat (Bartoshuk, 1965; Bartoshuk and Pfaffmann, 1965).

The contingent-water-response view differs markedly from the view of Zotterman and his colleagues. However, attention to the adapting conditions of Zotterman's experiments permits a simple reconciliation of the two views. Zotterman rinsed the tongue with Ringer's solution in most of his experiments. Ringer's solution is about 0.15 M NaCl. We now believe that Zotterman's water fibers correspond to the fibers sensitive to water follow-

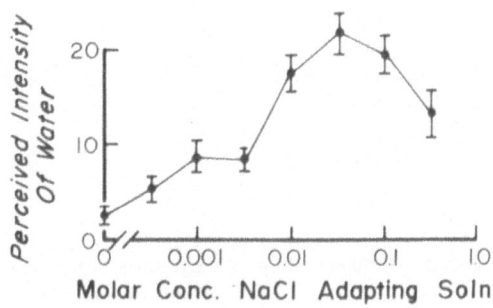

Figure 3. The total intensity of the taste of water following NaCl as a function of the concentration of NaCl. The predominant quality reported was bitter. Sour was the second most prominent quality. (Data collected by Keith Kishiyama as part of an undergraduate project at Yale University.)

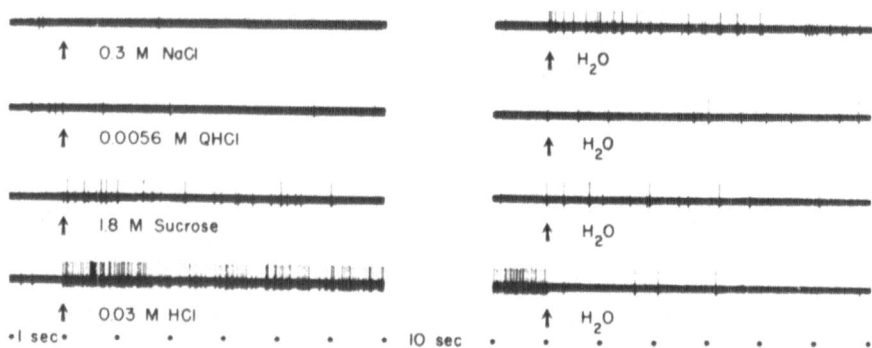

Figure 4. Responses from a single fiber in the chorda tympani taste nerve of the cat. Stimuli were applied in the order shown from left to right and from top to bottom. Stimulus onset is marked with an arrow. (From Bartoshuk *et al.*, 1971. Copyright 1971 by the American Association for the Advancement of Science.)

ing NaCl in our experiments. Both were insensitive to NaCl (Cohen *et al.*, 1955; Bartoshuk, 1965; Bartoshuk *et al.*, 1971), both responded to the removal of the chloride ion as tested with an anion–cation series (Cohen *et al.*, 1955; Bartoshuk, 1965), and both tended to respond to quinine and HCl (Cohen *et al.*, 1955; Bartoshuk and Frank, 1972).

The tendency of fibers sensitive to water following NaCl to be sensitive to QHCl and/or to HCl is particularly interesting. Frank has recently shown that taste fibers in the hamster chorda tympani (1973, 1974), the squirrel monkey chorda tympani (1974), and the rat glossopharyngeal nerve (1975) tend to fall into four fiber types according to the stimulus to which the fiber is most sensitive. Figure 7 shows the four fiber types found in the rat glossopharyngeal nerve (Frank, 1974).

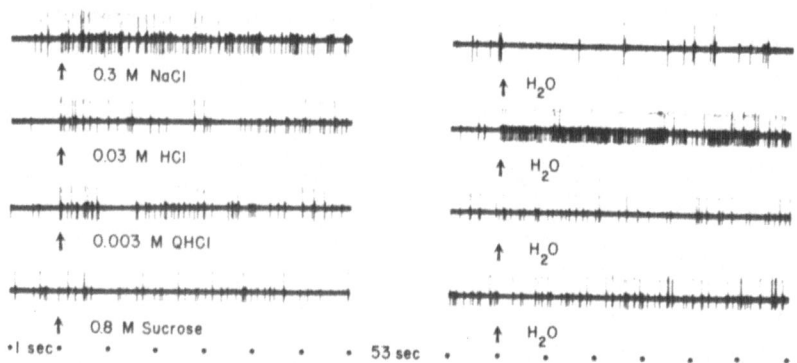

Figure 5. Responses from a single fiber in the chorda tympani taste nerve of the squirrel monkey (similar to Figure 4).

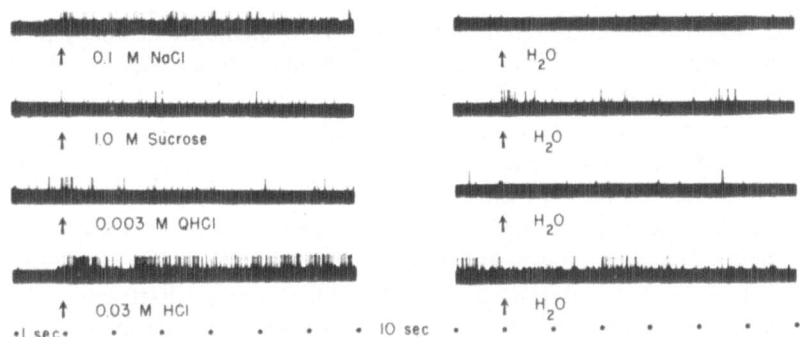

Figure 6. Responses from a single fiber in the chorda tympani taste nerve of the rat (similar to Figure 4).

The fiber types show peak sensitivities to stimuli representing the four basic tastes so familiar in human taste experience. The sucrose-best and the quinine-best fibers are relatively more specific than the NaCl-best and the HCl-best fibers. This finding suggests why this classification was not apparent earlier. Most of the research leading to belief in the lack of fiber types in taste was done on the rat chorda tympani, a nerve that is very sensitive to NaCl and HCl but insensitive to sucrose and quinine. Frank began her work on the hamster, a species quite sensitive to sucrose. It should be noted that although Cohen *et al.* (1955) proposed fiber types for the cat, Frank's classification suggests quite a different view of types. The types of Cohen *et al.* (1955) responded to a particular collection of stimuli not related in any obvious way. Frank's types respond to stimuli that have been ordered. That is, each fiber type has a maximal sensitivity to one stimulus and then responds with decreasing sensitivity with distance from the peak

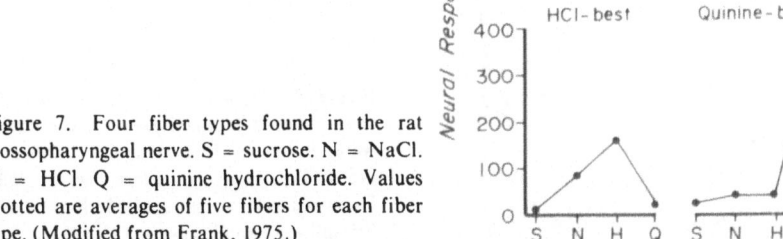

Figure 7. Four fiber types found in the rat glossopharyngeal nerve. S = sucrose. N = NaCl. H = HCl. Q = quinine hydrochloride. Values plotted are averages of five fibers for each fiber type. (Modified from Frank, 1975.)

stimulus along her ordering. This is reminiscent of the trichromacy theory of the coding of color, in which three receptor types with three peak sensitivities can code all color sensations. Frank's classification is completely consistent with the cross-fiber patterning theory of Erickson (1963), even though Erickson's original data suggested that there were no fiber types in taste. The only modification required is that the patterns that code qualities be generated across four fiber types rather than across the whole population of fibers. However, Frank's work has given rise to an alternate theory of taste coding known as the *labeled-line theory* (Pfaffmann, 1974). According to this view, each fiber codes only a single quality, that associated with its peak response. The fiber does respond to other stimuli, since the types are not totally specific, but the taste evoked is always the same. At present, both these coding theories remain logical possibilities.

We do not know exactly how water tastes fit into the debate over the coding of taste quality. However, the tendency of fibers sensitive to water following salt to be sensitive to quinine and/or HCl as well is suggestive. We know that water following NaCl tastes bitter–sour to man. It seems reasonable to conclude that water following NaCl produces the neural input normally produced by bitter–sour stimuli.

The fact that the actual stimulus for the response to water following NaCl is the removal of the chloride ion fits with the fact that the water response occurs to all NaCl concentrations less than the adapting concentration. The integrated whole-nerve recordings shown in Figures 8, 9,

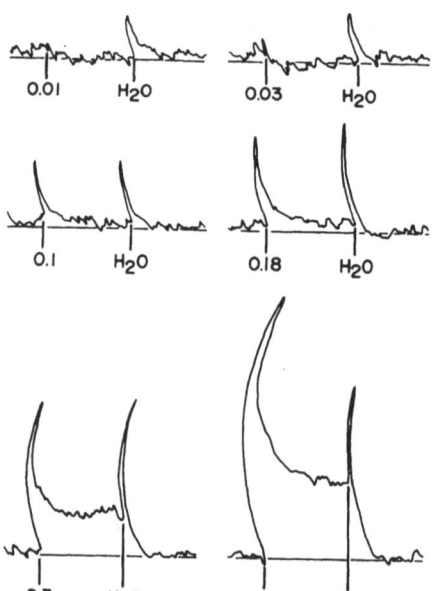

Figure 8. Integrated responses from the whole chorda tympani nerve of the cat to water following different concentrations of NaCl. The records read from left to right. Each stimulus was applied for 6 sec. Stimuli were applied every 3 min.

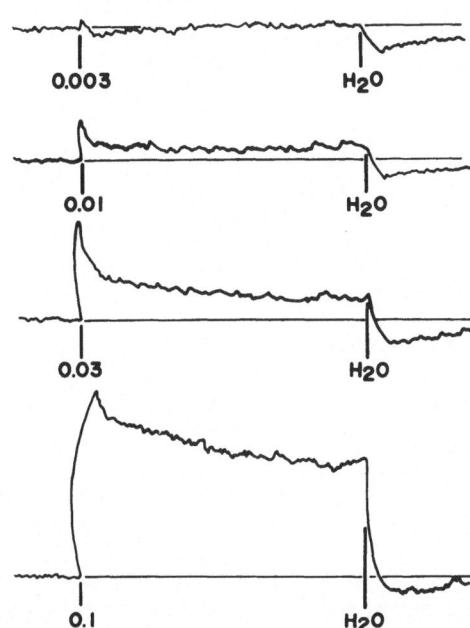

Figure 9. Similar to Figure 8, but recordings are from the rat. Stimuli were applied for 15 sec. Stimuli were applied every 5 min. (Modified from Pfaffmann and Powers, 1964.)

10, and 11 illustrate the relationships between responses from the fibers sensitive to water following NaCl and those sensitive to NaCl. Figure 8 shows responses to water following various concentrations of NaCl. Note that as the concentration of NaCl increases, the initial transient response to NaCl fails to return to baseline. That is, the NaCl stimulus produces a new steady-state response that is maintained until the NaCl is rinsed away with

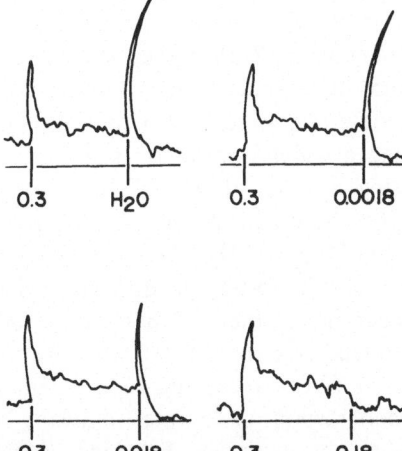

Figure 10. Integrated responses from the whole chorda tympani nerve of the cat to water and NaCl after 0.3 M NaCl. Responses shown were selected from a longer series. Water and NaCl concentrations were applied in ascending order of concentration alternately with 0.3 M NaCl. The NaCl concentrations that preceded 0.3 M NaCl in the records above were H_2O, 0.001 M, 0.01 M, and 0.1 M, respectively from left to right and from top to bottom.

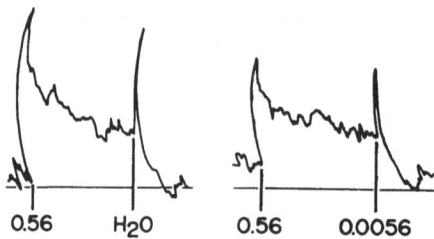

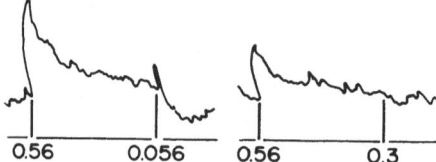

Figure 11. Similar to Figure 10, with 0.56 M NaCl used as the "rinse" instead 0.3 M NaCl. The NaCl concentrations that preceded 0.56 M NaCl in the records above were H_2O, 0.003 M, 0.03 M, and 0.1 M, respectively, from left to right.

water. Also note that water produces a transient increment followed by a decrement back to the original water baseline. Figure 9 shows responses of the whole chorda tympani of the rat to similar stimuli. Note that the rat nerve has a lower threshold for NaCl and that water does not produce a transient increment but only a decrement. Figure 10 shows responses to water and selected concentrations of NaCl applied following a 0.3 M NaCl rinse. Figure 11 shows similar responses with a 0.56 M NaCl rinse. Note that a comparison of Figures 8 and 11 shows that 0.056 M NaCl signals "water" when applied following 0.56 M NaCl but signals "salt" when applied following water.

The sensitivities of the "water" fiber and the "salt" fiber in the study of Cohen et al. (1955) are shown in the upper part of Figure 12. The concentrations stimulating these two fibers are not absolute, as suggested by Cohen et al. Rather, they were determined by the Ringer's solution used to rinse the tongue between stimuli. We obtained the U-shaped function in the middle of Figure 12, which is similar to the function produced by overlapping a water fiber and a salt fiber, from our cat whole-nerve preparation using a rinse of 0.1 M NaCl, which is slightly less concentrated than Ringer's. When stimuli are applied immediately after a water rinse (see bottom of Figure 12B), there is no response to water at all.

Since Zotterman and his colleagues did not vary the rinse in their experiments, they did not see the other adaptation contingencies that can produce responses to water. The rat chorda tympani nerve, which lacks water fibers according to Zotterman, actually seems to have few if any fibers responsive to water following NaCl but does have fibers responsive to other contingencies. The lack of the fibers responsive to water following

NaCl may be related to the great sensitivity of rat chorda tympani fibers to NaCl. The rat glossopharyngeal nerve is less sensitive to NaCl and does have fibers responsive to water following NaCl (Frank, 1968).

The apparent discrepancies between the studies on the cat done by Pfaffmann (1941) and by Cohen *et al.* (1955) can also be explained by the dependence of the water responses on the preceding substance. In Pfaffmann's study, the tongue was rinsed with water, and so water used as a stimulus was ineffective. When water followed acid, it did produce a response.

Man appeared to lack water fibers for a different reason. The electrophysiological experiments on human subjects were done with a water rinse (Diamant, Funakoshi, Ström, and Zotterman, 1963). Water after water would not be expected to produce a response. However, the application of each conventional tastant in the human experiments was followed by a distilled-water rinse. Some of these water rinses should have produced responses. Inspection of the published records shows that they did.

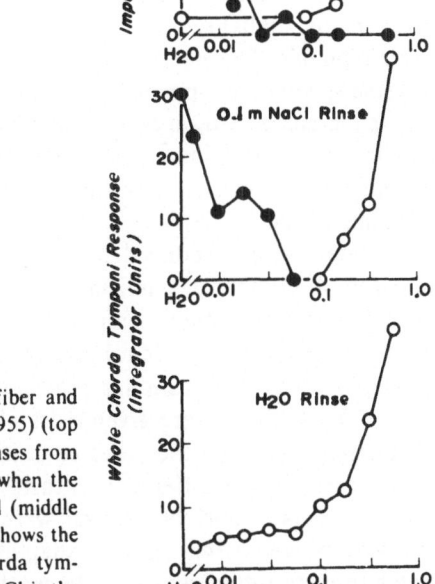

Figure 12. Characteristics of the "water" fiber and the "salt" fiber described by Cohen *et al.* (1955) (top part of figure) compared to integrated responses from the whole chorda tympani nerve of the cat when the cat tongue has been rinsed with 0.1 *M* NaCl (middle part of figure). The lower part of the figure shows the integrated responses of the same whole chorda tympani nerve when water rather than 0.1 *M* NaCl is the rinse.

However, Diamant *et al.* (1965) argued that these responses to water were probably due to cooling and/or touch. We cannot be certain now whether these responses were artifacts or were at least in part taste responses to water.

There is another reason why some early studies did not show water responses. The responses to water following NaCl appear to result from the removal of chloride. We do not know whether or not the other water responses are also produced by the removal of solutes, but this is a reasonable possibility. If the removal of a solute is the actual stimulating event, then the rate of removal—that is, the flow rate of water—would be expected to be of critical importance. In Pfaffmann's early studies on cats (1941, 1955), stimuli were applied with brushes, while in Cohen *et al.*'s (1955) study with cats, stimuli were flowed across the tongue at a rate of 5 ml/sec. These very different stimulation procedures would be expected to produce very different responses to water.

4. Water Responses Contingent on Saliva

4.1. Man

Whole-mouth saliva is a mixture of the salivas secreted primarily from three paired salivary glands: the sublingual, the submaxillary, and the parotid glands. The electrolyte content of the saliva secreted by some glands in some species can approach isotonicity (about 0.15 M) when salivary flow rate is increased. However, resting mixed saliva is hypotonic. Resting mixed human saliva contains about 0.008–0.06 M sodium. Chewing paraffin to increase flow rate can raise this value to about 0.02–0.13 M (Altman and Dittmer, 1961).

Water tastes produced by adaptation to saliva can complicate taste research. A number of published thresholds for NaCl may represent water-taste thresholds rather than solute thresholds (Bartoshuk, 1974). For example, Richter and MacLean (1939) measured the human detection threshold with a two-stimulus procedure that has become a standard technique. The subjects were given pairs of containers, one member of the pair containing NaCl and the other containing water. They were asked whether or not there was a difference in taste between the members of a pair. No rinses were used, and the subjects were permitted to sample the solutions in any way in which they chose. Adaptation to saliva could have caused the water to taste bitter–sour and the weak NaCl to be tasteless. In fact, some subjects did note that some stimuli tasted bitter. Figure 1 shows the detection thresholds for NaCl and water under adaptation to 0.01 M NaCl, an approximation to the NaCl in human saliva. Since detection thresholds involve only the detection of a taste and not recognition of any particular quality, the subjects could not discriminate the difference between these two thresholds.

Patients in some clinical studies that did not include rinses before each stimulus may have provided water-taste thresholds rather than NaCl thresholds. For example, cystic fibrosis has been reported to elevate salivary sodium and chloride (diSant'Agnese, Grossman, Darling, and Denning, 1958). In one study (Henkin and Powell, 1962), children with cystic fibrosis were reported to have unusually low detection thresholds for NaCl. In two other studies (Wotman, Mandel, Khotim, Thompson, Kutscher, Zegarelli, and Denning, 1964; Hertz, Cain, Bartoshuk, and Dolan, 1975), children with cystic fibrosis had normal thresholds. In one of these studies (Wotman *et al.*, 1964), the threshold was a recognition threshold for the taste of saltiness, which would not be confused with a water-taste threshold. The other study (Hertz *et al.*, 1975) used a detection criterion with a water rinse, so a water taste could not have occurred. That is, when the experimental design ensured that the thresholds measured were for the taste of NaCl rather than water, children with cystic fibrosis did not have thresholds below normal.

The contents of saliva vary not only across individuals but also within individuals. We know very little about how this variation might affect the taste and thereby the palatability of tap water.

4.2. Cat

Fibers responsive to water following NaCl are relatively common in the cat, which means that to the normally behaving cat, water may have an even stronger taste than it does to man. It seemed to us that this possibility might explain a curious contradiction in the literature on sugar preference in the cat. Frings (1951) showed that cats prefer dilute milk sweetened with sucrose over dilute milk itself. However, Carpenter (1956) failed to find any preference for water sweetened with sucrose over water in a standard Richter two-bottle preference test. We suspected that the taste of the water solvent might have been sufficiently strong to overwhelm the weaker sucrose taste in Carpenter's experiment. We decided to test this idea by dissolving sucrose in 0.03 M NaCl. Our electrophysiological data showed that 0.03 M NaCl was very near threshold for the fibers sensitive to NaCl, so that it would not taste "salty" to the cat. We did not know the exact concentration of NaCl in the cat's resting saliva; however, even if it were much higher than man's, 0.03 M NaCl would not be enough weaker than the NaCl in cat saliva to produce a large water response (see Figures 10 and

11). That is, we could feel relatively sure that 0.03 *M* NaCl would be a nearly tasteless solvent for the cat. The results of our experiment are shown in Figure 13. The cat was indifferent to sucrose dissolved in water but avidly consumed sucrose dissolved in 0.03 *M* NaCl. The NaCl content of the dilute milk used by Frings was probably about 0.006 *M* (Diem, 1962). This concentration would have suppressed the water taste, at least in part.

4.3. Rat

The sodium content of whole-mouth saliva in the rat is probably about 0.005–0.01 *M* (Hainsworth and Stricker, 1971). Although the rat chorda tympani appears to have few, if any, fibers sensitive to water following NaCl, the rat glossopharyngeal nerve does have such fibers (Frank, 1968). An ingenious psychophysical study of the rat gives some insight into the taste of water to the normal, behaving rat. Morrison (1967) trained food-deprived rats to use the taste of a solution as a cue in order to press the correct bar for a food reinforcement. The rats were trained on pairs of stimuli.

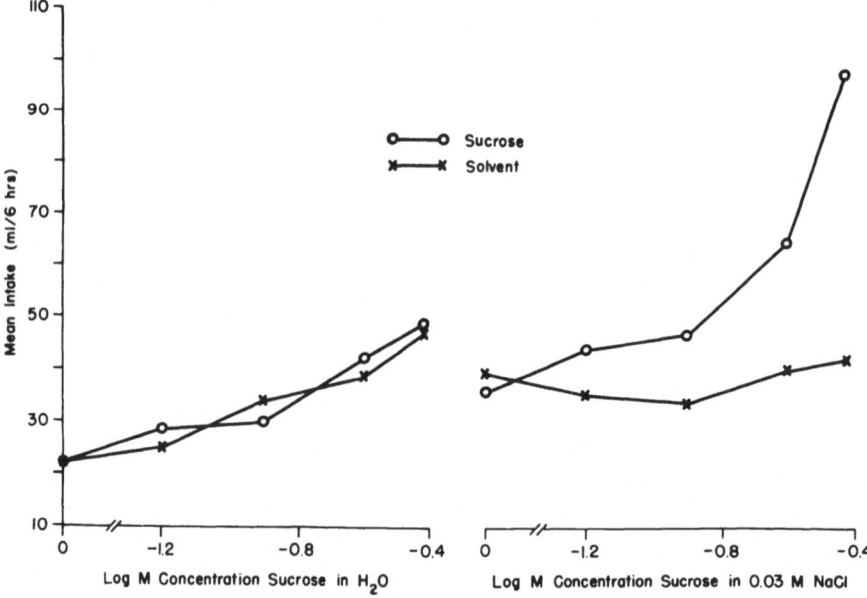

Figure 13. Two-bottle preference data from the cat. The functions on the left show the intake of water (*X*s) versus sucrose dissolved in water (circles). The functions on the right show the intake of 0.03 *M* NaCl (*X*s) versus sucrose dissolved in 0.03 *M* NaCl (circles). (Modified from Bartoshuk, Harned, and Parks, 1971. Copyright 1971 by the American Association for the Advancement of Science.)

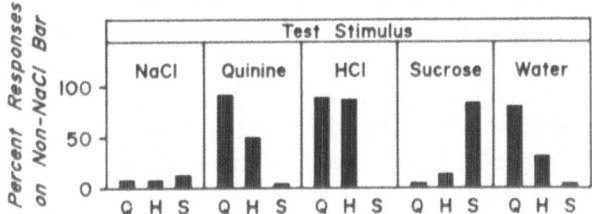

Figure 14. Profiles of the taste qualities of NaCl, quinine (Q), HCl (H), sucrose (S), and water for the rat adapted to its own saliva. Similar profiles suggest similar qualities. That is, water tastes predominantly like quinine to the rat adapted to its own saliva. (Modified from Morrison, 1967. Copyright 1967 by the American Psychological Association. Reprinted by permission.)

For example, one group of rats was presented with either 0.1 M NaCl or 0.2 M sucrose. On trials when the NaCl was the cue solution, one of the bars would provide a food reinforcement. On trials when the sucrose was the cue solution, the other bar would provide the reinforcement. When a new solution was introduced, the rats would then "describe" the similarity of its quality to NaCl and sucrose by choosing the appropriate bar. Morrison trained a second group on 0.1 M NaCl versus 0.0005 M quinine sulphate and a third group on 0.1 M NaCl versus 0.01 M hydrochloric acid (HCl). By combining the responses of each group to the non-NaCl bar, Morrison could construct profiles of taste quality. Figure 14 shows the profiles for NaCl, quinine, HCl, sucrose, and water. There were few responses on the non-NaCl bar when NaCl was the test stimulus. Most responses were to the sucrose bar when sucrose was the test stimulus, and most responses were to quinine when quinine was the test stimulus. When HCl was the test stimulus, the rats responded to both the quinine and the HCl bars, suggesting that HCl is somewhat similar to quinine. When water was the test stimulus, the rats produced a profile very similar to that produced by quinine. That is, the rat appears to be similar to man in that water following saliva is predominantly bitter.

There are a number of experiments in which salivation is experimentally disrupted (see Chapter 7). In order to determine exactly what effects this should have on the taste of water and of solutes, we would need to know what is on the tongue of the desalivated animal. However, some general observations about the probable effects of desalivation are possible, given what is known about taste adaptation. The concentrations of solutes in normal saliva are relatively low. These concentrations are important for water taste and for threshold concentrations of taste substances but are not very important for moderate and high concentrations of taste substances because of the recruitment that occurs with adaptation (McBurney, 1966). *Recruitment* refers to the fact that the effects of adaptation disappear as the concentration tested increases above the adapting concentration. Figure 1 shows recruitment. Note that adapting to 0.01 *M* NaCl affects the threshold for NaCl dramatically but that the function for the saltiness of NaCl concentrations above 0.01 *M* is hardly changed. Thus, unless desalivation were to markedly increase the concentration of solutes on the tongue, we would not expect this procedure to change the tastes of moderate and high concentrations of taste substances very much.

Variations in rat saliva probably do not affect taste substances of moderate and high intensity because of recruitment, as described above. However, certain variations could have very important effects on water taste and thresholds. Changes in the taste of water would be particularly important in experiments in which water intake is to be measured. Simply increasing the flow rate of saliva increases the concentration of Na and Cl (Schneyer and Schneyer, 1967). A variety of drugs are known to affect salivation as well. If experiments include operations that could affect salivation and if water intake or taste thresholds are of interest, then the possibility of taste effects induced by the salivary changes must be considered.

5. Implications for Research: How to Control for Water Tastes

Water tastes can complicate taste experiments in a variety of ways. As mentioned above, some published thresholds for NaCl appear to represent water-taste thresholds rather than solute thresholds (Bartoshuk, 1974). Detection thresholds for substances other than NaCl can also be affected by water tastes, since water is the solvent in which the threshold substances are dissolved.

Water tastes have also complicated the study of taste contrast. This phenomenon was first studied in the 19th century. We now believe that the early data represented not true contrast but rather water tastes. *Contrast* refers to the intensification of a given taste quality by prior exposure of the tongue to a different taste quality. We now call this phenomenon *cross-enhancement* because of its similarity to cross-adaptation. The following is an example of cross-enhancement. Tasting acid makes subsequently tasted sucrose solutions taste sweeter than normal. However, one cannot conclude from this that the sweetness of sucrose itself has been intensified. The tasting of the acid makes water taste sweet and the intensification of the sweet taste of the sucrose solution appears to represent the addition of the sweet water-taste to the sweet sucrose-taste rather than intensification of the sweet taste of sucrose. In a study controlling for water tastes, cross-enhancement did not occur (McBurney and Bartoshuk, 1973).

There are basically two ways to avoid incorrect inferences due to water tastes. First, when possible, the tongue can be rinsed with water prior to the tasting of the test solution. This rinsing ensures that any taste in the test solution comes from the solute. Unfortunately, rinsing is often difficult, especially when one is working with species other than man. The second basic way to eliminate errors due to water tastes is to know the adaptation state of the tongue and to allow for it. For example, one can measure NaCl thresholds accurately when the subject is adapted to saliva. One easy way is to use a recognition criterion so that solute and water tastes can be sorted out on the basis of quality. A detection criterion can also work, with a bit more effort. The actual contents of saliva can be measured so that the experimenter knows if the test stimuli are above or below the adaptation concentration. In addition, if a single test stimulus were presented on each trial, then there would be two thresholds found in ascending or descending series: the lower one would be that of water taste, the upper one that of NaCl taste. When a saliva baseline is used, the most important consideration is to make certain that the baseline is reestablished before *each* solution is tasted. Two minutes of rest following a thorough water rinse to remove the prior stimulus usually ensures a relatively stable saliva baseline.

The most confusing situation is produced when the adaptation state varies throughout the experiment. This confusion has been the problem with the Richter two-stimulus technique in the way it has been used in much research. The subject may be adapted to saliva, either stimulus, or a combination of these on each trial. The Richter procedure can be correctly used with a rinse or rest period before the tasting of each stimulus. However, there are also several single-stimulus techniques that provide easier control over the adaptation conditions. Morrison's procedure (1967) and the single-stimulus technique of Weiner and Stellar (1951) can both be used with appropriate delays to ensure a stable saliva baseline. Shaber, Brent, and Rumsey (1970) have devised a very useful procedure that requires a rat to sample several water stimuli before tasting a test stimulus. This kind of procedure offers a way to provide the adaptation condition of choice experimentally.

6. Summary

Water is not intrinsically tasteless. Water can taste sweet, sour, bitter, or salty, depending on the nature of the substance that precedes it. Taste responses to water in species other than man are also contingent on the nature of the preceding solution. The sweet taste of water that some indi- viduals notice after eating globe artichokes is an example of a water taste induced by exposure of the tongue to certain salts in the artichoke. A more commonly observed water taste is the bitter or flat taste produced by distilled water. This water taste is produced by adaptation to saliva. This saliva-induced water taste has important behavioral consequences for man as well as for other species. In some experiments, taste responses produced by water have been attributed to solutes dissolved in the water. One can avoid the confusions produced by water tastes either by the rinsing of saliva from the tongue before stimuli are tasted or by determining the contents of saliva and allowing for their effects.

References

Aducco, V., and Mosso, U., 1886, Richerche sopra la fisiologia de gusto. *G. Accad. Med. Torino* **34**:39–42.

Akaike, N., and Yamada, F., 1965, The depressant effect of calcium ions on the taste response of the frog, *Kumamoto Med. J.* **18**:169–170.

Altman, P. L., and Dittmer, D. S., 1961, "Blood and Other Body Fluids," Federation of American Societies for Experimental Biology, Washington, D.C.

Anderson, R. J., 1959, The taste of water, *Am. J. Psychol.* **72**:462–463.

Bartoshuk, L. M., 1965, Effects of adaptation on responses to water in cat and rat, doctoral dissertation, Brown University (University Microfilms, Ann Arbor, Mich., No. 65-13, 630).

Bartoshuk, L. M., 1968, Water taste in man, *Percep. Psychophys.* **3**:69–72.

Bartoshuk, L. M., 1974, NaCl thresholds in man: Thresholds for water taste or NaCl taste? *J. Comp. Physiol. Psychol.* **87**:310–325.

Bartoshuk, L. M., and Frank, M., 1972, Taste of water: Neural recordings from rat, hamster, cat, and squirrel monkey, paper presented at the meeting of the Society for Neuroscience, Houston, October 1972.

Bartoshuk, L. M., Harned, M. A., and Parks, L. H., 1971, Taste of water in the cat: Effects on sucrose preference, *Science* **171**:699–701.

Bartoshuk, L. M., Lee, C. H., and Scarpellino, R., 1972, Sweet taste of water induced by artichoke (*Cynara scolymus*), *Science* **178**:988–990.

Bartoshuk, L. M., McBurney, D. H., and Pfaffmann, C., 1964, Taste of sodium chloride solutions after adaptation to sodium chloride: Implications for the "water taste," *Science* **143**:967–968.

Bartoshuk, L. M., and Pfaffmann, C., 1965, Effects of pre-treatment on the water taste response in cat and rat, *Fed. Proc.* **24**:207.

Beidler, L. M., 1953, Properties of chemoreceptors of tongue of rat, *J. Neurophysiol.* **16**:595–607.

Beidler, L. M., 1955, The physiological basis of taste, *in* "Symposium on Physiological Psychology," Office of Naval Research, Pensacola, Fla., p. 1.

Brown, W., 1914, The judgment of very weak sensory stimuli, *U. Calif. Pub. Psychol.* **1**:199–268.

Camerer, W., 1870, Über die Abhängigkeit des Geschmacksinns von der gereizten Stelle der Mundhöhle, *Zeit. Biol.* **6**:440-452.

Carpenter, J. A., 1956, Species differences in taste preferences, *J. Comp. Physiol. Psychol.* **49**:139-144.

Cohen, M. J., Hagiwara, S., and Zotterman, Y., 1955, The response spectrum of taste fibres in the cat: A single fibre analysis, *Acta Physiol. Scand.* **33**:316-332.

Diamant, H., Funakoshi, M., Ström, L., and Zotterman, Y., 1963, Electrophysiological studies on human taste nerves, *in* "Olfaction and Taste," I (Y. Zotterman, ed.), Macmillan, New York, p. 193.

Diamant, H., Oakley, B., Ström, L., Wells, C., and Zotterman, Y., 1965, A comparison of neural and psychophysical responses to taste stimuli in man, *Acta Physiol. Scand.* **64**:67-74.

Diem, K. (ed.), 1962, "Scientific Tables" (6th ed.) Geigy Pharmaceuticals, Ardsley, N.Y., p. 516.

di Sant'Agnese, P., Grossman, H., Darling, R. C., and Denning, C. R., 1958, Saliva, tears and duodenal contents in cystic fibrosis of the pancreas, *Pediatrics* **22**:507-514.

Erickson, R. P., 1963, Sensory neural patterns and gustation, *in* "Olfaction and Taste," I (Y. Zotterman, ed.), Pergamon Press, New York, p. 205.

Frank, M., 1968, Single fiber responses in the glossopharyngeal nerve of the rat to chemical, thermal, and mechanical stimulation of the posterior tongue, doctoral dissertation, Brown University (University Microfilms, Ann Arbor, Mich., No. 69-9957).

Frank, M., 1973, An analysis of hamster afferent taste nerve response functions, *J. Gen. Physiol.* **61**:588-618.

Frank, M., 1974, The classification of mammalian afferent taste nerve fibers, *Chem. Sen. Flav.* **1**:53-60.

Frank, M., 1975, Response patterns of rat glossopharyngeal taste neurons, *in* "Olfaction and Taste," V (D. A. Denton and J. P. Coghlan, eds.), Academic Press, New York, p. 59.

Frawley, T. F., and Thorn, G. W., 1951, The relation of the salivary sodium-potassium ratio to adrenal cortical activity, *in* "Proceedings of the Second Clinical ACTH Conference," (J. R. Mote, ed.), Blakiston, New York.

Frings, H., 1951, Sweet taste in the cat and the taste-spectrum, *Experientia* **7**:424-426.

Gruner, O. C. 1930, "A Treatise on the Canon of Medicine of Avicenna," Luzac and Co., London.

Hahn, H., 1949, "Beiträge zur Reizphysiologie," Scherer, Heidelberg.

Hainsworth, F. R., and Stricker, E. M., 1971, Evaporation cooling in the rat: Differences between salivary glands as thermoregulatory effectors, *Can. J. Physiol. Pharmacol.* **49**:573-580.

Hammond, W. A., 1902, "Aristotle's *de Anima*," Macmillan, New York.

Hellekant, G., 1965a, Electrophysiological investigation of the gustatory effect of ethyl alcohol I, *Acta Physiol. Scand.* **64**:392-397.

Hellekant, G., 1965b, Electrophysiological investigation of the gustatory effect of ethyl alcohol II, *Acta Physiol. Scand.* **64**:398-406.

Henkin, R. I., and Powell, G. F., 1962, Increased sensitivity of taste and smell in cystic fibrosis, *Science* **138**:1107-1108.

Henle, J., 1880, Über den Geschmackssinn, *Anthrop. Vor.* **2**:1-24.

Hertz, J., Cain, W. S., Bartoshuk, L. M., and Dolan, T. F., 1975, Olfactory and taste sensitivity in children with cystic fibrosis, *Physiol. Behav.* **14**:89-94.

Kiesow, F., 1894, Beiträge zur physiologischen Psychologie des Geschmackssinnes, *Phil. Stud.* **10**:523-561.

Liljestrand, G., and Zotterman, Y., 1954, The water taste in mammals, *Acta Physiol. Scand.* **32:**291-303.

McBurney, D. H., 1966, Magnitude estimation of the taste of sodium chloride after adaptation to sodium chloride, *J. Exp. Psychol.* **72:**869-873.

McBurney, D. H., 1969, Effects of adaptation on human taste function, *in* "Olfaction and Taste," III (C. Pfaffmann, ed.), Rockefeller University Press, New York.

McBurney, D. H., and Bartoshuk, L. M., 1973, Interactions across stimuli with different taste qualities, *Physiol. Behav.* **10:**1101-1106.

McBurney, D. H., and Pfaffmann, C., 1963, Gustatory adaptation to saliva and sodium chloride, *J. Exp. Psychol.* **65:**523-529.

McBurney, D. H., and Shick, T. R., 1971, Taste and water taste of twenty-six compounds for man, *Percept. Psychophys.* **10:**249-252.

Morrison, G. R., 1967, Behavioral response patterns to salt stimuli in the rat, *Can. J. Psychol.* **21:**141-152.

Nagel, W. A., 1896, Über die Wirkung des Chlorsauren Kalis auf den Geschmackssinn, *Z. Psychol. Physiol. Sinnesorg.* **10:**235-239.

Nomura, H., and Ishizaki, M., 1972, Stimulation mechanism of water response in the frog: Roles of anions in the activity of a chemoreceptor, *Bull. Tokyo Dent. Coll.* **13:**21-52.

Nomura, H., and Sakada, S., 1965, On the "water response" of frog's tongue, *Jap. J. Physiol.* **15:**433-443.

Ogawa, H., Yamashita, S., Noma, A., and Sato, M., 1972, Taste responses in the Macaque monkey chorda tympani, *Physiol. Behav.* **9:**325-331.

Öhrwall, H., 1891, Untersuchungen über den Geschmackssinn, *Skand. Arch. Physiol.* **2:**1-69.

Pfaffmann, C., 1941, Gustatory afferent impulses, *J. Cell. Comp. Physiol.* **17:**243-258.

Pfaffmann, C., 1955, Gustatory nerve impulses in rat, cat, and rabbit, *J. Neurophysiol.* **18:**429-440.

Pfaffmann, C., 1974, Specificity of the sweet receptors in the squirrel monkey, *Chem. Sen. Flav.* **1:**61-67.

Pfaffmann, C., and Bare, J. K., 1950, Gustatory nerve discharges in normal and adrenalectomized rats, *J. Comp. Physiol. Psychol.* **43:**320-324.

Pfaffmann, C., Frank, M., Bartoshuk, L. M., and Snell, T. C., 1975, Coding gustatory information in the squirrel monkey chorda tympani, *in* "Progress in Physiological Psychology," VI (O. Sprague and A. Epstein, eds.), Academic Press, New York.

Pfaffmann, C., and Powers, J. B., 1964, Partial adaptation of taste, *Psychon. Sci.* **1:**41-42.

Richter, C. P., and MacLean, A., 1939, Salt taste threshold of humans, *Am. J. Physiol.* **126:**1-6.

Sarton, G., 1931, "Introduction to the history science," II, Williams and Wilkins, Baltimore, Md., p. 29.

Schneyer, C. H., and Schneyer, C. A., 1967, Inorganic composition of saliva, *in* "Handbook of Physiology" (Sect. 6. Alimentary Canal, Vol. 2. Secretion), (W. Heidel, ed.), American Physiological Society, Washington, D.C.

Shaber, G. S., Brent, K. L., and Rumsey, J. A., 1970, Conditioned suppression taste thresholds in the rat, *J. Comp. Physiol. Psychol.* **73:**193-201.

Skramlik, E. von, 1922, Mischungsgleichungen im Gebiete des Geschmackssinnes, *Z. Psychol. Physiol. Sinnesorg.* **53:**36-78.

Smith, D. V., and Frank, M., 1972, Cross-adaptation between salts in the chorda tympani nerve of the rat, *Physiol. Behav.* **8:**213-220.

Weiner, I. H., and Stellar, E., 1951, Salt preference of the rat determined by a single-stimulus method, *J. Comp. Physiol. Psychol.* **44:**394-401.

Wotman, S., Mandel, I. D., Khotim, S., Thompson, R. H., Kutscher, A. H., Zegarelli, E. V., and Denning, C. R., 1964, Salt taste thresholds and cystic fibrosis, *Am. J. Dis. Child.* **108**:372–374.

Yamamoto, T. and Kawamura, Y., 1974, An off-type response of the chorda tympani nerve in the rat, *Physiol. Behav.* **13**:239–243.

Zotterman, Y., 1949, The response of the frog's taste fibres to the application of pure water, *Acta Physiol. Scand.* **18**:181–189.

Zotterman, Y., 1956, Species differences in the water taste, *Acta Physiol. Scand.* **37**:60–70.

Zotterman, Y., and Diamant, H., 1959, Has water a specific taste? *Nature* **183**:191–192.

9

Schedule-Induced Polydipsia: The Role of Orolingual Factors and a New Hypothesis

William J. Freed, Ronald F. Zec, and Joseph Mendelson

1. Introduction

1.1. Schedule-Induced Polydipsia (SIP)

When a food-deprived animal is intermittently fed small amounts of food, it rapidly develops a tendency to ingest water immediately following the ingestion of each morsel (Falk, 1961, 1969). An appropriate choice of various parameters can lead these animals to ingest greatly excessive quantities of water. In one experiment, female rats with a mean free-feeding weight of 264 g were deprived of food until their body weights dropped to 70–80% of normal. When they were given the opportunity to bar-press for 45-mg dry food pellets on a variable-interval 1-min schedule, they ingested a mean of

Willam J. Freed • The Laboratory of Clinical Psychopharmacology, National Institute of Mental Health, William A. White Building, St. Elizabeth's Hospital, Washington, D.C. *Ronald F. Zec and Joseph Mendelson* • Department of Psychology, University of Kansas, Lawrence, Kansas. The authors' research reported in this paper was supported in part by grants to Joseph Mendelson from the National Institute of Mental Health, U.S. Public Health Service, U.S. Department of Health, Education, and Welfare (MH-14410 and MH-21955), the National Science Foundation (GB7370), and the University of Kansas General Research Fund (3080-5038 and 3582-5038) and by a Biomedical Sciences Support Grant (RR-07037) to the University of Kansas from the U.S. Public Health Service.

92.5 ml of water per 3.17-hr session (Falk, 1961). Such quantities of intake are greatly excessive whether compared to the normal 24-hr intake or to the amount of water that would be ingested if the rats were allowed to eat the same amount of food *ad libitum*. This phenomenon, which is referred to as *schedule-induced polydipsia* or by the acronym *SIP* (Falk, 1964), can be explained neither in terms of traditional behavioral phenomena, such as adventitious reinforcement or timing behavior (Falk, 1969), nor in terms of water-regulatory variables, such as impaired renal concentrating ability or *de facto* water deprivation (Falk, 1969; Stricker and Adair, 1966).

1.2. The Dry-Mouth Theory

One theory has maintained that mouth dryness plays a crucial role in generating SIP (*cf.* Chapman and Epstein, 1970; Hawkins, Schrot, Githens, and Everett, 1972; Kissileff, 1973; Stein, 1964; Stricker and Adair, 1966; Teitelbaum, 1966). Teitelbaum (1966) expressed one "dry mouth" explanation as follows: "Since feeding causes dehydration by pulling water into the stomach from the tissues, this should produce a state of thirst accompanied by a dry mouth. Chewing and swallowing a dry pellet may be difficult for a thirsty animal, so he drinks to help wash each pellet down" (p. 589). A variant of this notion was proposed by Stricker and Adair (1966): "a given amount of fluid is required to eliminate the postprandial oral effects of each pellet (a 'gargle effect' of drinking)" (p. 453). Neither of these explanations can provide a full account of SIP; Falk (1969, 1971) has discussed evidence that argues against such notions, and other investigators (McKearny, 1973; Mendelson and Chillag, 1970; Murphy and Brown, 1975) who have tested these hypotheses have not found support for them. For example, liquid reinforcers sustain SIP, preloading the animal with moderately large amounts of water does not attenuate SIP, and the effects of variations in the feeding schedule are inconsistent with a dry-mouth interpretation (Falk, 1969; also *cf.* Sections 2.3, 4.1, 5.1, 6.3, and 8.1).

Thus, it would appear that there is little support for the notion that SIP is reinforced either by a mouth-moistening or by a washing effect of water. However, many investigators have suggested that SIP is influenced by orolingual stimulation that is provided by either water or food (Carlisle and Laudenslager, 1976; Chillag and Mendelson, 1971; Falk, 1969; Freed and Mendelson, 1977; Gilbert, 1974c; Jacobs, 1969; Mendelson and Chillag, 1970). The purpose of this chapter is to examine the evidence pertaining to these suggestions, specifically, to examine the role of orolingual sensory stimulation in schedule-induced polydipsia.

1.3. Synopsis and Orientation

Explanations of SIP in terms of dryness of the mouth and subsequent moistening of the mouth by water have previously been dismissed (Falk, 1969). But other orolingual effects, many of which have been systematically studied, might influence the SIP phenomenon. In this chapter, we review the literature on this topic.

Section 2 of this chapter briefly reviews the general characteristics of SIP. In Section 3, we discuss the question of whether the durations or the individual bouts of drinking are regulated by the nature of the sensory feedback obtained from drinking. In subsequent sections, we examine the effects of desalivation, tongue denervation, properties of the drinking solution, and orolingual cooling in terms of the maintenance of drinking and licking behaviors. Section 7 consists of a review of orolingual factors related to food reinforcers, such as the dryness of the food, the sugar content of the food, and the possibility of the substitution of nonorolingual stimuli for food. In Section 8, we outline a new theory of SIP that suggests that excessive insulin secretion is conditioned to peripheral stimuli associated with intermittent food reinforcement. We then discuss how this theory can account for various heretofore-unexplained features of the SIP phenomenon. In conclusion, we summarize the current state of the literature on the control of SIP in terms of the way in which peripheral variables may function as reinforcers, regulators, predisposers, motivators, and initiators of schedule-induced polydipsia.

2. *The General Characteristics of SIP*

2.1. Food Deprivation and Hunger

There is a direct relationship between the degree of food-deprivation–induced deficit in body weight and the degree of schedule-induced drinking or airlicking. Falk (1969, 1971) reported that as the weight of food-deprived rats was gradually allowed to increase to levels above 80% of their free-feeding weight, SIP was essentially invariant until 95%; then, as body weight increased beyond 95%, there was a gradual decrease in SIP until little SIP was evidence at 105%. This decrement occurred even though bar pressing for 45-mg Noyes pellets on a fixed-internal 90-sec schedule of reinforcement did not diminish substantially until the animals reached 104–109% of their initial free-feeding weights. E. X. Freed and Hymowitz (1972) found that the acquisition and maintenance of SIP, under a fixed-time (FT) 30- or 60-

sec schedule with 45-mg Noyes pellets, were greatly enhanced in animals maintained at 60% or 70% as compared to 80% or 90% of their free-feeding weights. In another study (Chillag and Mendelson, 1971), it was found that schedule-induced airlicking declined substantially when rats' body weights were allowed to rise from 80% to 90% of their original free-feeding levels, with 45-mg Noyes pellets delivered noncontingently on an FT 1-min schedule (also see Chapter 4). Thus, the magnitude of schedule-induced drinking and airlicking is an increasing function of the severity of food-deprivation–induced deficit in body weight.

We have recently obtained evidence indicating that the effect of hunger level on SIP is not mediated through its influence on (1) the strength of the feeding response or (2) the reinforcing potency of feeding (Bruce-Wolfe, W. J. Freed, and Mendelson, 1976; Bruce-Wolfe, 1976). Feeding-inducing hypothalamic stimulation was found to engender SIP in satiated rats only when it was delivered within several seconds following the ingestion of each food pellet; such stimulation delivered only during the ingestion of each pellet or more than 14 sec thereafter was found to be ineffective in inducing drinking. It is clear that in the context of certain intermittent-feeding situations, brain-stimulation hunger directly motivates SIP. Whether deprivation-induced hunger engenders SIP in a similar fashion is an open question.

2.2. Type of Reinforcement Schedule

SIP is effectively engendered by several schedules of reinforcement, that is variable-interval (VI; Falk, 1961), fixed-interval (FI; Falk, 1966c), fixed-ratio (FR; Carlisle, 1971; Burks, 1970), differential reinforcement-of-low-rate (DRL; Segal and Deadwyler, 1965; Githens, Hawkins, and Schrot, 1973), multiple (Jacquet, 1972), and tandem (McLeod and Gollub, 1976) schedules. It is the food reinforcement itself, rather than its contingency upon bar pressing, that is important for the emergence of SIP, as fixed-rate and variable-rate noncontingent food-delivery schedules (i.e., fixed-time and variable-time, or FT and VT, schedules) are similarly effective (Falk, 1964, 1971; Hawkins *et al.*, 1972; Segal, Oden, and Deadwyler, 1965).

2.3. The Degree of Reinforcement Intermittency

A number of investigators have examined the relationship between the rate of delivery of food pellets and the degree of SIP, virtually all of them having observed that an intermediate range of feeding intermittencies is most effective in engendering SIP. High rates of delivery of food pellets are ineffective (Falk, 1966c; Keehn and Colotla, 1971), possibly because the feeding response competes successfully with drinking under these conditions (Cohen, 1975). Very low rates of pellet ingestion are also less effective in terms of the amount of water ingested per pellet (Bond, 1973; Falk, 1966c, 1971; Flory, 1971; Hawkins *et al.*, 1972; Keehn and Colotla, 1971), but there is some dispute about the minimal rate of pellet delivery that no longer engenders SIP. Falk (1966c, 1971) and others (Hawkins *et al.*, 1972; Keehn and Colotla, 1971) found that SIP was drastically reduced or eliminated when the rate of pellet delivery was decreased to one per 5–6 min, but a significant amount of SIP was observed by Flory (1971) even when pellets were delivered at a rate as low as one per 8 min. However, Flory (1971) did observe that SIP was substantially less when pellet-delivery rates were one per 8 min than when they were one per 1–2 min.

Falk (1969) has suggested that the effects of rate of pellet delivery are actually attributable to effects of consummatory rate, because increases in the quantity of food reinforcements are associated with decreased SIP when high rates of reinforcement are used but are associated with increased SIP when low rates are used. However, subsequent investigations have not entirely supported this suggestion (Bond, 1973; Flory, 1971).

In essence, a number of investigations have demonstrated that it is a cardinal feature of SIP that very low or very high rates of food-pellet delivery induce less polydipsia than do intermediate rates.

2.4. The Temporal Locus of Drinking

The observation that SIP is almost entirely postpellet in its temporal locus (Falk, 1971) left open the possibility that there might be something peculiar about the immediate postpellet interval that is effective in predisposing drinking. However, subsequent research has revealed that this is not the case; when water is unavailable in the immediate postpellet period, similar amounts of drinking will occur when water becomes available later on in the interpellet interval (IPI; Flory and O'Boyle, 1972; Gilbert, 1974c).

2.5. Drinking as Reinforcement during Intermittent Food Reinforcement

Falk (1966a) demonstrated that rats that were bar-pressing for food on a VI 1-min schedule would concurrently bar-press for 0.1 ml portions of water on FR schedules. One of two animals studied by Falk continued to emit as many as 50 successive responses and the other rat went to FR 20.

Subsequently, Cohen (1975) found that rats prefer a schedule component with water available to a component during which no water is available.

2.6. Species Generality

As well as in the rat, SIP has been demonstrated in the mouse (Ogata, Ogata, Jack H. Mendelson, and Mello, 1972; Palfai, Kutscher, and Symons, 1971), in the rhesus monkey (Porter and Kenshalo, 1974; Schuster and Woods, 1966), and in one pigeon (Shanab and Peterson, 1969). Guinea pigs do not exhibit SIP, but these animals exhibit robust schedule-induced oral palpation of cold metal (see Chapter 4 and Section 6.2). SIP was not evident in food-deprived hamsters while they were bar-pressing for 20-mg food reinforcements on a VI 1-min schedule (Wilson and Spencer, 1975). S. H. Hobbs (personal communication, 1976[1]) also failed to observe SIP in hamsters maintained at 75%, 80%, or 90% of *ad libitum* weights and given 45-mg Noyes pellets on FT 90-sec or FI 90-sec schedules. Although the animals occasionally pouched rather than ate the pellets, the use of mals occasionally pouched rather than ate the pellets, the use of powdered meal to eliminate pouching did not result in the emergence of SIP.

Kachanoff, Leveille, McLelland, and Wayner (1973) observed that two of seven schizophrenic patients, intermittently reinforced with pennies, occasionally drank moderately large amounts of water (about 400–600 ml/ 30-min session for one of the subjects). Although this is hardly enough water consumption to be termed polydipsia, it does indicate that a phenomenon similar to SIP may be demonstrable in humans.

2.7. Acquisition

SIP does not appear immediately upon the introduction of a rat to an SIP-engendering situation but develops gradually over the course of a few testing sessions (Reynierse and Spanier, 1968). Hymowitz, E. X. Freed, and Lester (1970) have found that SIP develops gradually even if bar pressing for food has been allowed to reach asymptotic levels before water is introduced into the testing situation. Therefore, the developmental phase of SIP does not correspond to the time required for adaptation to the schedule of feeding or to the learning of the contingencies of reinforcement.

Hymowitz and Koronakos (1968) found that the acquisition of SIP was essentially normal in animals that had been raised without ever having simultaneous access to food and water. Thus, developmental experience with prandial drinking is not essential for SIP.

---

[1] S. H. Hobbs, Department of Psychology, Augusta College, Augusta, Ga.

3. Is SIP Controlled by Sensory Feedback Associated with Water Ingestion?

Are the durations of bouts of drinking during SIP controlled by water-associated (primarily exteroceptive) or movement-associated (primarily proprioceptive) sensory feedback, or are they independent of feedback and merely the consequence of the discharge of a fixed amount of motor activity? The bouts of drinking during each SIP session tend to be of rather constant duration (*cf.* W. J. Freed, Mendelson, and Bramble, 1976). Accordingly, three possibilities immediately present themselves as candidates for the control of individual drinking-bout durations: (1) drinking is terminated by a certain amount of that portion of the sensory feedback obtained as a consequence of the fact that licking movements are successful in bringing a certain volume of water into the organism, for example, by water-associated exteroceptive feedback; (2) drinking is terminated by a certain amount of feedback from the *in vacuo* motor response, that is, the feedback that would be obtained if the motor response were run off in the absence of an appropriate reinforcer like water or an airstream; and (3) the postpellet bout of drinking is a centrally patterned motor phenomenon that is run off whenever the appropriate releasing conditions (water and the scheduling conditions) are present. (See Hinde, 1970, pp. 26–49, for a detailed discussion of the distinction between these latter two possibilities.)

Various theories of SIP might give different answers to this question. On the one hand, theories that emphasize the motor aspects of the SIP phenomenon (Cook and Singer, 1976; Singer, Wayner, Stein, Cimino, and King, 1974; Wayner, 1970, 1974a,b; Wayner and Greenberg, 1973; Wayner, Greenberg, Fraley, and Fischer, 1973), emotional induction (Segal, 1972), arousal (Killeen, 1975), or frustration (Denney and Ratner, 1970; Kissileff, 1973; *cf.* Stricker and Adair, 1966) might stress the importance of central programming or feedback from the muscles. However, since schedule-induced licking does not take place *in vacuo* (Stein, 1964) and rats will bar-press for water during SIP (Falk, 1966a), it is necessary also to implicate water or an airstream as a releaser for the performance of reinforcing licking behavior. One example of such a theory is the following:

> A rat drinks not because of a need for water or thirst or because some drinking center in the brain is activated but initially due to a general increase in motor excitability mediated via the LH [lateral hypothalamus] and the incidental chance occurrence of making contact with the drinking device and the first burst of licking. The sensory feedback from this interaction seems essential for drinking to continue and to occur again under similar circumstances in the future. (Wayner, 1974b, p. 392)

However, Carlisle and Laudenslager (1976) have found that water temperature affects the duration of postpellet drinking bouts; the reinforcing and
satiating value of water may depend on its temperature (Chapter 6;
Ramsauer, Mendelson, and W. J. Freed, 1974), and this finding might
therefore indicate that drinking-bout durations are regulated by sensory
feedback associated with water ingestion.

We attempted to conduct an additional test that could rule out at least
one of the three possible controlling factors by introducing variability into
the rate of water ingestion during SIP through the use of drinking tubes
with orifices of different diameters. It had previously been found that
diminishing the diameter of drinking-tube orifices to 1 mm was effective in
reducing rats' water-ingestion rates as compared to their rates of ingestion
from drinking tubes bearing 4-mm apertures (Wike and Barrientos, 1957).
Usually SIP is characterized by the occurrence of postpellet bouts of drinking that are relatively constant in duration. We sought to determine whether
reductions in the rate of water ingestion that are brought about by a
decrease in the diameters of drinking-tube orifices would lead to increases
in the durations of postpellet bouts of drinking. We expected one of three
possible results:

1. Postpellet drinking-bout durations might be unaffected by ingestion
   rate, so that ingested volumes would be proportional to ingestion
   rate. In this case (perfect-time regulation), we would be able to conclude that each drinking bout is either a centrally patterned motor
   phenomenon or is terminated by a type of water-volume-inde-
   pendent sensory feedback.
2. The duration of postpellet bouts of drinking might be inversely proportional to the rate of water ingestion, so that ingestion volumes
   are maintained at constant levels. In this case (perfect-volume regulation), we would be able to conclude that the magnitude of each
   drinking bout is entirely under the control of feedback that is
   obtained by virtue of the fact that a certain volume of water is being
   consumed.
3. Some combination of 1 and 2.

The results of these experiments were overwhelmingly in support of the
view that SIP is regulated by sensory feedback from water ingestion (W. J.
Freed *et al.*, 1976; Freed and Mendelson, 1977). Each of the animals tested
regulated volume of intake more precisely than it regulated time spent
drinking (Figure 1). The animals maintained fairly precise degrees of
accuracy in their adjustments of drinking durations in order to maintain
constant volumes of fluid intake. The overall mean-volume regulation was
calculated to be 103%, while the time regulation was 1.3%. Reductions in

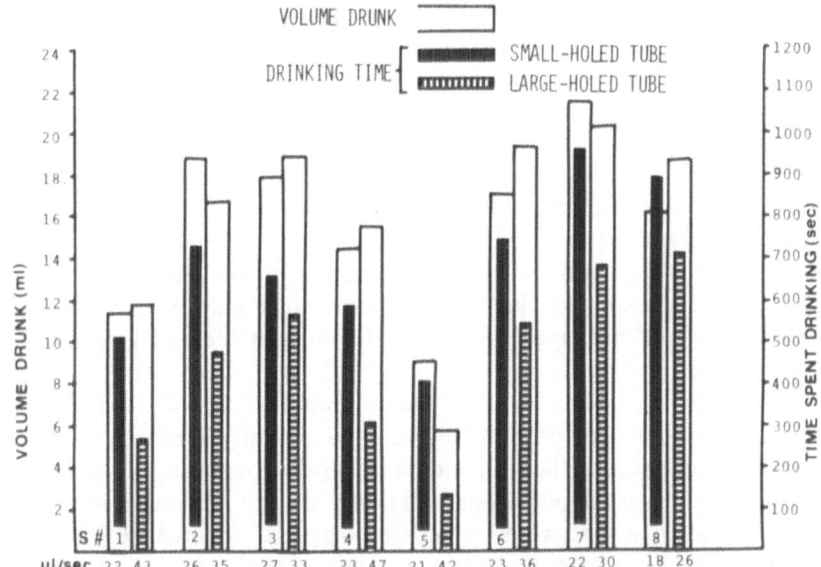

Figure 1. Mean volume ingested (open bars) and time spent drinking (black and striped bars) per 45-min SIP session for each rat. Testing sessions with the larger-holed drinking tubes available are represented by striped drinking-time bars, sessions with the smaller-holed drinking tubes by solid black drinking-time bars. Each rat's mean rate of water ingestion for each tube is indicated beneath the bar; the rat's identity number is indicated on the bottom portion of each pair of bars. (From W. J. Freed, Mendelson, and Bramble, 1976. Copyright 1976 by Academic Press, Inc. Reprinted by permission.)

rates of ingestion led to increases in drinking durations in 90% of the sessions but led to decreases in intake volumes in only 43% of the testing sessions.

The increases in drinking durations that were brought about by decreases in rates of ingestion were usually manifest in terms of increases in the durations of individual drinking bouts, but in half of the animals (those that drank during less than 90% of the IPIs), significant increases in the proportion of IPIs during which drinking occurred were noted when the drinking tubes with small apertures were used. Overall, about 81% of the total regulation of intake volumes was due to increases in the durations of individual drinking bouts, while the remainder was due to increases in the number of IPIs during which some drinking occurred. The accuracy of regulation of intake volumes during SIP was similar to that observed after water deprivation with no food available during drinking but more precise than that observed after water deprivation when food was simultaneously available during testing.

Several observations suggest that orolingual feedback was involved in the regulation of intake volumes by these animals: (1) differences in drinking-bout durations were noticeable even at the beginnings of sessions and immediately after drinking-tube apertures were changed halfway through the sessions; and (2) when diminished-aperture and normal drinking tubes were switched after each drinking bout, or in a nonsystematic fashion, adjustments in drinking-bout durations were still evident, although the adjustments observed under these conditions seemed to be in excess of the adjustments required to maintain constant volumes of intake. Therefore, we think that sensory feedback is involved and that it is relevant to the control of each bout of drinking. This feedback seems to be manifest so rapidly that it is probably of orolingual or hypothalamic (Carlisle and Laudenslager, 1976) origin.

A finding by Chapman (1969) lends additional support to our view of SIP as a phenomenon controlled by volume-correlated feedback. Chapman observed that "wetting the rat's mouth post-pellet by means of an intraoral tube does not stop the polydipsia unless the amount administered approximates the amount that would have been ingested" (quoted in Falk, 1972, p. 165). Thus, if the volume-correlated sensory feedback normally associated with ingestion of water during SIP is provided without the necessity of the animal's engaging in licking behavior, the amount of postpellet licking behavior will be zero. This finding further indicates that rats are not specifically motivated to engage in any licking activity during SIP-inducing conditions but are motivated only to ingest a certain amount of water.

Just as Chapman found that imposition of the requisite amount of volume-correlated feedback eliminates postpellet drinking behavior, it appears that reduction in intake-volume–correlated sensory feedback greatly prolongs bouts of postpellet drinking or licking. An airstream sustains robust licking behavior but provides sensory feedback that is less satiating than that provided by water licking (see Chapter 4). Animals do engage in schedule-induced airlicking, a behavior that is similar in most respects to SIP. The most striking dissimilarity to SIP is that under similar testing conditions, animals engage in schedule-induced airlicking for much longer durations than they engage in SIP (Mendelson and Chillag, 1970; also see Section 6.3 and Chapter 4). Thus, the satiating capacity of an airstream is less than that of water for schedule-induced licking just as it is for water-deprivation–induced licking. Presumably, this finding indicates that airlicking provides only limited amounts of a kind of volume-correlated feedback that plays a role in terminating both water-deprivation–induced licking and bouts of schedule-induced licking. However, an airstream provides sufficient sensory feedback to initiate and sustain schedule-induced licking, while certain other stimuli, such as empty drinking tubes do not (see Section 6.1).

All of the available evidence is quite consistent with the notion that SIP is regulated by volume-correlated sensory feedback; SIP is not a centrally patterned motor phenomenon, nor is it under the control of feedback from the muscles (*cf.* Hinde, 1970). However, the possibility that there is a regulatory motor component to the SIP phenomenon that has heretofore escaped detection cannot be ruled out. It is also possible that the performance of licking movements and their ingestion-independent feedback contribute to the reinforcement obtained during SIP, even though these factors may contribute only minimally to the satiation of the schedule-induced licking tendency (see Section 6.3).

4. The Effects of Surgical Intervention

4.1. Desalivation

Murphy and Brown (1975) have recently reported that desalivated rats (DSs) develop SIP very rapidly, within a single 1-hr session. However, over the course of extended testing, the SIP increased only slightly and nonsignificantly, so that it was exceeded by that of the controls after three to five sessions of testing. Murphy and Brown suggested that the developmental phase of SIP for normal rats corresponds to the period during which prandial-drinking behavior is being acquired. Since the DSs used in their experiment had already acquired prandial-drinking behavior in their home cages, SIP was manifest on the first session of testing. In a subsequent experiment, these investigators found that in animals that were desalivated on the day before the start of SIP testing, SIP developed at the same rate as in normal animals (Murphy and Brown, 1976). Thus, at least for DSs, the learning phase of SIP coincides with the acquisition of prandial drinking.

Similar findings had been reported by Chapman and Richardson (1974), who employed surgical desalivation in one group of rats and atropine methyl nitrate to induce a state of functional desalivation in another group. The investigators found initially that presession intragastric loads of water prevented the development of SIP in normal animals. After surgical desalivation, relatively large intakes of water developed during a single session despite presession loading with water. When functional desalivation was induced by the atropine compound, large intakes of water again developed in one session despite preloading with intragastric water. At the end of the experiment, the investigators administered one session without atropine and found that the large intake volumes persisted or even increased.

Since Murphy and Brown (1975) found that the water intake of the DSs was high both in their home cages under pretest *ad libitum* feeding conditions and during the first SIP test session they suggested that SIP might not actually have been produced in the DSs. To test for this possibility, the investigators changed the schedule of food presentation from fixed-time 1 min (FT1) to FT2 and also employed 97-mg food reinforcements in place of the customary 45-mg pellets. Although the water intake of the DSs remained below that of the controls throughout the manipulations, they responded to each parametric change in a parallel fashion. Thus, the putative SIP found in DSs is probably true SIP in that it is similarly sensitive to parametric manipulations. Although the behavioral changes induced by these manipulations were fairly small, Murphy and Brown (1976) later compared the putative SIP that they observed in DSs with the drinking that these animals exhibited when given all of their session pellets at once, at the start of the session. A large reduction in drinking was observed under these conditions, which strongly supports the conclusion that the apparent SIP of desalivates is true SIP, akin to that observed in normal animals, rather than the simple prandial drinking which is characteristic of all desalivates.

It is very difficult to understand the DSs' decreased asymptotic level of SIP in terms of the characteristics of SIP, if one considers that during intermittent-feeding sessions DSs are probably drinking considerable amounts of water just to wash down the dry food pellets. If DSs exhibit SIP as well as their usual exaggerated prandial drinking, one would expect them to ingest more, rather than less, water during intermittent-feeding sessions. That the DSs actually drink less suggests that another factor is involved. Murphy and Brown (1975, 1976) interpreted the DSs' decreased asymptotic level of SIP in terms of the frustration (Denny and Ratner, 1970) and emotionality theories of SIP (they mistakenly ascribed the emotionality theory to Falk, 1969, 1971, who actually avoided any reference to mediating internal states or psychological variables). According to Murphy and Brown (1975), DSs should experience less emotional upset and frustration than do normal animals, because:

> Under ad-lib feeding conditions the desalivate rat displays prandial drinking, that is, short periods of eating interrupted frequently by bouts of drinking. It is precisely this behavior pattern, though exaggerated, that is elaborated by the intact rat during SIP. It could be argued thus that desalivate rats experience less emotional upset during SIP because of the similarity of ad-lib and session feeding patterns. Consequently, lower levels of SIP are observed relative to control animals. In the intact rat the respective patterns are markedly dissimilar, producing considerable emotional upset. (p. 315)

However, Murphy and Brown would also have to predict, from this theoretical position, that SIP would tend to decrease in normal rats over the course of extended testing, since the animals should not become "emotionally upset" by the SIP feeding schedule once they are accustomed to its temporal characteristics. This did not occur even in Murphy and Brown's own experiments. Thus an emotional-upset interpretation of reduced SIP in DSs is untenable, and an alternative interpretation is necessary.

One possible explanation is that the reduced SIP in DSs could be an instance of the reduced responsivity of such animals to various dipsogenic stimuli. It has been reported that DSs drink less water than controls in response to water deprivation (Vance, 1965), complete food deprivation (Vance, 1965), isoproterenol and diazoxide (Falk and Bryant, 1973), and hypovolemia (Gutman, Livneh, and Pietrokovski, 1970). Although the decreased responsivity to water deprivation reported by Vance (1965) may have been an artifact of the testing conditions (Mendelson, Zec, and Chillag, 1972), the other findings are as yet unchallenged. The reason that desalivation induces decrements in water ingestion in these various circumstances is not clearly understood; Falk and Bryant (1973) concluded that: "The decreased drinking response of the desalivate rat to certain dipsogenic physiological and pharmacological stimuli cannot at this time be assigned to a ready interpretation. An altered oral sensory environment or some endocrine output from these glands may participate in the normal response to various dipsogenic stimuli" (p. 210). Thus, Murphy and Brown (1975) may simply have demonstrated that schedule induction can be added to the list of dipsogenic stimuli to which the responsivity of DSs is decreased.

4.2. Denervation of the Tongue

Several investigators (see Section 1) have suggested that the orolingual sensory stimulation provided by water is important in SIP (see Section 1.2). There are several reasons to suspect that this is the case. For example, systemic overhydration occurs during SIP (Stricker and Adair, 1966), and moderate water preloads do not diminish established SIP (Falk, 1969), which suggests that systematic signals are not important reinforcers for schedule-induced licking behavior. Stimuli that are primarily of orolingual origin—for example, those provided by a stream of air (Mendelson and Chillag, 1970) and by the thermal properties of water (Carlisle and Laudenslager, 1970; also see Chapter 6)—have effects on schedule-induced licking.

As a result of such considerations, Mendelson and Chillag (1970) suggested that "schedule-induced drinking should be particularly sensitive to experimental manipulations which interfere with the transmission of normal feedback from the tongue. For example, sensory denervation of the tongue might induce large decrements in schedule-induced drinking although it leaves normal drinking largely unaffected" (p. 604). Carlisle and Laudenslager (1976) suggested a possible alternative, the basal forebrain, for the locus through which water temperature affects SIP. Airlicking occurs during intermittent feeding schedules, and the latter investigators had previously found that airlicking causes decreases in the temperature of the preoptic area of the hypothalamus (Carlisle and Laudenslager, 1975; also see Chapter 4). This finding suggests a possible role for hypothalamic cooling in the reinforcement that maintains schedule-induced licking.

It has been reported that total sensory denervation of the tongue has little effect on airlicking motivated by water deprivation (Weijnen, 1976; also see Chapter 4). The reinforcing effects of airlicking are based on its cooling properties; thus, the cooling of the extralingual oral tissues or the hypothalamus must be a sufficient reinforcer to maintain airlicking behavior motivated by water deprivation (see Chapter 4). We have also obtained evidence indicating that sensory feedback from the anterior tongue is not essential for schedule-induced drinking and schedule-induced airlicking.

We tested the hypothesis that lingual sensory feedback is essential for the reinforcement of schedule-induced licking behavior by severing the sensory nerves innervating the tongue. We sectioned the lingual and chorda tympani nerves in five animals and the medial branch of the glossopharyngeal nerve in three animals and combined both treatments in six animals. Schedule-induced drinking and airlicking were evaluated in these animals in terms of the percentage of IPIs during which some licking occurred. Because some of the denervated animals required more time for pellet consumption so that they were left with less time to drink after each pellet, volume-ingestion figures were not directly comparable among the groups.

The acquisition of schedule-induced drinking and airlicking was somewhat diminished or delayed in the animals with multiple denervations. Animals with only glossopharyngeal denervation exhibited SIP that was comparable to that of sham-operated controls (see Figure 2).

Two additional animals were pretrained on SIP before multiple denervations were performed. Deficits in SIP were noted (see Figure 2), which, however, faded away in one of the animals after additional SIP experience (not shown in the figure). It is of interest that the performance of the most severely denervated animals was independent of whether or not the animals had been allowed to acquire SIP before the denervations.

When an airstream was substituted for water, the schedule-induced airlicking of the animals with the most complete denervations was similar to that of the sham-operated controls. However, the frequency of postpellet airlicking for the animals with medial glossopharyngeal denervations decreased precipitously as compared to SIP (from 90% to 40% of IPIs) and remained at low levels for two of the three animals that were tested. The percentage of IPIs during which licking occurred for the animals with

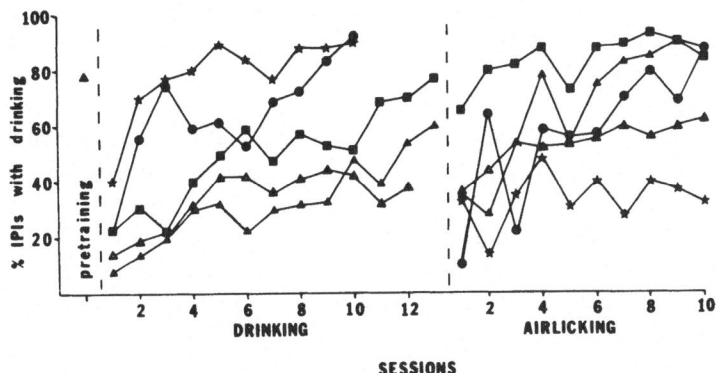

Figure 2. The development of schedule-induced polydipsia and the subsequent development of schedule-induced airlicking in rats with bilateral denervations of the lingual plus chorda tympani nerves (filled squares; five animals) or the medial branch of the glossopharyngeal nerve (filled stars; three animals), or both treatments combined (filled triangles; six animals), and sham-operated controls (filled circles; three animals). Two additional animals (open triangles) were trained to exhibit SIP, and then the combined treatments (denervations of the lingual, chorda tympani, and medial glossopharyngeal nerves) were administered. The median of the last three preoperative training sessions for these latter animals is labeled "pretraining." Following surgery, this group was tested in the same sequence as the other animals.

combined lingual amd chorda tympani lesions continued to increase gradually during tests for schedule-induced drinking and airlicking. Thus, during schedule-induced airlicking tests, they licked for greater percentages of IPIs than did the sham-operated controls or the animals with the more severe denervation.

The differential effects of the various denervation regimes on schedule-induced drinking and airlicking may simply be related to changes in the stimuli provided by the ingestion of food or to a decrease in the tendency of dry food to elicit drinking. However, it is of interest that medial glosso-pharyngeal denervations only attenuated schedule-induced airlicking, while all other types of denervations seemed to adversely affect only SIP, not schedule-induced airlicking. The deficits in SIP but not in schedule-induced airlicking may have been caused by a difficulty these animals had in obtaining water; they had bitten off portions of their tongues but could nonetheless activate the lick-contingent airstreams easily. We are at a loss to explain the deficits restricted to schedule-induced airlicking in the animals with medial glossopharyngeal denervations particularly since these deficits were not manifest by animals with multiple denervations including the glossopharyngeal nerve.

Thus, although denervations of the tongue in some circumstances do interfere with SIP, drinking or airlicking was observed during a mean of

80% or more of the IPIs in all of the groups of denervated animals at some time during testing. Many individual animals with each type of denervation drank or airlicked during 90% or more of the IPIs in at least some sessions. It therefore may be concluded that although partial denervation of the tongue has some deleterious effects on schedule-induced drinking or airlicking, sensory feedback from the anterior tongue is not essential for the emergence of schedule-induced licking behavior. Thus, the reinforcement of SIP must be accomplished, at least in part, by the activation of receptors that are not located on the anterior tongue. These receptors are likely to be located in either the posterior tongue, extralingual oral tissues, or the hypothalamus.

5. The Effects of Fluid Properties on SIP

5.1. Temperature

Carlisle (1971) found that the temperature of the rostral hypothalamic area is elevated during intermittent reinforcement by food and suggested that SIP is caused by this hypothalamic heating. This is an attractive hypothesis because it proposes that the cause of SIP is a concrete physiological mechanism that can be readily measured. This hypothesis also may account for certain data that are not easily explained by most other SIP theories (see Section 3; Carlisle and Laudenslager, 1976).

Carlisle (1971) reported that rostral hypothalamic temperature was elevated by a few tenths of a degree, to as much as 39.6°C, during fixed-ratio food-reinforcement schedules that engendered SIP. Although this temperature level was slightly above that previously observed in eating, drinking, or sleeping rats (Carlisle, 1968), substantially higher hypothalamic temperatures had been found to be necessary to induce drinking in goats (Andersson and Larsson, 1961). However, peak hypothalamic temperatures were consistently a few tenths of a degree higher when water was not available than when it was available on the same reinforcement schedules. Also, Carlisle (1971) observed that bar pressing for food was associated with increases in hypothalamic temperature, while drinking and pauses in bar pressing were associated with decreases. He concluded that while rostral hypothalamic temperature increased during SIP-engendering food-reinforcement schedules, these temperature increases were insufficient in magnitude to induce polydipsic drinking. Notwithstanding this negative evidence, Carlisle (1971) left open the possibility that "drinking served to mitigate an increase in body temperature" (p. 21). This possibility is the more attractive if one considers that if SIP were caused by a hypothalamic warming effect, the effect could be localized in a more restricted location than the rostral hypothalamic area in general, or the heating effect could be centered in an area that is at some distance from the rostral hypothalamus. In either case, the increase in temperature of the rostral hypothalamus that is correlated with drinking behavior would certainly be somewhat less than the amount of artificially induced warming that is required to induce drinking behavior. Thus, it may be premature to conclude that SIP is not mediated by hypothalamic warming, although there is not yet any evidence that would positively support such a suggestion.

Carlisle (1973) subsequently found that SIP is greatly attenuated by cooling of the rostral area of the hypothalamus. When preoptic hypothalamic temperature was maintained at 33°C by means of implanted thermodes, water ingestion was reduced by a mean of about 70-80% as compared to that of control subjects. When hypothalamic temperature was no longer controlled, the subjects with implanted thermodes quickly developed SIP, but their drinking never reached the level of the control group, possibly because of damage from the implanted thermode assembly (control animals were not implanted). After SIP had developed in the animals with an implanted thermode, subsequent cooling of the hypothalamus reduced the amount of water ingested during SIP by only 22% as compared to the no-cooling condition. Thus, hypothalamic cooling reduced SIP by about 73% prior to the acquisition of normal levels of SIP, but by only 22% after SIP had developed. Carlisle (1973) concluded that "Once polydipsia has developed, it is less amenable to the influence of hypothalamic cooling" (p. 218). This is reminiscent of reports that presession stomach-loads of water prevent the development (Chapman and Richardson, 1974) but not the maintenance (Falk, 1969) of SIP. Since hypothalamic cooling and stomach preloading also suppress drinking induced by water deprivation (Andersson, Larsson, and Persson, 1960; Sundsten, 1969), it appears that some manipulations that reduce normal drinking can prevent the development of SIP but only slightly affect the maintenance of SIP once it has developed. Interestingly enough, desalivation, which *facilitates* normal prandial drinking, seems to *accelerate* the development of SIP (see Section 4.1). The results of these experiments support the suggestion by Falk (1971) that normal, low base rates of drinking behavior in SIP-like situations are critical to the development of SIP.

In later studies, Carlisle (1973) and Carlisle and Laudenslager (1976) found that SIP is affected by water temperature and ambient temperature in ways that are similar to the effects of these variables on water-deprivation–induced drinking (the latter findings are discussed in Chapter 6). Carlisle (1973) found that when water temperature co-varied with ambient temperature, SIP was greatest at about 30°C; when water temperature was varied independently, the greatest degree of SIP was observed at warm water temperatures (33°C). In a later study, Carlisle and Laudenslager (1976) conducted investigations of the effects on SIP of ambient temperature and water temperature varied independently. They found that ambient temperature had little effect but that water temperature had a large effect on SIP. Peak volumes of intake were obtained when the ingested water was at a temperature near 30°C, with intakes decreasing when water temperatures were above or below that level (about 35% less at 5°C and 24% less at 40°C). Thus, the effects of ambient temperature reported by Carlisle (1973) were probably due to the co-varying temperature of water (see Chapter 6).

The effects of water temperature on SIP occurred largely through an effect on the volume of water ingested per drinking bout, but at temperatures above 30°C, there was also a drop in the frequency of drinking bouts (Carlisle and Laudenslager, 1976). The combined water- and ambient-temperature variable also affected SIP through increasing or decreasing drinking-bout duration in the Carlisle (1973) study. In contrast, hypothalamic cooling depressed SIP by decreasing both the frequency and the duration of postpellet drinking bouts (Carlisle, 1973). Carlisle (1973) attributed this difference to the control of drinking-bout duration and frequency by different classes of thermal receptors, peripheral receptors purportedly playing a role in volume ingested per drink, and central (hypothalamic) receptors controlling frequency of drinking bouts or both frequency and duration of drinking bouts. But an alternative possibility is that the difference is due to the involvement of motivational versus satiation factors. Hypothalamic cooling might be expected to diminish the motivation to drink, thus decreasing the frequency of drinking bouts. Water temperature might be expected to affect only the satiation aspects of SIP, since the influence of water temperature becomes manifest only after drinking has been initiated. If the effects of water temperature on SIP are also modulated through an effect on hypothalamic thermal receptors, we would expect hypothalamic cooling through implanted thermodes to decrease both the duration and the frequency of bouts of drinking.

Carlisle and Laudenslager (1976) suggested that water temperature might affect SIP via temperature receptors located in lingual, oral extralingual, or hypothalamic tissues. Although the available evidence does not support the possibility that overheating of the hypothalamus causes SIP (Carlisle, 1971), there is every reason to entertain the possibility that either orolingual cooling or hypothalamic cooling plays a role in reinforcing SIP. Cool water may engender less volume per drink because it satiates the drinking tendency more effectively. However, the reason that warm water induces greater drinking volumes might be that it more slowly cools orolingual or hypothalamic thermal receptors. Very warm water (above 30°C) may not be effective as a cooling agent, which might explain the fact that it is less effective both in sustaining drinking bouts and in initiating them. Airlicking occurs during SIP-engendering conditions (see Chapter 4) and is reinforcing because of its cooling effects (Chapter 4). However, although air may cool orolingual tissues (Chapter 4) as well as the hypothalamus (Carlisle and Laudenslager, 1975), it may not be as effective a cooling agent as is water, which might explain why schedule-induced airlicking bouts are longer than are schedule-induced water-licking bouts (Mendelson and Chillag, 1970). If cooling reinforces SIP, this would explain the finding that animals accurately regulate intake volumes during SIP (see Section 3). One would expect animals to compensate for a reduced intake rate by drinking for longer durations in order to effect a given amount of hypothalamic or oral cooling.

The question of whether the activation of hypothalamic thermal receptors modulates the effects of water temperature on SIP or even reinforces SIP could be investigated in several experiments. First, one could test the hypothesis that hypothalamic thermal receptors play a role in the reinforcement of SIP by determining whether animals will press a lever to cool their hypothalamic areas during SIP-engendering conditions. Clamping the temperature of the hypothalamus at 39°C should not directly induce drinking (Andersson and Larsson, 1961). However, clamping the hypothalamic temperature at this level should increase the duration of bouts of schedule-induced drinking or airlicking. A third prediction is that rapid cooling of the hypothalamus during individual bouts of schedule-induced airlicking should cause the cessation of those airlicking bouts. One could test the converse hypothesis, that SIP is reinforced through the activation of orolingual or other thermal receptors, in part by determining whether schedule-induced drinking or airlicking is diminished by denervation of the sensory nerves from the tongue and/or the mouth (see Section 4.2).

5.2. Saline

Rats ingest larger volumes of saline than water during SIP sessions, provided that the concentration of the saline does not exceed 0.9% (Falk, 1964, 1966a). The optimal concentration is 0.3%. Hypertonic saline induces decrements in SIP that are correlated with the degree of hypertonicity (Falk, 1964, 1966a). The finding that 0.3% NaCl increases fluid consumption during SIP has been confirmed by Segal and Deadwyler (1965), using a DRL 30-sec schedule of delivery of 45-mg Noyes pellets. The NaCl acceptance–rejection functions observed during SIP (Falk, 1964, 1966a) were similar in shape to those obtained after 21 and 69 hr of water deprivation;

however, peak levels of intake in water-deprived animals were observed at
0.8-1.0% NaCl concentrations, as compared to 0.3% during SIP.

An additional finding of interest was that the decrease in ingested
volumes of hypertonic saline appeared to be caused by a decrease in the
number of IPIs during which drinking occurred. Postpellet periods with no
drinking first became evident when the concentration of NaCl reached or
exceeded 0.9-1.2%, and gradually increased in number as NaCl concentration was further increased. This finding suggests that strong NaCl concentrations are relatively ineffective in initiating bouts of drinking. The mean
length of all non-zero-length drinking bouts was similar for weak (0.9-1.2%)
and for strong (2.1-3.0%) NaCl concentrations but was somewhat longer
(40-100%) at intermediate (1.2-2.1%) NaCl concentrations. Thus, once
drinking was initiated, it was sustained by all concentrations of NaCl at
least as well as by water. It is of interest to note, however, that the increase
in total-session intake at NaCl concentrations around 0.3% could not be
accounted for either by increases in the number of bouts of drinking *or* by
increases in the length of individual bouts of drinking (Falk, 1964, 1966c).
Thus, it might be suspected that instantaneous ingestion rates (e.g., ml/sec)
were greater for those NaCl solutions that were ingested in greater volumes.

One question regarding NaCl adulteration during SIP remains unan-
swered: Are larger volumes of 0.3% NaCl ingested during SIP because 0.3%
NaCl is more preferred or because 0.3% NaCl is less "satiating"? For
normal thirsty animals, one school of thought views increases in the
consumption of hypotonic saline by water-deprived animals to be a result of
"dilution" of the water, making it less effective in terms of its satiating
capacity (J. A. Deutsch and D. Deutsch, 1973, pp. 612-618). An opposing
view argues that greater volumes of hypotonic saline are ingested at least in
part because the taste of hypotonic saline is preferred to water (Chiang and
Wilson, 1963; Falk and Titlebaum, 1963; Young and Falk, 1956). It would
be interesting to determine whether the effects of saline on intake volumes
during SIP are mediated by preference factors or by a satiation process.
One could study this problem by measuring preferences for water versus
various concentrations of saline during SIP. The findings that SIP
gradually declines during the course of single testing sessions (Hymowitz
and E. X. Freed, 1972) and that large stomach preloads of water (6% of
body weight) somewhat reduce SIP (Falk, 1969) suggest that overhydration
limits SIP to some degree. Thus, one might suspect that 0.3% saline is
ingested in larger amounts during SIP because it disturbs the homeostatic
balance less than does pure water. If the postingestional consequences of
ingestion of large amounts of water were more aversive than those of 0.3%
saline, one would expect that animals would prefer 0.3% saline to water during SIP. On the other hand, animals ingest precisely regulated volumes of
water during SIP (see Section 3), suggesting the involvement of what we

have called a *pseudosatiation mechanism* (see Section 6.1) in terminating individual bouts of drinking. Dilute (0.3%) saline might be less effective in activating the pseudosatiation mechanism than pure water, leading to the ingestion of larger volumes.

If 0.3% saline is preferred to water during SIP, one would suspect that the increased ingestion of saline is mediated by taste factors, whereas if water is preferred, one would expect that a satiating osmotic factor mediates the enhanced SIP produced by 0.3% saline. One test of this question would be to determine whether the effects of saline on SIP are abolished by denervation of taste receptors. Another test would be to determine whether saline preferences can be demonstrated through the use of a forced-choice procedure. During every other IPI, the rat would be allowed access only to the solution that it had not chosen during the previous IPI. The small amounts ingested during single IPIs, combined with a rapid succession of IPIs during which different solutions are ingested, would not allow the animal to learn to associate the postingestional consequences of either solution with the taste of that solution; thus, it could be determined which of the tastes (0.3% saline or water) is preferred by rats during SIP in the absence of any learning based on the postingestional consequences of either solution.

5.3. Saccharin

Saccharin adulteration of the available drinking fluid increases the volume ingested during SIP sessions (Colotla and Keehn, 1975; Keehn, Colotla, and Beaton, 1970; Samson and Falk, 1974b; Segal and Deadwyler, 1966; Wayner and Greenberg, 1972; Wayner, Greenberg, Fraley, and Fischer, 1973). However, Keehn and his colleagues (Keehn, 1972; Keehn *et al.*, 1970) have argued that saccharin does not enhance SIP *per se* but instead enhances concurrent non-schedule-induced drinking that takes place in addition to SIP during testing sessions. These investigators (Keehn *et al.*, 1970) found that postpellet durations of drinking were not enhanced by 0.4% saccharin but that additional bouts of drinking appeared later in the IPIs. Such additional nonpostpellet bouts of drinking occurred only rarely when water was the only available drinking fluid. Presatiation with saccharin tended to decrease these extra, interpellet bouts of drinking but did not decrease postpellet drinking. Also, interpellet drinking declined markedly within sessions, while postpellet drinking remained relatively stable. Since the enhancement of SIP by saccharin was essentially eliminated by presatiation with saccharin, it is certainly not advisable to neglect the issue of presatiation when studying the effects of saccharin on SIP.

5.4. Ethanol, Acetone, Quinine, and Conditioned Taste Aversions

Increasing the concentration of ethanol in the drinking solution decreases the volume of fluid consumed during SIP in a dose-related fashion (Falk and Samson, 1976; Gilbert, 1974b; Githens, Hawkins, and Schrot, 1973; Holman and Myers, 1968; Meisch and Thompson, 1972; Schrot, Hawkins, and Githens, 1971). However, the decreases in volume of intake are less than proportional to the increases in concentration, so that the amount of absolute ethanol that is actually consumed increases with ethanol concentrations (Falk and Samson, 1976) of up to 32% in some circumstances (Githens, Hawkins, and Schrot, 1973; Meisch and Thompson, 1972). Although it has been suggested that ethanol adulteration attenuates SIP because of the aversive taste of ethanol (Holman and Myers, 1968), it would seem more plausible to blame a pharmacological effect of the ethanol (*cf.* Gilbert, 1975; Meisch and Thompson, 1972). Gilbert (1973) has found that SIP is attenuated by intraperitoneal doses of ethanol that have little effect on food-maintained lever pressing. Furthermore, schedule-induced ethanol drinking declines within single sessions (Colotla and Keehn, 1975; E. X. Freed and Lester, 1970; E. X. Freed, Carpenter, and Hymowitz, 1970; Gilbert, 1974a; Githens, Hawkins, and Schrot, 1973; Meisch and Thompson, 1972) in a dose-related fashion (Meisch and Thompson, 1972) to a far greater extent than does SIP (Hymowitz and E. X. Freed, 1972). Although the intake of ethanol solutions is generally somewhat higher than that of water during the first few minutes of SIP sessions for concentrations as high as 16% (Meisch and Thompson, 1972), the rate of intake begins to drop after about 10 min, and this drop continues until, after 2–3 hr, the rate of ethanol ingestion is only a small fraction of that of water (Meisch and Thompson, 1972). Therefore, the decreased consumption of fluid during SIP that is brought about by increased ethanol concentration is probably due to within-session decrements in SIP; these decrements seem to be related to the pharmacological effects of ethanol. However, we cannot entirely dismiss the possibility that an aversive taste of ethanol contributes to this phenomenon; rats might be able to tolerate only limited amounts of the aversive ethanol taste before this aversive taste becomes unbearable.

Others have observed that rats drink substantial amounts of such aversive fluids as quinine and acetone solutions during SIP. Segal and Deadwyler (1965) found that 0.008% and 0.016% quinine had little effect on the amount of fluid consumed by rats during a DRL 30-sec schedule, and Wayner and Greenberg (1973) found that 0.008% quinine decreased fluid consumption during SIP by about 40% but that mean intakes were still 18 ml/hr. E. X. Freed and Lester (1970) observed within-session decrements in schedule-induced drinking of acetone solutions, although water was ingested in preference to 5.6% acetone (E. X. Freed, 1974). Thus, it must be concluded that even relatively aversive fluids are ingested during SIP; indeed, even water would probably otherwise be somewhat aversive in the amounts that are consumed.

Samson and Falk (1974a) reported that ethanol-dependent animals prefer 5.0% ethanol to water and even to 3.0% dextrose during SIP, even though they prefer that dextrose solution to 5.0% ethanol in the home cages. This finding may be contrasted with the finding that nondependent animals, during SIP-inducing conditions, prefer water to ethanol solutions in concentrations that are preferred to water in the home cages (Keehn and Coulson, 1975). Also, Colotla and Keehn (1975) found that similar amounts of water and 5% alcohol were consumed during SIP sessions and that drinking occurred during about the same percentage of IPIs for water and ethanol solutions. However, the latency to drink following pellet consumption was greater for ethanol than for water; when water was available, virtually all bouts of drinking began less than 10 sec after pellet delivery. When an ethanol solution was available, only about half of the bouts of drinking were postpellet, and the remaining half occurred later in the IPIs i.e., they began at least 10 sec after pellet delivery (Keehn and Colotla, 1971). Thus, fluid composition does influence SIP in some respects. In this connection, it would be interesting to compare quinine-rejection thresholds during SIP with those obtained during food deprivation (perhaps with sucrose added to the quinine, so that quinine is associated with caloric value) and with those obtained when all pellets are delivered simultaneously at the start of the sessions. Postpellet drinking latencies should be concurrently measured in such an experiment.

It is difficult to induce animals to become dependent upon ethanol by voluntary ingestion (Mello, 1973; Myers and Veale, 1972). When ethanol solutions are given as the sole source of drinking fluid, animals do not become dependent unless the ethanol solutions also contain 1.7% saline (Falk, Samson, and Tang, 1973). It has been suggested that this limitation is related to the formation of conditioned aversions to the taste of ethanol (Eckardt, 1975, 1976; Lester and E. X. Freed, 1973). Evidence for this suggestion includes the finding that small doses of ethanol paired with the taste of saccharin cause the saccharin subsequently to become aversive (Berman and Cannon, 1974; Eckardt, 1975, 1976; Eckardt, Skurdal, and Brown, 1974; Lester, Nachman, and LeMagnen, 1970). In addition, we (V. Bruce-Wolfe and J. Mendelson, unpublished experiment, 1974) have obtained indirect support for the suggestion that the formation of conditioned taste aversions may actually function to limit ethanol ingestion. When rats can obtain highly rewarding electrical stimulation of the hypothalamus by licking an ethanol solution, they ingest only moderate amounts (Martin and Myers, 1972). However, when we applied this procedure to animals that had never been exposed to ethanol, very substantial amounts of 20% ethanol were ingested during the first few sessions of testing. Sometimes as much as 5 g/kg of ethanol was ingested during the 20-min sessions, which resulted in states of extreme intoxication, including ataxia and sometimes anesthesia. However, in subsequent sessions, only moderate amounts were ingested (see Figure 3). This finding suggests that although the taste of ethanol was not sufficiently aversive to limit ingestion during the first exposure, the taste of

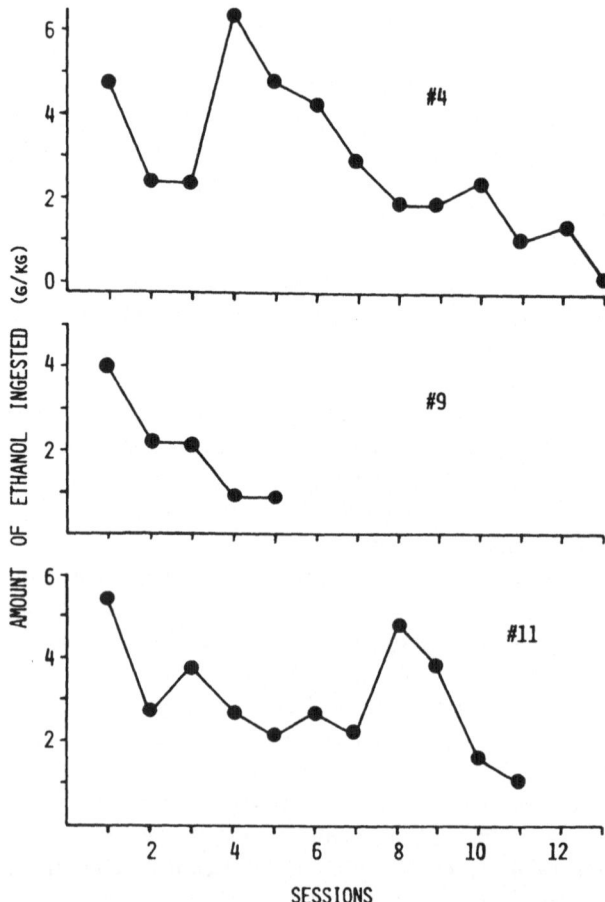

Figure 3.   The gradual decline of ethanol ingestion over the course of several sessions of testing for three rats. The animals obtained reinforcing electrical brain stimulation by licking a drinking tube that contained a 20% v/v ethanol solution. Volumes ingested during the 20-min testing sessions were measured by the animals' being weighed before and after testing. Bipolar stainless-steel electrodes were implanted in the lateral hypothalamus at Krieg coordinates 1.0 mm posterior to bregma, 1.5 mm lateral from the midline, and 8.2 mm below the superior surface of the skull. After each lick, a constant-current 60-Hz sine-wave pulse 0.5 sec in duration was delivered to one of the electrodes.

ethanol, through the establishment of a conditioned aversion, subsequently *became* so aversive that it was not ingested even when paired with highly rewarding brain stimulation. It is notable that after the first few sessions, the animals made persistent attempts to obtain lateral hypothalamic stimulation by licking without ingesting fluid, for example, by licking at the sides of the drinking tubes. When such subversive attempts to obtain brain stimulation were made impossible, licking ceased altogether. Previously, it had been reported that ethanol preference is decreased following forced drinking of 10% ethanol during water deprivation (Carey, 1972). This latter finding is also consistent with the possibility that learned taste aversions function to limit animals' drinking of ethanol.

Two experimental regimens that have been successfully used to induce rats to voluntarily ingest sufficient quantities of ethanol solutions to become physically dependent are (1) hunger: when ethanol is mixed with nutritive solutions that provide the only available source of calories, ethanol dependence develops, usually with a large (up to 33%; Freund, 1969) loss of body weight (Freund, 1969; Hunter, Riley, Walker, and Freund, 1975; Walker and Freund, 1971); and (2) schedule induction: when 5% ethanol is presented during six equally spaced daily SIP sessions, ethanol dependence develops over the course of several weeks (Falk and Samson, 1976; Falk *et al.*, 1973; Falk, Samson, and Winger, 1972; Samson and Falk, 1974a). There is also some evidence that both SIP and unusually intense hunger can increase rats' resistance to the development of conditioned taste aversions. Roll, Schaeffer, and Smith (1969) found that it was necessary to repeatedly pair saccharin with ionizing radiation in order to induce a decrement in schedule-induced drinking of saccharin solutions. In contrast, even single pairings of saccharin with smaller doses of radiation (54 as opposed to 100–200 R) had previously been found to result in large decrements in saccharin consumption motivated by 48 hr of water deprivation (Garcia and Kimeldorf, 1960). Thus, SIP-engendering conditions apparently induce a motivational state sufficient to overcome conditioned taste aversions, at least partially.

Rats' resistance to conditioned taste aversions may also be increased by certain forms of intense hunger. Lesions of the ventromedial hypothalamus increase food motivation under certain testing circumstances (e.g., Kent and Peters, 1973; King and Gaston, 1973; Marks and Remley, 1972). Similar lesions and testing circumstances impair the acquisition and accelerate the extinction of conditioned taste aversions to saccharin (Gold and Proulx, 1972). Thus, it may be that unusually strong hunger is also a sufficient motivational condition to overcome conditioned taste aversions, at least partially.

These considerations suggest that ethanol dependence can be induced by strong hunger motivation and by SIP because both of these conditions are capable of partially overcoming or preventing the formation of conditioned taste aversions. An interesting speculation is that hunger and SIP are governed by the same motivational state; for example, SIP is mediated by the induction of excess "hunger," if one defines hunger to be the internal state that is brought about by food deprivation. Although no direct evidence is available to support this latter hypothesis, it is consistent with a finding that quinine in concentrations of 0.008% and 0.016% does not substantially reduce SIP (Segal and Deadwyler, 1965), although even less-concentrated solutions reduce the fluid intake of water-deprived rats by 50% or more (Johnson and Fischer, 1973; also *cf.* Wade and Zucker, 1970).

There are other data in the literature that may be interpreted in terms of an ability of SIP to suppress conditioned taste aversions. Samson and Falk (1974b) found that when 5% ethanol was available as the only source of drinking fluid in rats' home cages, the addition of 0.25% saccharin did not enhance its consumption. However, when 5% ethanol was available during SIP, its consumption was significantly increased by the addition of 0.25% saccharin. These findings may be readily interpreted in terms of the putative conditioned-aversion–attenuating properties of SIP. Saccharin adulteration of an ethanol solution made available during a SIP session would increase drinking simply by increasing the palatability of the available fluid, much as saccharin enhances the ingestion of plain water during SIP (see Section 5.3). However, the pairing of ethanol with saccharin in the home cages would result in the formation of a conditioned taste aversion to saccharin as well as to ethanol, both taste aversions having been induced by the aversive consequences of ethanol ingestion. Thus, saccharin would not be expected to remain palatable after repeated pairings with ethanol, and saccharin would not be expected to enhance the consumption of ethanol when it is the only available fluid. It would be of interest to run a similar experiment in which ethanol would be mixed with the only source of calories rather than with the source of water in the home cage. Would saccharin adulteration enhance ethanol consumption? We predict that ethanol consumption would be enhanced under these hunger-motivated conditions, although perhaps not to as great an extent as it is during SIP.

6. What Will Sustain Schedule-Induced Licking Behavior Other Than Fluids?

6.1. Empty Metal Drinking Tubes

The fact that animals drink precise volumes of water during SIP (see Section 3) means that there must be a volume-sensitive mechanism or process that determines when drinking will be terminated, that is, a pseudosatiation mechanism. There seem to be certain stimuli that are effective in sustaining schedule-induced licking but which activate the hypothetical pseudosatiation mechanism less effectively than does water. For example, the increases in the duration of schedule-induced licking that are evident when 0.3% saline (see Section 5.2), an airstream (see Chapter 4 and Section 6.3), or warm water (see Chapter 6 and Section 5.1) are available may be caused by a decreased efficacy of these stimuli in activating the pseudosatiation mechanism.

Conversely, intermittent food-reinforcement schedules do not seem to cause licking movements *in vacuo*; some minimal sensory feedback from a liquid, an airstream, or a cold object (for guinea pigs; see Section 6.2) must be obtained by the licking movements in order for schedule-induced licking behavior to be initiated and sustained. There are also some lick-produced stimuli that are relatively ineffective in initiating or sustaining schedule-induced licking behavior. One example of such a subadequate stimulus may be saline solutions at concentrations above 0.9–1.2%, which are less effective in initiating postpellet drinking bouts (see Falk, 1966c, and Section 5.2).

There is some doubt as to whether an empty metal drinking tube, which is certainly less effective than water, is capable of supporting schedule-induced licking. Stein (1964) found that when an empty drinking tube was substituted for a full one during SIP, licking frequency dropped to very low levels but did not cease entirely. Recently, Cook and Singer (1976) observed that while the substitution of an empty drinking tube for a water-filled one caused schedule-induced licking rates to drop by about 70% (to about 1500 "licks" per 30-min session), contacts with the drinking tube did not cease entirely, and there occurred a concomitant increase in the number of attacks on the drinking tube. These investigators did not substract the number of lickometer counts resulting from drinking-tube attacks from the total number of drinkometer counts; thus, it is not possible to evaluate the actual rate of dry-tube licking on the basis of the information contained in their report. However, one would not expect the rate of drinking-tube attacks to have approached the residual 50/min drinking-tube contact rate. Stein (1964) did not record drinking-tube attacks; thus, the few dry-tube contacts that he reported may have been a consequence of attacks rather than licks.

Deadwyler and Segal (1965) tested three rats for schedule-induced empty-metal-tube licking by allowing drinking bottles to run dry midway through SIP sessions (DRL 30 sec, 45-mg Noyes pellets). In one of the animals, substantial, although greatly reduced, licking of the drinking tube persisted after it had run dry; the behavior persisted during 20 consecutive sessions of testing. This animal also engaged in some licking when an empty tube was presented at the start of one session. From cumulative records presented by Deadwyler and Segal (1965), it did not appear that this empty-metal-tube licking was exclusively a postpellet phenomenon.

In conclusion, it seems that a dry, empty drinking tube is a relatively ineffective stimulus for supporting schedule-induced licking behavior. At present, it is not clear whether a dry metal tube is less effective in initiating postpellet bouts of licking, in sustaining licking bouts, or in both. It also is not known to what extent the licking of a dry metal tube is confined to the immediate postpellet period.

6.2. Cold Metal Objects

Food-deprived guinea pigs given 45-mg Noyes food pellets on an FT 1-min schedule did not develop SIP (see Chapter 4). However, these same animals rapidly developed schedule-induced cold-licking behavior when a cold metal rod was made available. This schedule-induced cold-licking was postpellet in locus, was diminished in most animals by warming of the metal rod, and was eliminated when the food pellets were delivered 15 at a time at 15-min intervals. All of the guinea pigs preferred a cold metal rod to a warm one and preferred a cold metal rod to water when given a choice.

We tested four experienced and four naïve rats for schedule-induced cold-licking behavior during an FT 1-min schedule of delivery of 45-mg Noyes pellets (R. F. Zec, unpublished experiment, 1972). Robust cold licking emerged only in one of the experienced animals. This rat had been used in previous experiments (described in Sections 4.2 and 6.3), and thus the fact that it engaged in this behavior may have been related to this extensive experience. The schedule-induced metal-rod licking of this animal was not eliminated by warming of the metal rod to room temperature or to near body temperature. However, this animal preferred a cold metal rod to a warm one, a room-temperature rod to a warm rod, and a cold metal rod to a room-temperature rod (see Figure 4). Typically, this rat would lick both the sides and the tip of the rod with its tongue. A second experienced animal engaged in moderate amounts of schedule-induced cold licking; during the first five sessions, it licked for a mean of 9.13 min of the 90-min sessions and engaged in cold licking during 54% of the IPIs. The other six animals did not, however, engage in significant amounts of schedule-induced cold licking.

Thus, there are some conditions under which at least some rats will lick dry metal objects during intermittent feeding schedules, and this behavior is not eliminated when the metal is warmed to near body temperature. This phenomenon may be similar to the licking of empty metal drinking tubes observed by other investigators (see Section 6.1).

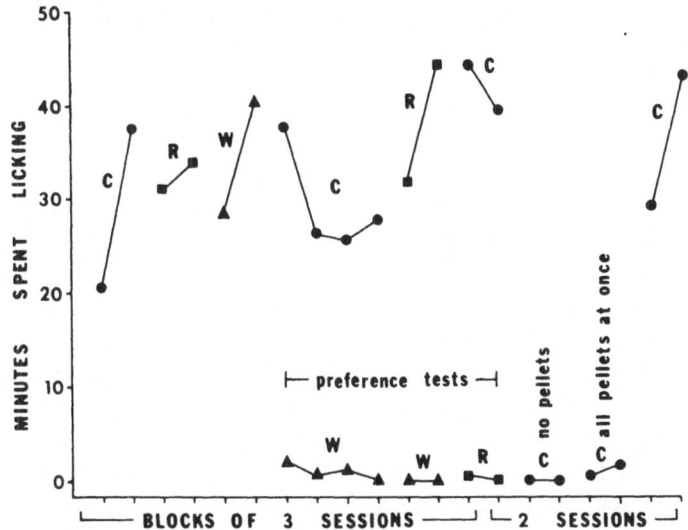

Figure 4. Time spent licking per 90-min testing session for rat MG 1. Each point represents the median of successive blocks of three sessions (first 13 points) or two sessions (last 7 points). A solid-brass rod with a rounded end protruded from a thermos bottle into the test chamber. Solid circles (marked by the letter *C*) indicate the presence of ice water in the thermos bottle (rod temperature was about 17°C). Squares (marked by the letter *R*) indicate the presence of room-temperature water in the thermos bottle, while triangles (marked by the letter *W*) indicate the presence of warm water in the bottle (rod termperature was 38–43°C). This animal was deprived of food to 80% of its *ad libitum* weight, and during testing, it was given one 45-mg Noyes food pellet every minute. During preference tests, two metal rods with different temperatures were available. Control sessions with no food present, or with all 90 pellets delivered at the start of the sessions, generated only minimal amounts of licking behavior and thus demonstrated that the behavior was, in fact, schedule-induced.

6.3. Streams of Air

Thirsty rodents engage in airlicking and in cold-metal licking because both of these behaviors effect cooling of the orolingual or, perhaps, of the hypothalamic tissues (see Chapter 4). Mendelson and Chillag (1968, 1970; also see Chapter 4) have found that schedule-induced airlicking occurs during intermittent reinforcement schedules that also produce SIP. In most respects, this schedule-induced airlicking was quite similar to schedule-induced water licking; it was postpellet in locus, occurred during a large proportion of IPIs, was directly related to the degree of body-weight deficit (Chillag and Mendelson, 1971; also see Section 2.1), and was dependent on schedule manipulations (Mendelson, Zec, and Chillag, 1971). Schedule-induced airlicking was a robust phenomenon and persisted throughout many sessions of testing. This behavior did, however, differ from SIP in one important respect: on the average, animals engaged in schedule-induced airlicking for 1.5–2 times as long as they drank during previous or subsequent SIP sessions.

When water and an airstream were made available simultaneously during SIP-inducing conditions (FT 1-min, 45-mg Noyes pellets) for six experienced rats (used in schedule-induced airlicking experiments) and nine naïve rats (R. F. Zec and S. M. Ramsauer, unpublished experiment, 1971), five of the experienced animals engaged in both schedule-induced airlicking and water-licking behavior (see Table 1). Of the naïve animals, only four engaged in substantial amounts of airlicking. Of those animals that did engage in some airlicking behavior, most displayed a characteristic pattern of postpellet licking. Typically, the ingestion of a food pellet would be followed by a bout of water drinking. A bout of airlicking would generally follow this bout of drinking; on many occasions, these bouts of airlicking were longer in duration than the bouts of drinking (see Table 1). Thus, some animals, particularly those that have had extensive experience with schedule-induced airlicking, engage in frequent bouts of schedule-induced airlicking even when water is also available. Usually each airlicking bout followed a bout of schedule-induced drinking.

Table 1.   Water-Licking and Airlicking Behavior of Food-Deprived Rats[a]

Animal	D (sec)	AL (sec)	Pattern of behavior during each interpellet interval
Experienced animals:			
AL35	345	2366	D or AL, not both
AL39	876	2140	D or AL, not both
MG1	564	2559	D then AL, or AL then D then AL
MG3	1607	29	D
MG5	1355	988	D then AL
DC5	923	279	D; D then AL on some occasions
Naïve animals:			
AL41	1718	793	D then AL
AL42	2292	124	D then 2-sec burst of AL
AL43	1981	104	D
AL44	1397	20	D
AL45	1881	71	D
AL47	1175	2336	D then AL
AL48	1496	71	D
AL49	852	1547	D then AL
AL50	1322	382	D, then AL on about half of the occasions

[a] Rats were tested with an airstream and water simultaneously available during SIP-inducing conditions. The animals were deprived of food to 80% of their free-feeding body weights and during 90-min sessions were given 45-mg Noyes pellets on a noncontingent FI 1-min schedule. Results shown are the mean times spent in each activity for the last four sessions of testing and the general pattern of behavior observed during each interpellet interval. D = drinking and AL = airlicking.

As discussed in Section 3, all of the available evidence is consistent with the notion that the magnitude of SIP is regulated by volume-correlated feedback. However, it may be that drinking motor activity itself makes a substantial contribution to the reinforcement obtained from drinking during SIP sessions. It has been reported that water-deprived rats tend to maximize durations of drinking (Wike, 1965; Wike and Barrientos, 1957). It was found that rats develop a preference for the side of a T-maze in which a fixed amount of water is available from a drinking tube with a small orifice, to the other side of the maze, where the same amount of water is available from a tube with a large orifice (Wike and Barrientos, 1957). This tendency to maximize consummatory behavior cannot be extended to rats that are choosing between 10 sec of access to water and 5 min of access to an airstream (Chapter 4). The airlicking observed during scheduled feeding sessions even with water concurrently available could be explained in terms of a tendency for the animals to maximize the duration of consummatory activity by beginning to drink immediately after consuming each food portion, but by switching to airlicking prior to the moment at which the drinking tendency would become satiated through the ingestion of a particular quantity of water.

7. Effects of Alterations in or Elimination of the Orolingual Consequences of Feeding

7.1. Sweetened Dry-Food Pellets

A fairly consistent finding regarding SIP has been that increasing the sugar content of dry-food reinforcements decreases SIP. Falk (1967) initially found that SIP was reduced by 61% when sucrose pellets were substituted for standard Noyes pellets and by 74% when dextrose pellets were used. Another study (Christian and Schaeffer, 1973) reported that SIP (FT 1-min schedule) was inversely related to the sugar content of food-pellet reinforcements. Whereas standard Noyes food pellets (containing 7.5% glucose) caused a mean increase of 29 ml in water intake above baseline conditions (100 of the appropriate type of pellets delivered at once; 100-min sessions), sugarless pellets caused a mean 36-ml increase. When the sugarless pellets were adulterated with sucrose, the mean increase in intake above baseline was depressed: 4% sucrose caused a 24-ml increase above baseline, 16% sucrose a 19-ml increase, and 32% sucrose caused a mean increase of only 13 ml above that observed during baseline conditions.

Christian (1976) later reported that 45-mg 90%-dextrose food pellets reduced water intake to levels similar to those observed during baseline conditions (all pellets delivered at once), although bar-pressing rates (FI 1-min) were similar to those obtained for sugarless pellets. In this study the water intake of two rats was reduced to levels that were less than those observed during baseline conditions, while in a third rat the use of 90%-dextrose pellets reduced intake levels by a mean of 15.2 ml, to a level that was still 15.5 ml above that observed during baseline conditions. The finding that intakes were reduced to below baseline levels might be related to the fact that standard Noyes pellets were used for baseline measurements; perhaps this finding would not be duplicated if SIP intake were compared with a baseline condition using 90%-dextrose pellets, like the comparison made in the Christian and Schaeffer (1973) study.

The finding that levels of SIP are inversely related to sugar content of dry food-pellet reinforcements has been replicated in several other investigations (Christian, 1975; Christian, Reister, and Schaeffer, 1973; Christian and Schaeffer, 1975; Colotla and Keehn, 1975). One additional finding regarding the composition of dry reinforcements is that the substitution of sweetened nonnutritive pellets for standard Noyes pellets reduced mean water intake during SIP (E. X. Freed, 1971). In this study, the delivery of nonnutritive pellets at random on half of the reinforcement occasions reduced SIP by 24% (from 15.3 ml/100 g body weight to 11.6 ml/100 g during a 45-min session). The exclusive use of nonnutritive pellets reduced mean intake by 75%, to 3.77 ml/100 g body weight. It was reported that on the 50% nonnutritive-pellet schedule, "drinking bursts generally followed each reinforcement but were seemingly not as long (or as efficient) as when all pellets were food. Generally, all the pellets delivered were eaten" (E. X. Freed, 1971, p. 368). The authors did not report whether all bouts of drinking were reduced to a similar degree when half of the food pellets were nonnutritive, or whether those drinking bouts that followed nonnutritive pellets were diminished to a greater degree than those that followed nutritive pellets.

A variety of experiments can be designed to separate oral (e.g., sweetness) versus postingestional (e.g., blood-glucose elevation) factors in the suppression of SIP by sugar reinforcements. A third effect of sugar pellets—that they may be completely dissolved in the mouth and thus swallowed differently—could also be investigated experimentally. For example, nutritive food pellets could be sweetened with saccharin or other nonnutritive sweeteners, or food pellets could be composed of the slowly metabolized sugars, mannitol and sorbitol. Pellets composed of small-molecular-weight starches, proteins, nonsweet amino acids, or peptides such as glycylglycine that are soluble and provide calories but do not taste sweet could be used to duplicate the "swallowability" characteristics of sugar pellets. To duplicate experimentally the short-latency effects of sugar on blood glucose while eliminating the solubility characteristics of sugar pellets, intravenous glucose injections could be given concurrently with the ingestion of nonnutritive pellets.

7.2. Liquid Food

Several attempts have been made to obtain SIP through the use of liquids rather than food pellets as reinforcers (see Table 2). Failures have been reported with 0.15-ml portions of 33% Borden's sweetened condensed milk (Stein, 1964), 0.06-ml portions of Wesson oil (Stricker and Adair, 1966), 0.08-ml portions of 30% sucrose solution (Falk, 1967), and 0.02-ml or 0.08-ml portions of Metrecal (Falk, 1967). In contrast, standard liquid monkey diet (Nutritional Biochemicals) has proved to be highly effective (Falk, 1967; Hawkins *et al.*, 1972). Falk (1967) found that drinking engendered by 22-mg portions of this diet was comparable to that obtained when 45-mg Noyes pellets were used. Larger (63-mg and 353-mg) portions of the diet were much less effective, probably because they caused excessive weight gains during the long (3.33 hour) sessions. Later, Hawkins *et al.* (1972) reported that substantial SIP was produced by the delivery of 90-mg portions of liquid monkey diet during shorter (50-min) sessions that used an FT 1-min schedule. The SIP observed by Hawkins and his colleagues developed more slowly and reached a lower asymptote than that obtained when 45-mg Noyes pellets were used. Of 10 animals tested with liquid diet, 4 drank a mean of only about 0.2 ml of water per food portion, an amount considerably less than the mean of about 0.4 ml/portion drunk by the worst of 10 animals receiving 45-mg Noyes pellets. The SIP of the Noyes-pellet

Table 2.   *Effectiveness of Liquid Foods in Producing SIP*

Food	Amount	Result	Reference
30% sucrose	0.08 ml	No SIP	Falk, 1967
Metrecal	0.02 ml	No SIP	Falk, 1967
Metrecal	0.08 ml	No SIP	Falk, 1967
Wesson Oil	0.06 ml	No SIP	Stricker & Adair, 1966
33% Borden's sweetened condensed milk	0.15 ml	No SIP	Stein, 1964
Liquid Monkey Diet	22 ml	Similar to that with 45-mg pellets	Falk, 1967
Liquid Monkey Diet	63 ml	No SIP[a]	Falk, 1967
Liquid Monkey Diet	353 ml	No SIP[a]	Falk, 1967
Liquid Monkey Diet	90 ml	63% of that with 45-mg pellets[a]	Hawkins *et al.*, 1972

[a] See text.

group stabilized after about 10 days of testing, while the liquid-diet group stabilized after about 13 days. The mean asymptotic amount of water ingested per food reinforcement was about 0.73 ml for the Noyes-pellet group and about 0.46 ml for the liquid-diet group.

The decrements in SIP in animals receiving liquid reinforcers may be understood in terms of two factors: (1) the fact that some liquid is provided by the reinforcer itself and (2) the possibility that the sweetness or the sugar content of the liquid diet attenuates SIP as does sucrose when added to dry food pellets (see Section 7.A). As prepared by Falk (1967), 90 mg of the liquid diet would contain only about 30 mg of water; the ingestion of this amount of water could account for only about 10% of the decrement in SIP that was brought about by the use of liquid reinforcers by Hawkins *et al.* (1972). However, the diet also contained 51% sucrose in terms of dry weight, or about 31% sucrose in terms of the wet weight. The use of dry food pellets containing 32% sucrose attenuates SIP by about 32% (i.e., about 0.23 ml/reinforcement) as compared to standard Noyes pellets delivered on the same reinforcement schedule (Christian and Schaeffer, 1973). This rate is very close to the mean attenuation of 0.27 ml/reinforcement brought about by the use of a 30% sucrose diet by Hawkins *et al.* (1972).

There is currently no satisfactory explanation for the fact that other liquid foods, such as Metrecal, 30% sucrose in water, oil, and milk, do not generate SIP. We would like to suggest that these other liquid foods do not generate SIP because they are more-or-less completely ineffective in generating water-drinking behavior—prandial or otherwise—in normal home-cage or non-SIP situations. Falk (1971) has suggested that the scheduling condition prevalent during SIP (a "drive operation") exaggerates behavior that would otherwise have a low but nonzero probability in the context of feeding. Falk (1967) also conducted an experiment during which animals, water-satiated but deprived of food to 80% of their free-feeding body weights, were given rations of various diets while water intake was recorded for the subsequent 3.5 hr. When a ration of Noyes pellets was used for several sessions, the animals ingested a mean of about 8.0–11.2 ml of water. Sucrose and dextrose pellets engendered somewhat less drinking, about 6.3 and 7.7 ml respectively. The amount of drinking following standard liquid monkey diet was still less, about 5.6 ml. However, the volume of water ingested during the 3.5 hr after the presentation of either Metrecal or 39% sucrose was only about 0.5 ml. Thus, the two liquid foods that failed to engender drinking under non-SIP conditions also failed to engender SIP, but liquid monkey diet, which engendered drinking under non-SIP conditions, did engender SIP. This finding supports the conclusion of Cohen and Mendelson (1974) that, "It seems that the reinforcer itself must be capable of engendering a certain base level of drinking when available on an *ad lib.*

basis, otherwise it will fail to promote schedule-induced drinking" (p. 28). This conclusion is similar to the suggestion of Falk (1971) that SIP involves an increase in the rate of a behavior that is already somewhat probable in the situation and suggests specifically that the normal tendency of a food to elicit drinking is a necessary ingredient for the production of SIP. This property of a food is, however, not a sufficient condition, because sugar pellets did not cause SIP even though they engendered food-associated drinking (Falk, 1967). Thus, the reason that sugar pellets are ineffective in producing SIP is probably not because of factors related to their effectiveness in producing normal prandial drinking.

7.3. Is Food Essential?

7.3.1. Can Conditioned Stimuli Substitute for Food?

Several investigators have examined the question of whether a neutral stimulus that has been repeatedly paired with food is capable of substituting for the food pellet as a stimulus to initiate schedule-induced drinking. These investigations have employed second-order, or percentage-reinforcement schedules (Ferster and Skinner, 1957); following the completion of each fixed-interval (FI) component, an auditory and/or visual stimulus (the conditioned stimulus) is presented, but a food-pellet reinforcement is delivered after only some fraction (percentage) of the fixed intervals. In the initial study of this type, Rosenblith (1970) delivered a brief light flash and a click after the completion of each fixed-interval component of the schedule and simultaneously reinforced every third FI component with a food pellet. She consistently observed drinking in the intervals following the flash–click alone, but this drinking was weaker than that observed after pellet delivery, in that: (1) it required many more sessions of training to develop than did drinking after pellet delivery; (2) it was frequently interrupted by bar pressing, and (3) it was not consistently initiated immediately after the presentation of the conditioned stimuli. Rosenblith interpreted her findings in terms of an analogy to the classical-conditioning paradigm; the food pellets served as an unconditioned stimulus, and the flash–click stimulus served as a conditioned stimulus.

Five other studies have investigated SIP during similar second-order reinforcement schedules:

1. Falk and Bryant (1971, as cited in Falk, 1971) observed "relatively little post-buzzer drinking" (p. 584) when presenting a 3-sec buzzer after each FI 1-min component and a food pellet after half of the FI components.

2. Wuttke and Innis (1972) employed a procedure very similar to that used by Rosenblith (1970)—including some of the same animals—and obtained similar results.

3. Porter and Kenshalo (1974) used rhesus monkeys and paired the offset of a 3-sec white noise with the delivery of food during training (until lever-pressing behavior stabilized) and throughout testing, but after training was completed, they reinforced only 80% of the FI components with food pellets. For four of the seven animals, drinking after the noise offset alone was similar in quantity to that observed after food-pellet delivery, but for the other three animals, drinking after nonreinforced intervals was only 39%, 24%, and 1% of that which occurred after food-pellet delivery.

4 and 5. In two other studies using rats, Porter, Arazie, Holbrook, Cheek, and Allen (1975) and Allen, Porter, and Arazie (1975) did not observe drinking following light-plus-noise stimuli (Porter *et al.*, 1975) or light stimuli alone (Allen *et al.*, 1975) except in one of three (Porter *et al.*, 1975) or one of six (Allen *et al.*, 1975) animals that were tested. In both studies, it was established that the conditioned stimuli had discriminative control over pauses in bar pressing. Extensive pretraining (30–33 sessions) with 100% of intervals reinforced was used in both studies. In one study (Porter *et al.*, 1975), 90% of the intervals were reinforced during all test sessions, and in the other study (Allen *et al.*, 1975) 10%, 30%, 50%, 70%, and 90% of the intervals were reinforced on different test sessions.

The conclusion from these experiments is that drinking can only sometimes be conditioned to occur after the presentation of previously neutral stimuli that have been repeatedly paired with food delivery. This phenomenon is relatively weak: it develops only when certain experimental procedures are used, it is not easy to replicate, it can be observed only in some animals, and even when drinking does occur it is less robust than drinking engendered by food pellets. An additional characteristic of drinking following conditioned stimuli is that such drinking takes much longer to develop than does SIP, which suggests the involvement of an additional conditioning process. Thus, even though schedule-induced drinking engendered by food pellets requires some conditioning, drinking following nonorolingual stimuli requires far more.

An additional interesting finding reported by Allen *et al.* (1975) is that although no drinking was evident following intervals lacking food reinforcement, the amount of drinking that occurred following food-reinforced intervals was greater when a smaller percentage of intervals was reinforced. Thus, the mean amount of water ingested per pellet was about 0.58 ml when 100% of the intervals were reinforced, but a mean of about 1.85 ml was ingested after each pellet when 10% of the intervals were reinforced with food pellets on a random basis. All intervals were followed by a light stimulus (house light off and a cue light over the response lever turned on for 1 sec). Allen *et al.* (1975) pointed out that this and related findings (Jacquet, 1972; Porter *et al.*, 1975) are analogous to the positive-contrast phenomenon (Reynolds, 1961), in which response rates during unchanged reinforced components of multiple schedules are increased after the imposition of extinction on the other schedule component. However, when one considers the relationship between their findings and the bitonic function relating rate of pellet delivery to amount of water ingested during SIP (see Section 2.3), an additional consideration arises. If the light-change stimuli in the Allen *et al.* (1975) study were entirely ineffective, one would expect that drinking would be greatly reduced, rather than greatly enhanced, by the omission of 90% of the reinforcers. When the rate of food-pellet delivery is reduced from 1/min to 1/5–8 min, SIP is generally reduced or eliminated (see Section 2.3). In the Allen *et al.* (1975) study, however, a mean rate of pellet delivery of 1/10 min resulted in increases in the amount of water ingested per pellet. Thus, the light-change stimuli were apparently capable of eliminating, even reversing, the SIP-attenuating effect of long IPIs. If we may speak of mechanisms for the moment—that is, the mechanism that is responsible for mediating the SIP-engendering effect of intermediate IPI durations—conditioned light-change stimuli must be capable of effectively influencing this mechanism in a manner similar to, or with the same result as, orolingual or postingestional stimuli provided by food. Nonetheless, conditioned light-change stimuli are either completely or relatively ineffective as compared to food in terms of initiating bouts of drinking. They merely extend the effective interval during which the previous food reinforcer can interact with the next food reinforcer (or the anticipation thereof) to induce drinking. In this sense they contribute to the maintenance of the postfeeding motivational state which engenders SIP.

In view of these considerations, an alternative explanation of the Allen *et al.* (1975) "positive-contrast" finding can be made in terms of the fact that animals ingest precise volumes of water during SIP (see Section 3). It may be that the presentation of the conditioned stimulus during most of the FI components is not able to initiate drinking behavior but it seems to reset the satiation mechanism in such a way that greater quantities of water are ingested when the animal finally does drink in response to the feeding stimulus that initiates drinking. Thus, when food is finally ingested the animals drink in excess, as if they are compensating for water that was not ingested after previous nonreinforced IPIs. We have also observed this kind of "compensatory drinking" in an experiment in which some IPIs were shortened so that the next pellet was delivered before the rats were able to ingest their customary volume. The rats drank more water following the short IPI but reverted to their usual volume after the next standard-length IPI (W. J. Freed and Mendelson, unpublished experiment, 1974). We are left with the task of determining the character of the motivational mediator that (1) is maximal at intermediate levels of schedule intermittency; (2) can sometimes be maintained by conditioned stimuli; and (3) is responsible for determining the excessive volumes of water ingested during SIP sessions.

7.3.2. Can Electrical Stimulation of the Brain Substitute for Food?

As a substitute for food pellets, several investigators have attempted to employ electrical stimulation of the lateral hypothalamic area of the brain (ESLH), which induces feeding and may mimic its reinforcing effects (Hoebel, 1968), in order to demonstrate SIP. In one such study, Atrens (1973) reinforced five rats working on FI or VI schedules with the opportunity to bar-press for five consecutive trains of ESLH of up to 2.5-sec duration, separated by at least 0.5 sec. Atrens reported that during the interreinforcement intervals two of the rats drank excessive quantities of water while three of them ate excessive amounts of wet mash. The maximum quantities (21–32 g) of water or food consumed were substantial compared to the baseline levels of drinking and eating which "never exceeded 5 g and were usually in the range of 1–2 g" (p. 322). However, these excessive intakes declined very rapidly and disappeared entirely after four or five sessions of testing. The excessive intakes could not then be reinstated by food deprivation (water but not food available), changes in ESLH intensity, or changes in the schedule of reinforcement. This finding is in striking contrast to the permanence of SIP during intermittent food-reinforcement schedules.

In another study, Wayner et al. (1973; also cf. Wayner, 1974b) allowed rats working on an FI 30-sec schedule to develop SIP with food reinforcements and then substituted 3-sec trains of ESLH for food. Only one of the four animals exhibited post-ESLH drinking, and this drinking was considerably less than postpellet drinking. This animal drank after 98% of the food pellets and ingested 18 ml of water during a pellet-delivery session, but it drank only after a mean of 27% of the ESLH reinforcements and ingested 68% less water. Also, post-ESLH drinking was tested for only two sessions and therefore might have disappeared during extended testing, as did the post-ESLH drinking reported by Atrens (1973). Thus, although it seems to be possible under some conditions to substitute ESLH for food reinforcement and still obtain some postreinforcement drinking, the drinking thus obtained is a much weaker and less permanent phenomenon than that engendered by the eating of food.

Cohen and Mendelson (1974) demonstrated the relative weakness of ESLH in this regard even more convincingly. These investigators employed "simultaneous" VI schedules of delivery of 45-mg Noyes food pellets and ESLH trains similar to those used by Atrens (1973), with the lever-pressing rates for food and ESLH roughly equated. A single response on a second lever effected a changeover from the ESLH schedule to the food schedule and vice versa. (Because there was no stimulus to indicate which reinforcer was available at any particular time, this schedule is known as a "simultaneous" rather than a "concurrent" one; Catania, 1966). These investigators found that while drinking was observed consistently following the delivery of food pellets, it occurred only rarely after the delivery of ESLH. Thus, a highly reinforcing intracranial stimulation (ICS) was not effective in inducing postreinforcement drinking even though postpellet drinking was occurring during the same sessions. These investigators concluded that in order to be effective in promoting schedule-induced drinking, a reinforcer must be capable of engendering some drinking when available on an *ad libitum* basis. Additional support for this hypothesis can be found in the literature on the effectiveness of liquid food reinforcers in producing SIP (see Sections 7.2 and 5.1). ESLH provides another instance in which a nonfood stimulus fails to engender substantial amounts of SIP.

8. A Theoretical Proposal

8.1. Can Orolingual Factors Fully Account for SIP?

In searching for a theoretical understanding of the SIP phenomenon, some investigators have ascribed crucial roles to the mouth-moistening properties of water or to the thirst-inducing properties of dry food (see Section 1.2). Although the close temporal relationship between the ingestion of food and the bouts of drinking (also *cf.* Section 4.2) certainly suggests that an orolingual factor is important in the phenomenon, the fact that long IPI's do not engender SIP effectively (see Section 2.3) indicates that orolingual factors cannot provide a full explanation for the phenomenon: any characteristic of food pellets that would tend to cause drinking should be just as effective even when IPI's are very long. The fact that the amount of water consumed per pellet is a bitonic function of IPI length makes difficulties for any theory of SIP that suggests that orolingual stimulation by food is sufficient for engendering SIP.

One of the recent theoretical suggestions regarding SIP has been that "the rat drinks a certain fixed amount of water after every bite, independently of the size of that bite" (Lotter, Woods & Vasseli, 1973, p. 478). These authors (Lotter *et al.*, 1973) suggest that SIP is simply an "artifact" of the large number of bites (i.e., reinforcements) eaten during a session of SIP. However, this hypothesis, which we will also refer to as the "many-meals" hypothesis, cannot account for the bitonic relation between magnitude of SIP and length of IPI. In fact, Falk (1964) and Stein (1964) initially proposed similar explanations for SIP: "In my experiment meal size is defined for the animal. I don't have any evidence and this is speculation but the defined meal size is 45 mg and the animal drinks after he eats as if it were a regular meal. I keep serving him meals and he drinks as if he had 180 meals" (Falk, 1964, p. 115). Stein in 1964 independently proposed the same explanation. However, Falk was forced to discard this hypothesis after he discovered the bitonic function relating SIP to fixed-interval length (Falk, 1966b, p. 37).

As a consequence of the bitonic relationship between IPI length and water consumed per pellet, a *mediating factor* must be interposed to account for the excessive drinking that is observed at moderate IPI lengths. This mediating factor must be such as to augment the normal tendency for animals to engage in postprandial drinking; this augmentation must occur most strongly with moderate IPI lengths (30 sec to 3 min) but not with long IPIs (greater than 5–8 min). The hypothetical mediator may or may not be decreased at very short IPIs; Cohen (1975) has suggested that the decreases in drinking under these conditions may merely be an effect of time limitations (i.e., a kind of "floor effect").

There have been several explicit or implicit suggestions as to the nature of the motivational mediators of SIP. These include:

1. Emotion or frustration in various forms (Segal, 1972; Denny and Ratner, 1970; Kissileff, 1973; Stricker and Adair, 1966; Thomka and Rosellini, 1975).
2. Excitation of the lateral hypothalamus that induces behavioral excitability (Wayner, 1970, 1974a,b; Wayner *et al.*, 1973; Cook and Singer, 1976).
3. Arousal (Killeen, 1975).
4. Hypothalamic warming (Carlisle, 1971, 1973).

However, none of the proponents of these candidates for a mediator have provided satisfactory arguments as to why long IPIs do not effectively engender SIP. It must simply be assumed, for example, that long IPIs are just not as "frustrating" as moderate ones.

Another theoretical suggestion (Falk, 1971) is that SIP is an instance of a stable displacement behavior brought about by the repeated thwarting of feeding tendencies as a consequence of the intermittent schedule of feeding. This theory is on a different explanatory level than the others, because thwarting is used in the sense of an interruption or blocking of an activity that would be otherwise continued. Thus, Falk's theory steers clear of mediating psychological states. Although it might be said that long IPIs simply do not engender stable displacement behaviors, a fully satisfactory explanation will eventually require the identification of the motivational mediator that is at a peak level at those IPI lengths that most effectively engender SIP. This mediator may very well have a psychological name (e.g., "arousal" or "frustration"), but to be fully satisfactory, an independent measure of the psychologically named mediator will have to be identified. This measure would also have to be a widely accepted measure of the psychological construct and one that is valid in situations other than SIP. At any rate, it is certainly not enough simply to say that SIP is mediated by "frustration," for example, without providing any evidence that those parameters that most effectively engender SIP also most effectively engender "frustration."

8.2. Precedents for Our Theoretical Position

Our theoretical position regarding SIP is an outgrowth of an earlier suggestion that SIP is related to two factors: (1) the normal tendency for low base rates of drinking to occur in most *ad libitum* feeding situations (Falk, 1971, 1972); and (2) the involvement of a "drive operation," for example, "an operation that changes the effectiveness of a stimulus as a reinforcer" (Catania, 1968, p. 334, as cited by Falk, 1971, p. 586).

Falk (1971, 1972) has considered adjunctive behavior, of which SIP is a prominent example, to be "an increase in behavior already present in the situation at a lower base rate" (Falk, 1971, p. 584). One datum supporting this suggestion is that decreasing the base rate of drinking in the SIP-engendering situation to zero prior to the acquisition of SIP, by stomach water preloads, prevents the acquisition of SIP (Chapman and Richardson, 1974), while similar preloads given to animals with established SIP have little effect on their drinking (Falk, 1969). Although Falk (1971, 1972) did not attempt to identify the factors that produce the base rates of drinking, he specified that the schedule conditions "probably do not generate such behavior as new responses, but as large increases in the base rate of a behavior already present as a response to the current situation" (Falk, 1971, p. 584).

Our theory involves a novel motivational state, for example, the drive operation invoked by Falk (1971, 1972). The purpose of the theory that we outline below is to suggest a specific basis for the drive-operation characteristics of SIP that have been referred to by Falk. We not only suggest a specific physiological basis for this curious motivational state, but we outline specific characteristics of the motivational state in terms of a possible similarity to other naturally occurring motivational conditions.

8.3. Is SIP an Insulin-Dependent Phenomenon?

In this section, we develop the hypothesis that SIP is related to an excessive secretion of insulin, which is induced by certain intermittent schedules of food reinforcement. Specifically, we suggest that SIP-engendering feeding schedules cause excessive insulin secretion, which in turn causes hypoglycemia and an exaggerated motivational state. This exaggerated motivational state is sometimes manifest in terms of drinking; in particular, we suggest that it accentuates the normal tendency of animals to engage in feeding-associated postprandial drinking. We discuss here the theoretical basis for the insulin hypothesis in terms of data regarding conditioned insulin secretion in feeding-associated contexts and the motivational effects of insulin, and we show how the insulin hypothesis can readily account for many findings that pose difficulties for other theories.

It is known that insulin injections can cause feeding, obesity, and increased motivation to obtain food (Booth and Pitt, 1968; Booth and Brookover, 1968; Panksepp, 1975; Jacobs, 1958; Woods, Decke and Vasseli, 1974), effects that presumably are secondary to its influence on blood-glucose levels (*cf.* LeMagnen, 1971; Woods *et al.*, 1974; Steffens, 1969). These findings and a series of reviews by Woods and Porte (1974) and by Woods *et al.* (1974) provide the basis for our theory. Woods and his colleagues have concluded that excessive secretion of insulin and consequent hypoglycemia can be conditioned to occur in response to a wide range of food-associated stimuli. For example, animals exhibit severe hypoglycemia when they ingest saccharin, which presumably provides a sweet stimulus that has been associated with food (R. Deutsch, 1974; Tepperman, 1975; Thompson and Mayer, 1959). Insulin secretion can be conditioned to occur following the sight or the smell of food (Parra-Corarrubias, Rivera-Rodriguez, and Almarez-Ugalde, 1971; Penick, Prince, and Hinkle, 1966), the suggestion of food (Goldfine, Abraira, Gruenwald, and Goldstein, 1970), and the time of day associated with feeding (Balagura, 1968; R. Deutsch, 1974; Woods, Alexander, and Porte, 1972; Woods, Hutton, and Makous, 1970). Following the onset of a meal, a biphasic insulin-secretion response has been observed, consisting of one peak of blood insulin between 5 and 15 min following the onset of a meal, and another peak at 30–60 min (Fischer, Hommel, Ziegler, and Michael, 1972; Hommel Fischer, Retzlaff, and Knofler, 1972; Steffens, 1970; Tepperman, 1975). Animals fed intragastrically show only the second peak (Hommel *et al.*, 1972), which suggests that the first peak is elicited by external food-related cues, while the second peak is related to the postingestional consequences of eating (also see Tepperman, 1975). It has even been suggested that "Insulin may also prime feeding control systems in anticipation of a meal" (Panksepp, 1975, p. 664). In summary, animals secrete insulin in response to external cues that are associated with food ingestion prior to the absorption of the food and even in anticipation of its presentation. Thus, it is possible that when an animal is receiving one small food morsel every minute, the combined effects of intermittent priming by food and by stimuli associated with the testing situation could lead to an amount of insulin secretion that is excessive in relation to the amount of food that is actually ingested. Since at least some amount of new conditioning must be involved in this process, the fact that SIP requires the presentation of several hundred reinforcements to develop fully (see Section 2.7) is entirely consistent with our hypothesis.

Our hypothesis is in one respect a reversal of the many-meals hypothesis of Lotter and her colleagues (1973), and it is in reality a "one-meal" hypothesis. We suggest that SIP is not a consequence of the rat's treating each of the 180 45-mg food pellets that are dispensed over a 3-hr period as many little meals but that SIP is a consequence of the rat's treating the 180 individual food pellets as one large, extended meal. Our hypothesis is thus an explanation for the exaggerated amount of fluid that is ingested during SIP. The facts that (1) drinking occurs at all and (2) SIP drinking is usually postpellet in locus may be an example of the well-known postprandial drinking effect, while the excessive aspects of SIP may be best explained by the insulin hypothesis. Thus the postpellet drinking phenomenon may be explainable in terms of the many-meals hypothesis, but the exaggerated levels of drinking are best explained by the "one drawn-out meal" hypothesis.

In other words, we believe that the amounts of insulin secreted during SIP sessions are similar to those that would be secreted if the animal were eating continuously throughout the session. If continuous food ingestion were sustained for this length of time, the hypoglycemia-inducing effects of the insulin would be neutralized by the glucose derived from the ingested foodstuffs. In the SIP situation, however, we suggest that the amount of food ingested is insufficient to neutralize the hypoglycemia but provides a sufficient unconditioned stimulus to maintain a large conditioned secretion of insulin in response to peripheral stimuli associated with the testing situation. Thus, the amount of insulin that is secreted during a session of SIP is excessive in relationship to the amount of food that is actually consumed, so that an extreme hypoglycemic state may develop during SIP. These abnormal physiological conditions may result in a heightened motivational state; it is likely that this motivational state is similar to extraordinarily severe hunger. This heightened motivational state would constitute the characteristic of SIP that has been referred to by Falk (1971) as the "drive operation" (*cf.* section 8.2). When combined with intermittent feeding and the availability of water this motivational state could potentiate SIP; when other stimulus objects are available, however, the motivational state could potentiate other adjunctive (*cf.* Falk, 1971) behaviors.

With the effective range of feeding intermittency that generates SIP, insulin levels may remain elevated during the interval between pellet deliveries. At longer IPIs, the individual pellets may function as separate meals so that the insulin level attained after the eating of a pellet may peak and begin to fall before the next pellet arrives. This progression would culminate in only moderate levels of circulating insulin; consequently, the elevated insulin levels that are a necessary condition for SIP would be absent. For shorter IPIs, SIP may be attenuated either because the rate of food ingestion is sufficient to neutralize the hypoglycemia or simply because the time available for drinking is limited (Cohen, 1975). In this way, the insulin hypothesis can account for the characteristic bitonic function relating SIP to FI length.

We believe that the heightened motivational state brought about by excessive insulin secretion during SIP may be similar to extreme hunger. There are some data in the literature that suggest that a heightened state of hunger is an accompaniment of SIP. For example, conditioned taste aversions are not easily demonstrated during SIP, which may also hold true for heightened states of hunger (see Section 5.4). Also, rats drink more alcohol during intermittent feeding schedules than when they are just food-deprived (Lester and E. X. Freed, 1972; Samson and Falk, 1974a); there have been suggestions that the caloric value of ethanol is involved in this phenomenon (E. X. Freed, 1972, 1974; Lester and E. X. Freed, 1972). This possibility may mean that intermittent feeding schedules increase rats' motivation to obtain calories via ethanol.

However, there is no direct evidence that hunger is increased during SIP to levels above that produced by mere food deprivation or that extraordinary hunger is the mediating motivational variable that causes SIP within the effective range of feeding intermittencies. The accentuated hunger state that may accompany SIP could be indexed in terms of measures that have been shown to be sensitive to food-deprivation levels (*cf.* Allen, Stein and Long, 1972; Miller, Bailey, and Stevenson, 1950). It is also possible, however, that the heightened motivational state that occurs during SIP is dissimilar to hunger in various respects, although still caused by insulin.

We should make clear that it is our position that excessive insulin release and the resultant hypoglycemia are necessary but not sufficient conditions for the production of SIP. Food deprivation and intermittent food delivery are also necessary conditions. There is substantial evidence, however, that insulin alone is sufficient to increase drinking (Booth and Brookover, 1968; Booth and Pitt, 1968; Novin, 1964). Also, the amounts of water ingested during food deprivation are excessive in relation to body needs for water (Baumann, Guyot-Jeannin, and Dobrowolski, 1964; Cumming and Morrison, 1960; Millar and Morrison, 1968; Morrison, 1967; Morrison, MacNay, Hurlbrink, Wier, Nick, and Millar, 1967). Furthermore, there is some normal tendency for rats to engage in drinking during and after feeding bouts (see Falk, 1971, 1972); this tendency is also probably a necessary condition for SIP.

SIP is greatly decreased in rats that have recovered from lateral hypothalamic lesions (Falk, 1964). Such animals show certain long-lasting deficits in drinking behavior (Stricker, 1976); these behavioral deficits might account for such a finding. However, such animals also do not eat in response to insulin injections (Epstein and Teitelbaum, 1967), and according to our theory, this finding would be sufficient to explain the absence of SIP in animals that have recovered from these lesions. Recently, Blass and Kraly (1974) have shown that lesions of the medial forebrain bundle cause a specific deficit in the response to glucoprivation without accompanying deficits in water regulation. In accordance with the insulin hypothesis, we predict that such lesions will also eliminate SIP.

Another effect that can be accounted for by the insulin hypothesis is the absence of or reduction in SIP when Noyes pellets are replaced by pellets containing large amounts of sucrose or dextrose (Falk, 1967; Christian, 1976; Christian, Reister, and Schaeffer, 1973). Previous investigators (Falk, 1971; Christian, 1976) have related this effect to a decreased preference for such sweetened pellets by food-deprived animals. However, sugar would be expected to alleviate the hypoglycemia caused by excessive secretion of insulin more effectively than would other substances, and, in fact, Booth and Pitt (1968) have shown that glucose, but not nonmetabolizable sugars, attenuates insulin-induced feeding when administered concurrently. Jacobs (1958) found that insulin injections induce strong preferences for glucose solutions and suggested as a reason the ability of glucose to remedy the resultant glucoprivation more effectively than can other foods. Thus, sucrose and glucose reinforcements may not effectively engender SIP because these substances more effectively neutralize hypoglycemia than do other foods.

The insulin hypothesis is also consistent with the findings regarding SIP during second-order reinforcement schedules. Although drinking can sometimes be demonstrated following the presentation of initially neutral stimuli that have been repeatedly paired with food, the phenomenon is relatively weak, develops slowly, and is difficult to obtain (see Section 7.3.1). These findings are entirely consistent with the data regarding conditioned insulin secretion (Woods and Porte, 1974). Although a short-latency insulin response can be elicited by orolingual stimulation associated with food (Woods and Porte, 1974; Tepperman, 1975), nonorolingual stimuli may also be conditioned to induce insulin secretion (Woods and Porte, 1974). Accordingly, nonorolingual stimuli such as lights and tones could be conditioned to induce insulin secretion during second-order reinforcement schedules and thereby to produce SIP. However, some additional conditioning would be necessary; insulin secretion has been conditioned to orolingual stimuli long before the start of SIP testing, but it has not been similarly conditioned to neutral stimuli such as lights and tones. Thus, the insulin hypothesis would predict a delayed development of drinking following conditioned stimuli. However, because of the absence of the orolingual stimuli that usually induce the secretion of insulin, as well as the absence of the normal tendency of eating to induce a small amount of drinking, one would expect the drinking response to conditioned stimuli to be relatively weak, as is the case.

Finally, the finding that hunger during pellet consumption is neither necessary nor sufficient for SIP but that hunger during the IPI is a necessary condition (Bruce-Wolfe *et al.*, 1976; also see Section 2.1) is consistent with the idea that a hungerlike state is the motivational factor that induces SIP. It is also interesting that ESLH, which was used to induce hunger artificially in the Bruce-Wolfe *et al.* (1976) study, itself causes short-latency insulin secretion (Kuzuya, 1962; Steffens, Mogenson and Stevenson, 1972); this may contribute to its effectiveness in the induction of SIP.

In conclusion, it is of interest to note that the insulin hypothesis is entirely consistent with Falk's (1971, 1972) interpretation of SIP. He considered the requisite intermittent feeding schedule to be a drive operation that exaggerates the normal tendency of animals to engage in drinking behavior during the SIP situation or in similar situations. The insulin hypothesis may lead to the identification of the physiological basis for this drive operation. The insulin hypothesis is also, in general, consistent with emotionality interpretations of SIP. There have been many reports in recent years that suggest that hypoglycemia is a cause or a correlate of emotional disorders (Ford, Bray, and Swerdloff, 1976). Our hypothesis that hypoglycemia is an accompaniment of intermittent feeding schedules with "thwarting" characteristics (*cf.* Falk, 1971) is compatible with interpretations of SIP in terms of emotional mediators.

9. Summary and Conclusions

That animals regulate volumes of ingested water during SIP suggests that schedule-induced drinking is reinforced by volume-correlated sensory feedback as obtained through the ingestion of water. Saccharin, quinine, and conditioned taste aversions do affect SIP to some degree, but in general, these effects are relatively minor as compared to the effects of these substances on drinking motivated by water deprivation alone. Ethanol may influence SIP, primarily through its pharmacological effects. Near-total sensory denervation of the tongue, although decreasing SIP somewhat, does not eliminate it, and thus most sensory feedback from the tongue is not essential for reinforcing SIP. Water temperature and saline do affect SIP substantially and in ways that roughly parallel the effects of these manipulations on drinking induced by water deprivation. Saline may affect SIP because of its taste, its osmolarity, or both. Since an airstream, which is reinforcing because of its cooling effects, sustains schedule-induced licking, while near-total sensory denervation of the tongue does not eliminate SIP or schedule-induced airlicking, we suggest that cooling of the extralingual oral tissues or of the hypothalamus contributes to the reinforcement of schedule-induced licking and probably mediates the efforts of water temperature on SIP.

It has been reported that rats will self-inject water intravenously in order to regulate body-fluid content (Nicolaidis and Rowland, 1974). Self-injection drinking demonstrates that rats will maintain their water balance "in the absence of all extrasystemic metering of reinforcement" (p. 12). In Nicolaidis and Rowland's study (1974), each lever press delivered 0.8 ml of fluid over a 2-min period. Since this delivery rate approximates the rate at which rats drink during SIP, the self-intravenous drinking technique could be used to determine whether peripheral feedback is a necessary condition for the occurrence of SIP. The tongue-denervation experiments were inconclusive with regard to this question (see Section 4). Based on the experiments reviewed above, which indicate the importance of sensory feedback in SIP, we would predict that rats will not display SIP when their only means of obtaining water is by self-intravenous injection.

The tendency of food reinforcers to engender drinking when available on an *ad libitum* basis predisposes the emergence of SIP and is therefore a requisite attribute of a food, or of any other stimulus that is to be considered a candidate for the induction of SIP. There are a number of liquid foods that are not effective in producing SIP.

One liquid food, liquid monkey diet, does engender SIP, but it is an unusual liquid food in that it has been shown to engender substantial amounts of drinking even when available *ad libitum*. Neutral stimuli that have been repeatedly paired with food are of limited effectiveness in inducing poststimulus drinking, as is brain stimulation that induces feeding. The SIP thus obtained occurs less consistently, and even when it is observed, it is less robust than SIP engendered by food reinforcers. The ineffectiveness of these stimuli cannot be attributed entirely to a relative weakness of these stimuli as reinforcers, since brain stimulation and sugar pellets are also relatively ineffective, even though they are strong reinforcers.

Food consumption is probably associated with certain stimuli that are essential to the induction of scheduled drinking. Although it is not now possible to specify the nature of these stimuli, it is likely that they are the same ones that cause food to engender prandial drinking outside of the SIP context. It is possible to test for the presence of this attribute of food by presenting it *ad libitum* and determining whether the subsequent drinking is similar to that observed after the *ad libitum* ingestion of dry food pellets. This attribute of food seems to be *necessary* for SIP, but it is not a guarantee that the food will be effective in producing SIP; sugar pellets engender prandial drinking when available *ad libitum*, but they do not engender SIP.

That SIP is greatly attenuated by the use of long IPIs suggests that neither oral factors, such as food dryness, nor the tendency for animals to drink certain volumes of water following meals can provide a full account of SIP. Falk (1971) suggested that the SIP-inducing food-reinforcement schedule is a drive operation that enhances the initial tendency for some drinking to occur in the SIP context. This enhancement might take the form of an increase in the probability with which food ingestion will be followed by drinking, plus an increase in the amounts of water that will be ingested during each postprandial bout of drinking.

We have attempted to provide a suggestion as to the physiological mechanisms that underlie this drive operation. Various considerations and experimental data regarding SIP led us to suspect that conditioned secretion of insulin (*cf.* Woods *et al.*, 1974) mediates the motivational aspect of SIP, that is, the drive operation. Conditioned stimuli associated with food ingestion and with the SIP context might enhance the secretion of insulin. The amounts of insulin thus secreted might be large, perhaps comparable to the amounts that would be secreted if food ingestion were maintained continuously throughout the SIP session. During SIP, however, the amounts of food that are ingested are small, and they provide insufficient glucose to neutralize the hypoglycemia that is a consequence of the insulin secretion. Thus, we suggest that hypoglycemia occurs during SIP to a degree that is rarely encountered in other situations. The hypothesized hypoglycemia is translated by the brain into a powerful motivational state that may be similar to extraordinarily severe hunger. It is this motivational state in combination with intermittent feeding that results in SIP.

These suggestions have yet to be verified experimentally. If the insulin hypothesis is correct, the implication would be that intermittent reinforcers other than food would be ineffective in causing SIP-like phenomena. On the other hand, the fact that schedule-induced behaviors other than drinking are demonstrable with the use of food reinforcers (*cf.* Falk, 1971) is not inconsistent with the insulin hypothesis. Thus, the generalizability of the adjunctive-behavior paradigm may be restricted in terms of reinforcer specificity but probably not in terms of the range of adjunctive behaviors that can be induced by scheduled feeding.

ACKNOWLEDGMENTS

The experiments described in Section 4.2 were conducted with the assistance of M. Paramesvaran. We thank L. M. Freed for consultation during the preparation of Section 5.4, V. Bruce-Wolfe for reading and commenting on Section 5.4, and Norman Hymowitz, Don Justesen, and John L. Falk for reading and commenting on an earlier version of this paper. The insulin hypothesis of schedule-induced polydipsia was originally conceived of by R. F. Zec in an unpublished manuscript submitted in April 1976 to partially fulfill requirements for a Ph.D. degree at the University of Kansas.

References

Allen, J. D., and Porter, J. H., 1975, Demonstration of behavioral contrast with adjunctive drinking, *Physiol. Behav.* **15**:511–515.

Allen, J. D., Porter, J. H., and Arazie, R., 1975, Schedule-induced drinking as a function of percentage reinforcement, *J. Exp. Anal. Behav.* **23**:223–232.

Allen, J. D., Stein, G. W., and Long, C. J., 1972, The effect of food deprivation on responding for food odor in the rat, *Learn. Motiv.* **3**:101–107.

Andersson, B., and Larsson, B., 1961, Influence of local temperature changes in the preoptic area and rostral hypothalamus on the regulation of food and water intake, *Acta. Physiol. Scand.* **52**:75–89.

Andersson, B., Larsson, S., and Persson, N., 1960, Some characteristics of the hypothalamic "drinking centre" in the goat as shown by the use of permanent electrodes, *Acta. Physiol. Scand.* **50**:140–152.

Atrens, D. M., 1973, Schedule-induced polydipsia and polyphagia in nondeprived rats reinforced by intracranial stimulation, *Learn. Motiv.* **4**:320–326.

Balagura, S. J., 1968, Conditioned glycemia responses in the control of food intake, *J. Comp. Physiol. Psychol.* **65**:30–32.

Baumann, J. W., Guyot-Jeannin, C., and Dobrowolski, J., 1964, Nutritional state and urine concentrating ability in the rat, *J. Endocrin.* **30**:147–148.

Berman, R. F., and Cannon, D. S., 1974, The effect of prior ethanol experience on ethanol-induced saccharin aversions, *Physiol. Behav.* **12**:1041–1044.

Blass, E. M., and Kraly, F. S., 1974, Medial forebrain bundle lesions: Specific loss of feeding to decreased glucose utilization in rats, *J. Comp. Physiol. Psychol.* **86**:679–692.

Bond, N., 1973, Schedule-induced polydipsia as a function of the consummatory rate, *Psychol. Rec.* **23**:377–382.

Booth, D. A., and Brookover, T., 1968, Hunger elicited in the rat by a single injection of bovine crystalline insulin, *Physiol. Behav.* **3**:439–446.

Booth, D. A., and Pitt, E., 1968, The role of glucose in insulin-induced feeding and drinking, *Physiol. Behav.* **3**:447–453.

Bruce-Wolfe, V. L., 1976, The role of hunger in schedule-induced polydipsia, unpublished doctoral dissertation, University of Kansas.

Bruce-Wolfe, V., Freed, W. J., and Mendelson, J., 1976, The role of hunger in schedule-induced polydipsia, *Bull. Psychon. Soc.* **7**:536–538.

Burks, C. D., 1970, Schedule-induced polydipsia: Are response-dependent schedules a limiting condition? *J. Exp. Anal. Behav.* **13**:351–358.

Carey, R. J., 1972, A decrease in ethanol preference in rats resulting from forced ethanol drinking under fluid deprivation, *Physiol. Behav.* **8**:373–375.

Carlisle, H. J., 1968, Initiation of behavioral responding for heat in a cold environment, *Physiol. Behav.* **3**:827–830.

Carlisle, H. J., 1971, Fixed-ratio polydipsia: Thermal effects of drinking, pausing, and responding, *J. Comp. Physiol. Psychol.* **75**:10–22.

Carlisle, H. J., 1973, Schedule-induced polydipsia: Effect of water temperature, ambient temperature, and hypothalamic cooling, *J. Comp. Physiol. Psychol.* **83**:208–220.

Carlisle, H. J., and Laudenslager, M. L., 1975, Inhibition of airlicking in thirsty rats by cooling the preoptic area, *Nature* **225**:72–73.

Carlisle, H. J., and Laudenslager, M. L., 1976, Separation of water and ambient temperature effects on polydipsia, *Physiol. Behav.* **16**:121–124.

Catania, A. C., 1966, Concurrent operants, *in* "Operant Behavior: Areas of Research and Application" (W. K. Honig, ed.), Appleton-Century-Crofts, New York, pp. 213–270.

Catania, A. C. (ed.), 1968, "Contemporary Research in Operant Behavior," Scott, Foresman, Glenview, Ill.

Chapman, H. W., 1969, Oropharyngeal determinants of nonregulatory drinking in the rat, unpublished doctoral dissertation, University of Pennsylvania, as cited by Falk (1971).

Chapman, H. W., and Richardson, H. M., 1974, The role of systemic hydration in the acquisition of schedule-induced polydipsia by rats, *Behav. Biol.* **12**:501–508.

Chiang, H. M., and Wilson, W. A., 1963, Some tests of the diluted-water hypothesis of saline consumption in rats, *J. Comp. Physiol. Psychol.* **56**:660–665.

Chillag, D., and Mendelson, J., 1971, Schedule-induced airlicking as a function of body-weight deficit in rats, *Physiol. Behav.* **6**:603–605.

Christian, W. P., 1975, Effects of the quality of dry food reinforcement on the rat's instrumental licking and drinking behavior, *Psychol. Rec.* **25**:237–242.

Christian, W. P., 1976, Control of schedule-induced polydipsia: Sugar content of the dry food reinforcer, *Psychol. Rec.* **26**:41–47.

Christian, W. P., Riester, R. W., and Schaeffer, R. W., 1973, Effects of sucrose concentrations upon schedule-induced polydipsia using free and response-contingent dry-food reinforcement schedules, *Bull. Psychon. Soc.* **2**:65–68.

Christian, W. P., and Schaeffer, R. W., 1973, Effects of sucrose concentrations upon schedule-induced polydipsia on a FI-60-sec dry-food reinforcement schedule, *Psychol. Rep.* **32**:1067–1073.

Christian, W. P. and Schaeffer, R. W., 1975, Motivational properties of fixed-interval reinforcement: A preliminary investigation, *Bull. Psychon. Soc.* **5**:143–145.

Cohen, I. L., 1975, The reinforcement value of schedule-induced drinking, *J. Exp. Anal. Behav.* **23**:37–44.

Cohen, I. L., and Mendelson, J., 1974, Schedule-induced drinking with food, but not ICS, reinforcement, *Behav. Biol.* **12**:21–29.

Colotla, V. A., and Keehn, J. D., 1975, Effects of reinforcer-pellet composition on schedule-induced polydipsia with alcohol, water, and saccharin, *Psychol. Rec.* **25**:91–98.

Cook, P., and Singer, G., 1976, Effects of stimulus displacement on adjunctive behavior, *Physiol. Behav.* **16**:79–82.

Cumming, M. C., and Morrison, S. D., 1960, The total metabolism of rats during fasting and refeeding, *J. Physiol.* **154**:219–243.

Deadwyler, S. A., and Segal, E. F., 1965, Determinants of polydipsia: VII. Removing the drinking solution midway through DRL sessions, *Psychon. Sci.* **3**:185–186.

Denny, M. R., and Ratner, S. C., 1970, "Comparative Psychology," Dorsey Press, Homewood, Ill.

Deutsch, J. A., and Deutsch, D., 1973, "Physiological Psychology," Dorsey Press, Homewood, Ill.

Deutsch, R., 1974, Conditioned hypoglycemia: A mechanism for saccharin-induced sensitivity to insulin in the rat, *J. Comp. Physiol. Psychol.* **86**:350–358.

Eckardt, M. J., 1975, Conditioned taste aversion produced by the oral ingestion of ethanol in the rat, *Physiol. Psychol.* **3**:317–321.

Eckardt, M. J., 1976, Alcohol-induced conditioned taste aversion in rats: Effect of concentration and prior exposure to alcohol, *J. Studies Alcohol* **37**:334–346.

Eckardt, M. J., Skurdal, A. J., and Brown, J. S., 1974, Conditioned taste aversion produced by low doses of alcohol, *Physiol. Psychol.* 2:89-92.

Epstein, A. N., and Teitelbaum, P., 1964, Severe and persistent deficits in thirst produced by lateral hypothalamic damage, *in* "Thirst in the Regulation of Body Water" (M. J. Wayner, ed.), Pergamon Press, Oxford, pp. 394-410.

Falk, J. L., 1961, Production of polydipsia in normal rats by an intermittent food schedule, *Science* 133:195-196.

Falk, J. L., 1964, Studies on schedule-induced polydipsia, *in* "Thirst: First International Symposium on Thirst in the Regulation of Body Water" (M. J. Wayner, ed.), Pergamon Press, New York, pp. 95-116.

Falk, J. L., 1966a, Analysis of water and NaCl solution acceptance by schedule-induced polydipsia, *J. Exp. Anal. Behav.* 9:111-118.

Falk, J. L., 1966b, The motivational properties of schedule-induced polydipsia, *J. Exp. Anal. Behav.* 9:19-25.

Falk, J. L., 1966c, Schedule-induced polydipsia as a function of fixed interval length, *J. Exp. Anal. Behav.* 9:37-39.

Falk, J. L., 1967, Control of schedule-induced polydipsia: Type, size, and spacing of meals, *J. Exp. Anal. Behav.* 10:199-206.

Falk, J. L., 1969, Conditions producing psychogenic polydipsia in animals, *Ann. N.Y. Acad. Sci.* 157:569-593.

Falk, J. L., 1971, The nature and determinants of adjunctive behavior, *Physiol. Behav.* 6:577-588.

Falk, J. L., 1972, The nature and determinants of adjunctive behavior, *in* "Schedule Effects: Drugs, Drinking, and Aggression" (R. M. Gilbert and J. D. Keehn, eds.), University of Toronto Press, Toronto, pp. 148-173.

Falk, J. L., and Bryant, R. W., 1973, Salivarectomy: Effect on drinking produced by isoproterenol, diazoxide, and NaCl loads, *Pharmacol. Biochem. Behav.* 1:207-210.

Falk, J. L., and Samson, H. H., 1976, Schedule-induced physical dependence on ethanol, *Pharmacol. Rev.* 27:449-464.

Falk, J. L., Samson, H. H., and Tang, M., 1973, Chronic ingestion techniques for the production of physical dependence on ethanol, *in* "Alcohol Intoxication and Withdrawal: Experimental Studies" (M. M. Gross, ed.), Plenum Press, New York, pp. 197-212.

Falk, J. L., Samson, H. H., and Winger, G., 1972, Behavioral maintenance of high concentrations of blood ethanol and physical dependence in the rat, *Science* 177:811-813.

Falk, J. L., and Titlebaum, L. F., 1963, Saline solution preference in the rat: Further demonstrations, *J. Comp. Physiol. Psychol.* 56:337-342.

Ferster, C. B., and Skinner, B. F., 1957, "Schedules of Reinforcement," Appleton-Century-Crofts, New York.

Fischer, U., Hommel, H., Ziegler, M., and Michael, R., 1972, The mechanism of insulin secretion after oral glucose administration: I. Multiphasic course of insulin mobilization after oral administration of glucose in conscious dogs. Differences to the behavior after intravenous administration, *Diabetologia* 8:104-110.

Flory, R. K., 1971, The control of schedule-induced polydipsia: Frequency and magnitude of reinforcement, *Learn. Motiv.* 2:215-227.

Flory, R. K., and O'Boyle, M. K., 1972, The effect of limited water availability on schedule-induced polydipsia, *Physiol. Behav.* 8:147-149.

Ford, C. V., Bray, G. A., Swerdloff, R. S., 1976, A psychiatric study of patients referred with a diagnosis of hypoglycemia, *Am. J. Psychiatry* 133:290-294.

Freed, E. X., 1971, Schedule-induced polydipsia with nutritive and nonnutritive reinforcers, *Psychon. Sci.* 23:367-368.

Freed, E. X., 1972, Alcohol polydipsia in the rat as a function of caloric need, *Q. J. Stud. Alc.* 33:504-507.

Freed, E. X., 1974, Fluid selection by rats during schedule-induced polydipsia, *Q. J. Stud. Alc.* **35:**1035–1043.

Freed, E. X., Carpenter, J. A., and Hymowitz, N., 1970, Acquisition and extinction of schedule-induced polydipsic consumption of alcohol and water, *Psychol. Rep.* **26:**915–922.

Freed, E. X., and Hymowitz, N., 1972, Effects of schedule, percent body weight, and magnitude of reinforcer on acquisition of schedule-induced polydipsia, *Psychol. Rep.* **31:**95–101.

Freed, E. X., and Lester, D., 1970, Schedule-induced consumption of ethanol: Calories or chemotherapy? *Physiol. Behav.* **5:**555–560.

Freed, W. J., and Mendelson, J., 1977, Water intake-volume regulation in the rat: Schedule-induced drinking compared to water-deprivation-induced drinking, *J. Comp. Physiol. Psychol.*, in press.

Freed, W. J., Mendelson, J., and Bramble, J. M., 1976, Intake-volume regulation during schedule-induced polydipsia in rats, *Behav. Biol.* **16:**245–250.

Freund, G., (1969), Alcohol withdrawal syndrome in mice, *Arch. Neurol.* **21:**315–320.

Garcia, J., and Kimeldorf, D., 1960, Some factors which influence radiation-conditioned behavior in rats, *Radiat. Res.* **12:**719–722.

Gilbert, R. M., 1973, Effects of ethanol on adjunctive drinking and bar-pressing under various schedules of reinforcement, *Bull. Psychon. Soc.* **1:**161–164.

Gilbert, R. M., 1974a, Schedule-induced ethanol polydipsia, *in* "Aportaciones al Analisis de la Conducta" (J. E. Diaz, E. Ribes, and S. Gomar, eds.), Trillas, Mexico, pp. 136–155.

Gilbert, R. M., 1974b, Schedule-induced ethanol polydipsia with restricted fluid availability, *Psychopharmacologia* **38:**151–157.

Gilbert, R. M., 1974c, Ubiquity of schedule-induced polydipsia, *J. Exp. Anal. Behav.* **21:**277–284.

Gilbert, R. M., 1977, Schedule-induced phenomena: Drug taking as excessive behavior, *in* "Behavioral Models of Drug Dependence" (R. Stretch, ed.), Raven Press, New York.

Githens, S. H., Hawkins, T. D., and Schrot, J., 1973, DRL schedule-induced alcohol ingestion, *Physiol. Psychol.* **1:**397–400.

Gold, R. M., and Proulx, D. M., 1972, Bait-shyness acquisition is impaired by VMH lesions that produce obesity, *J. Comp. Physiol. Psychol.* **79:**201–209.

Goldfine, J. D., Abraira, C., Gruenwald, D., and Goldstein, M. S., 1970, Plasma insulin levels during imaginary food ingestion under hypnosis, *Proc. Soc. Exp. Biol. Med.* **133:**274–276.

Gutman, Y., Livneh, P., and Pietrokovski, J., 1970, Role of salivary glands in response to thirst stimuli, *Isr. J. Med. Sci.* **6:**573–575.

Hawkins, T. D., Schrot, J. F., Githens, S. H., and Everett, P. B., 1972, Schedule-induced polydipsia: An analysis of water and alcohol ingestion, *in* "Schedule Effects: Drugs, Drinking, and Aggression" (R. M. Gilbert and J. D. Keehn, eds.), University of Toronto Press, Toronto, pp. 95–128.

Hinde, R. A., 1970, "Animal Behaviour: A Synthesis of Ethology and Comparative Psychology," McGraw-Hill, New York.

Holman, R. B., and Myers, R. D., 1968, Ethanol consumption under conditions of schedule-induced polydipsia, *Physiol. Behav.* **3:**369–371.

Hommel, H., Fischer, V., Retzlaft, K., and Knofler, H., 1972, The mechanism of insulin secretion after oral glucose administration: II. Reflex insulin secretion in conscious dogs bearing fistulas of the digestive tract by sham feeding of glucose or tap water, *Diabetologia* **8:**111–116.

Hunter, B. E., Riley, J. N., Walker, D. W., and Freund, G., 1975, Ethanol dependence in the rat: A parametric analysis, *Pharmacol. Biochem. Behav.* **3:**619–629.

Hymowitz, N., and Freed, E. X., 1972, Inconstancy of drinking bursts during schedule-induced polydipsia, *Psychon. Sci.* **28:**283–284.

Hymowitz, N., Freed, E. X., and Lester, D., 1970, The independence of barpressing and schedule-induced drinking, *Psychon. Sci.* **20**:45–46.

Hymowitz, N., and Koronakos, C., 1968, The effects of a controlled eating and drinking history on the development of schedule-induced polydipsia, *Psychon. Sci.* **13**:261–262.

Jacobs, H. L., 1958, Studies in sugar preference: I. The preference for glucose solution and its modification by injections of insulin, *J. Comp. Physiol. Psychol.* **51**:304 310.

Jacobs, H. L., 1969, Discussion following: Conditions producing psychogenic polydipsia in animals, J. L. Falk, *Ann. N.Y. Acad. Sci.* **157**:591.

Jacquet, J. H., 1972, Schedule-induced licking during multiple schedules, *J. Exp. Anal. Behav.* **17**:413–423.

Johnson, A. K., and Fischer, A. E., 1973, Tolerance for quinine under cholinergic versus deprivation induced thirst, *Physiol. Behav.* **10**:613–616.

Kachanoff, R., Leveille, R., McLelland, J. P., and Wayner, M. J., 1973, Schedule induced behavior in humans, *Physiol. Behav.* **11**:395 398.

Keehn, J. D., 1972, Schedule-dependence, schedule-induction, and the law of effect, *in* "Schedule Effects: Drugs, Drinking, and Aggression" (R. M. Gilbert and J. D. Keehn, eds.), University of Toronto Press, Toronto, pp. 65–94.

Keehn, J. D., and Colotla, V. A., 1971, Schedule-induced drinking as a function of interpellet interval, *Psychon. Sci.* **23**:69 71.

Keehn, J. D., Colotla, V. A., and Beaton, J. M., 1970, Palatability as a factor in the duration and pattern of schedule-induced drinking, *Psychol. Rec.* **20**:433–442.

Keehn, J. D., and Coulson, G. E., 1975, Schedule-induced choice of water versus alcohol, *Psychol. Rec.* **25**:325–328.

Kent, M. A., and Peters, R. H., 1973, Effects of ventromedial hypothalamic lesions on hunger-motivated behavior in rats, *J. Comp. Physiol. Psychol.* **83**:92–97.

Killeen, P., 1975, On the temporal control of behavior, *Psychol. Rev.* **82**:89 115.

King, B. M., and Gaston, M. G., 1973, The effects of pretraining on the bar-pressing performance of VMH-lesioned rats, *Physiol. Behav.* **11**:161–166.

Kissileff, H. R., 1969, Oropharyngeal control of prandial drinking, *J. Comp. Physiol. Psychol.* **67**:309–319.

Kissileff, H. R., 1973, Nonhomeostatic controls of drinking, *in* "The Neuropsychology of Thirst: New Findings and Advances in Concepts" (A. N. Epstein, H. R. Kissileff, and E. Stellar, eds.), V. H. Winston, Washington, D. C., pp. 163–198.

Kuzuya, T., 1962, Regulation of insulin secretion by the central nervous system: II. The role of the hypothalamus and the pituitary gland upon insulin secretion, *J. Jap. Soc. Internal. Med.* **51**:65–74.

LeMagnen, J., 1971, Advances in studies on the physiological control and regulation of food intake, *in* "Progress in Physiological Psychology," IV (E. Stellar and J. M. Sprague, eds.), Academic Press, New York, pp. 203–261.

Lester, D. and Freed, E. X., 1972, The rat views alcohol—Nutrition or nirvana? "International Symposium on the Biological Aspects of Alcohol Consumption," 27–29 September 1971, Helsinki, Finnish Foundation for Alcohol Studies **20**:51–57.

Lester, D., and Freed, E. X., 1973, Criteria for an animal model of alcoholism, *Pharmacol. Biochem. Behav.* **1**:103–107.

Lester, D., Nachman, M., and LeMagnen, J., 1970, Aversive conditioning by ethanol in the rat, *Q. J. Stud. Alc.* **31**:578–586.

Lotter, E. C., Woods. S. C., and Vasselli, J. R., 1973, Schedule-induced polydipsia: An artifact, *J. Comp. Physiol. Psychol.* **83**:478–484.

Marks, H. E., and Remley, N. R., 1972, The effects of type of lesion and percentage body weight loss on measures of motivated behavior in rats with hypothalamic lesions, *Behav. Biol.* **7**:95–111.

Martin, G. E., and Myers, R. D., 1972, Ethanol ingestion in the rat induced by rewarding brain stimulation, *Physiol. Behav.* **8**:1151–1160.

McKearny, J. W., 1973, Effects of methamphetamine and chlordiazepoxide on schedule-controlled and adjunctive licking in the rat, *Psychopharmacologia* **30**:375–384.

McLeod, D. R., and Gollub, L. R., 1976, An analysis of rats' drinking-tube contacts under tandem and fixed-interval schedules of food presentation, *J. Exp. Anal. Behav.* **25**:361–370.

Meisch, R. A. and Thompson, T., 1972, Ethanol intake during schedule-induced polydipsia, *Physiol. Behav.* **8**:471–475.

Mello, N. K., 1973, A review of methods to induce alcohol addiction in animals, *Pharmacol. Biochem. Behav.* **1**:89–101.

Mendelson, J., and Chillag, D., 1968, Schedule-induced air-licking in rats, *Am. Zool.* **8**:744.

Mendelson, J., and Chillag, 1970, Schedule-induced air licking in rats, *Physiol. Behav.* **5**:535–537.

Mendelson, J., and Zec, R., 1972, Effects of lingual denervation and desalivation on airlicking in the rat, *Physiol. Behav.* **8**:711–714.

Mendelson, J. Zec, R., and Chillag, D., 1971, Schedule dependency of schedule-induced airlicking, *Physiol. Behav.* **7**:207–210.

Mendelson, J., Zec. R., and Chillag, D., 1972, Effects of desalivation on drinking and air-licking induced by water deprivation and hypertonic saline injections, *J. Comp. Physiol. Psychol.* **80**:30–42.

Millar, F. K., and Morrison, S. D., 1968, Relation of tissue electrolyte losses to the relative polydipsia of early starvation in rats, *J. Nutr.* **94**:211–218.

Miller, N. E., Bailey. C. J., and Stevenson, J. A. F., 1950, "Decreased hunger" but increased food intake resulting from hypothalamic lesions, *Science* **112**:256–259.

Morrison, S. D., 1967, The adrenal cortex and the regulation of water exchange during food deprivation, *Endocrinology* **80**:835–839.

Morrison, S. D., MacKay. C., Hurlbrink, E., Wier, J. K., Nick, M. S., and Millar, F. K., 1967, The water exchange and polyuria of rats deprived of food, *Q. J. Exp. Physiol.* **52**:51–67.

Murphy, L. R., and Brown. T. S., 1975, Effects of desalivation on schedule-induced polydipsia in the rat, *J. Exp. Psychol.: Anim. Behav. Processes* **1**:309–317.

Myers, R. D., and Veale, W. L., 1972, The determinants of alcohol preference in animals, *in* "The Biology of Alcoholism (Vol. 2, Physiology and Behavior)" (B. Kissin and H. Begleiter, eds.), Plenum Press, New York, pp. 131–168.

Nicolaidis, S., and Rowland, N., 1974, Long-term self-intravenous "drinking" in the rat, *J. Comp. Physiol. Psychol.* **87**:1–15.

Novin, D., 1964, The effects of insulin on water intake in the rat, *in* "Thirst: First International Symposium on Thirst in the Regulation of Body Water" (M. J. Wayner, ed.), Pergamon Press, New York, pp. 177–184.

Ogata, H., Ogata, F., Mendelson, J. H., and Mello, N. K., 1972, A comparison of techniques to induce alcohol dependence and tolerance in the mouse, *J. Pharmacol. Exp. Ther.* **180**:216–230.

Palfai, T., Kutscher, C. L., and Symons, J. P., 1971, Schedule-induced polydipsia in the mouse, *Physiol. Behav.* **6**:461–462.

Panksepp, J., 1975, Hormonal control of feeding behavior and energy balance, *in* "Hormonal Correlates of Behavior" (B. E. Eleftheriou and R. L. Sprott, eds.), Plenum Press, New York, pp. 657–695.

Parra-Covarrubias, A., Rivera-Rodriguez, J., and Almarez-Ugalde, A., 1971, Cephalic phase of insulin release in obese adolescents, *Diabetes* **20**:800–802.

Penick, S. B., Prince, H., and Hinkle, L. E., 1966, Fall in plasma content of free fatty acids associated with the sight of food, *N. Engl. J. Med.* **275**:416-419.

Porter, J. H., Arazie, R. Holbrook, J. W., Cheek, M. S., and Allen, J. D., 1975, Effects of variable and fixed second-order schedules on schedule-induced polydipsia in the rat, *Physiol. Behav.* **14**:143-149.

Porter, J. H., and Kenshalo, D. R., Jr., 1974, Schedule-induced drinking following omission of reinforcement in the rhesus monkey, *Physiol. Behav.* **12**:1075-1077.

Ramsauer, S., Mendelson, J., and Freed, W. J., 1974, Effects of water temperature on the reward value and satiating capacity of water in water-deprived rats, *Behav. Biol.* **11**:381-393.

Reynierse, J. H., and Spanier, D., 1968, Excessive drinking in rats' adaptation to the schedule of feeding, *Psychon. Sci.* **10**:95-96.

Reynolds, G. S., 1961, Behavioral contrast, *J. Exp. Anal. Behav.* **4**:57-71.

Roll, D., Schaeffer, R. W., and Smith, J. C., 1969, Effects of a conditioned taste aversion on schedule-induced polydipsia, *Psychon. Sci.* **16**:39-41.

Rosenblith, J. Z., 1970, Polydipsia induced in the rat by a second-order schedule, *J. Exp. Anal. Behav.* **14**:139-144.

Samson, H. H., and Falk, J. L., 1974a, Alteration of fluid preference in ethanol-dependent animals, *J. Pharmacol. Exp. Ther.* **190**:365-376.

Samson, H. H., and Falk, J. L., 1974b, Schedule-induced ethanol polydipsia: Enhancement by saccharin, *Pharmacol. Biochem. Behav.* **2**:835-838.

Schuster, C. R., and Woods, J. H., 1966, Schedule-induced polydipsia in the rhesus monkey, *Psychol. Rep.* **19**:823-828.

Segal, E. F., 1972, Induction and the provenance of operants, *in* "Reinforcement: Behavioral Analyses" (R. M. Gilbert and J. R. Millenson, eds.), Academic Press, New York, pp. 1-34.

Segal, E. F., and Deadwyler, S. A., 1964, Water drinking patterns under several dry food reinforcement schedules, *Psychon. Sci.* **1**:271-272.

Segal, E. F., and Deadwyler, S. A., 1965, Determinants of polydipsia: VI. Taste of the drinking solution on DRL, *Psychon. Sci.* **3**:101-102.

Segal, E. F., Oden, D. L., and Deadwyler, S. A., 1965, Determinants of polydipsia: IV. Free-reinforcement schedules, *Psychon. Sci.* **3**:11-12.

Shanab, M. E., and Peterson, J. L., 1969, Polydipsia in the pigeon, *Psychon. Sci.* **15**:51-52.

Singer, G., Wayner, M. J., Stein, J., Cimino, K., and King, K., 1974, Adjunctive behavior induced by wheel-running, *Physiol. Behav.* **12**:493-495.

Steffens, A. B., 1969, Rapid absorbtion of glucose in the intestinal tract of the rat after ingestion of a meal, *Physiol. Behav.* **4**:829-832.

Steffens, A. B., 1970, Plasma insulin content in relation to blood glucose level and meal pattern in the normal and hypothalamic hyperphagic rat, *Physiol. Behav.* **5**:147-151.

Steffens, A. B., Mogenson, G. J., and Stevenson, J. A. F., 1972, Blood glucose, insulin, and free fatty acids after stimulation and lesions of the hypothalamus, *Am. J. Physiol.* **222**:1446-1452.

Stein, L., 1964, Excessive drinking in the rat: Superstition or thirst? *J. Comp. Physiol. Psychol.* **58**:237-242.

Stricker, E. M., 1976, Drinking by rats after lateral hypothalamic lesions: A new look at the lateral hypothalamic syndrome, *J. Comp. Physiol. Psychol.* **90**:127-143.

Stricker, E. M., and Adair, E. R., 1966, Body fluid balance, taste and postprandial factors in schedule-induced polydipsia, *J. Comp. Physiol. Psychol.* **62**:449-454.

Sundsten, J. W., 1969, Alterations in water intake and core temperature in baboons during hypothalamic thermal stimulation, *Ann. N.Y. Acad. Sci.* **157**:1018-1029.

Teitelbaum, P., 1966, The use of operant methods in the assessment and control of motivational states, *in* "Operant Behavior: Areas of Research and Application" (W. K. Honig, ed.), Appleton-Century-Crofts, New York, pp. 565–608.

Tepperman, J., 1974, Some aspects of the control of carbohydrate and fat metabolism, *in* "The Control of Metabolism" (J. D. Sink, ed.), Pennsylvania State University Press, Philadelphia, pp. 35 48.

Thomka, M. L., and Rosellini, R. A., 1975, Influence of prandial drinking vs. dry mouth in the attenuation of SIP in the desalivate rat, *Behav. Biol.* **17**:529 545.

Thompson, M. M., and Mayer, J., 1959, Hypoglycemic effects of saccharin in experimental animals, *Am. J. Clin. Nutr.* **7**:80 85.

Vance, W. B., 1965, Observations on the role of salivary secretions in the regulation of food and fluid intake in the white rat, *Psychol. Monogr.* **79**:1–22.

Wade, G. N., and Zucker, I., 1970, Hormonal modulation of responsiveness to an aversive taste stimulus in rats, *Physiol. Behav.* **5**:269–273.

Walker, D. W., and Freund, G., 1971, Impairment of shuttlebox avoidance learning following prolonged alcohol consumption in rats, *Physiol. Behav.* **7**:773–778.

Wayner, M. J., 1970, Motor control functions of the lateral hypothalamus and adjunctive behavior, *Physiol. Behav.* **5**:1319–1325.

Wayner, M. J., 1974a, The lateral hypothalamus and adjunctive drinking, *in* "Progress in Brain Research: Integrative Hypothalamic Activity" (D. F. Swaab and J. P. Schade, eds.), Elsevier, Amsterdam, pp. 371–394.

Wayner, M. J. 1974b, Specificity of behavioral regulation, *Physiol. Behav.* **12**:851 869.

Wayner, M. J., and Greenberg, I., 1972, Effects of septal lesions on palatability modulation of schedule-induced polydipsia, *Physiol. Behav.* **9**:663–665.

Wayner, M. J., and Greenberg, I., 1973, Schedule dependence of schedule-induced polydipsia and lever pressing, *Physiol. Behav.* **10**:965–966.

Wayner, M. J., Greenberg, I., Fraley, S., and Fisher, S., 1973, Effects of Δ 9-tetrahydrocannabinol and ethyl alcohol on adjunctive behavior and the lateral hypothalamus, *Physiol. Behav.* **10**:109–132.

Weijnen, J. A. W. M., 1975, Lingual stimulation and water intake, *in* "Control Mechanisms of Drinking, (G. Peters, J. T. Fitzsimons, and L. Peters-Haefeli, eds.), Springer-Verlag, New York, pp. 9 13.

Weijnen, J. A. W. M., 1976, Effects of denervation of the tongue on airlicking and current-licking behavior in the rat, paper presented at the Annual Meeting of the European Brain and Behaviour Society, Copenhagen, Denmark.

Wike, E. L., 1965, Comments on the Ison-Kniaz study of consummatory activity, *Psychol. Rep.* **12**:162.

Wike, E. L., and Barrientos, G., 1957, Selective learning as a function of differential consummatory activity, *Psychol. Rep.* **3**:255 258.

Wiley, J. J., and Leveille, G. A., 1970, Significance of insulin in the metabolic adaptation of rats to meal ingestion, *J. Nutr.* **100**:1073 1080.

Wilson, S., and Spencer, W. B., 1975, Schedule-induced polydipsia: Species limitations, *Psychol. Rep.* **36**:863 866.

Woods, S. C., Alexander, K. R., and Porte, D., Jr., 1972, Conditioned insulin secretion and hypoglycemia following repeated injections of tolbutamide in rats, *Endocrinology* **90**:227 231.

Woods, S. C., Decke, E., and Vasselli, J. R., 1974, Metabolic hormones and the regulation of body weight, *Psychol. Rev.* **81**:26 43.

Woods, S. C., Hutton, R. A., and Makous, W., 1970, Conditioned insulin secretion in the albino rat, *Proc. Soc. Exp. Biol. Med.* **133**:964 968.

Woods, S. C., and Porte, D., Jr., 1974, Neural control of the endocrine pancreas, *Physiol. Rev.* **54**:569–619.

Wuttke, W., and Innis, N. K., 1972, Drug effects upon behavior induced by second-order schedules of reinforcement. The relevance of ethological analyses, *in* "Schedule Effects: Drugs, Drinking, and Aggression" (R. M. Gilbert and J. D. Keehn, eds.), University of Toronto Press, Toronto, pp. 129–147.

Index